Shades of Green

An Environmental and Cultural History of Sitka Spruce

Ruth Tittensor

Ceremonial Potlatch Hat woven of Sitka spruce roots in the traditional Haida style by Debbie Young-Canaday of Juneau, Alaska. Photo: Marilyn Holmes

They that plant trees love others besides themselves.

Adapted from Thomas Fuller, *Gnomologia*, 1732

For Isabelle, Sebastian and Sylvie

Windgather Press is an imprint of Oxbow Books

Published in the United Kingdom in 2016 by
OXBOW BOOKS
10 Hythe Bridge Street, Oxford OX1 2EW

and in the United States by
OXBOW BOOKS
1950 Lawrence Road, Havertown, PA 19083

Paperback Edition: ISBN 978-1-909686-77-9
Digital Edition: ISBN 978-1-909686-78-6

A CIP record for this book is available from the British Library

Printed in Malta by Melita Press Ltd

For a complete list of Windgather titles, please contact:

United Kingdom
OXBOW BOOKS
Telephone (01865) 241249
Fax (01865) 794449
Email: oxbow@oxbowbooks.com
www.oxbowbooks.com

United States of America
OXBOW BOOKS
Telephone (800) 791-9354
Fax (610) 853-9146
Email: queries@casemateacademic.com
www.casemateacademic.com/oxbow

Oxbow Books is part of the Casemate group

Cover design by Susan Anderson, Eikon Design

Contents

Foreword

Richard Oram

Professor of Environmental and Medieval History, Centre for Environment, Heritage and Policy, University of Stirling, Scotland

In the intersecting fields of cultural and environmental history, the focus of scholarly study has been animal species, mineral commodities and rivers: cod and herring, salt and coal, the Forth and Severn, to name but a few.

Oddly, while trees and woodland generally have long been subjects of research, there has been no single tree species that has been treated to such close-focus research as a cultural entity around which a complex human system operates, rather than solely a living organism in a complex ecosystem. Given the English fixation with the oak as the symbol of everything from enduring royalty to the virility of their fighting men and as an essential element in everything from furniture-making and leather-tanning to the construction of warships – and to a lesser extent the elm and the yew – it is remarkable that there has been no study that has tried to draw these two strands – the cultural and the environmental – together in a truly interdisciplinary exploration. It is rather ironic, then, that the species which should be the subject of such a treatment is one whose very presence in these islands has been the focus of both censure and celebration since it was first planted on a large scale a century ago: *Picea sitchensis* – the Sitka spruce.

How can a tree that on one side of the globe was valued for its intrinsic worth as a bountiful source of manifold and versatile materials, and treated with spiritual reverence by the native peoples of America's Pacific North-West, become one of the most loathed and reviled organisms to so many people on the other? Known to Europeans since the later eighteenth century and grown as specimen trees in British arboreta, it burst into public awareness in the mid-twentieth century when vast swathes of Britain and Ireland's largely tree-less northern and western uplands began to be submerged beneath a tidal wave of dark-green saplings. It was loved by foresters and government planners who saw in this fast-growing and resilient species a strategic resource that could provide these islands with a secure, home-grown supply of commodities as diverse as timber and wood-spirit for future wars. Large-scale planting coincided with more intensive ecological study of the land on which they were planted, land long seen as almost valueless, and the intensification of an environmental consciousness that awakened the wider public to the destructive impact of humanity on the world around them. To many, however, it was the visual impact that was most harmful in a culturally transformative way: the distinct character of Britain's anthropogenic landscapes that had been shaped over millennia of human intervention into something familiar and loved, and around which so many aspects of rural life revolved, seemed to be swept away by something visually jarring and, simply, foreign.

For Sitka spruce the result was a popular loathing that was at once an ill-informed reaction to the shock-of-the-new usually associated with purely man-made features like electricity pylons or dams, reinforced by opponents of large-scale forestry planting. Often dismissed as alien vermin, the Sitka spruce plantations have been portrayed as the coniferous equivalent of the Dark Satanic Mills that scarred the face of Blake's green and pleasant land. Yet, like the output of those reviled mills, the products of the plantations have delivered to successive generations commodities upon which they depend.

It remains the most widely-planted tree species in these islands, but greater care is taken through strategic intermingling with stands of trees of other species – producing variation in height, colour, shape and texture.

In many ways, these opposed responses, of antipathy and dependence, of rejection and integration, reflect, often disturbingly, public attitudes to the presence of other exotica like roads, airports or foreigners. And it is here, in its exploration of the cultural significance of the tree in both its 'home' and its 'host' lands, that Ruth Tittensor's study of this quite remarkable tree transcends the traditional nature/culture opposition still found in so many natural or environmental history books. In her discussion of still deeply-entrenched public attitudes towards the Sitka spruce she holds up a mirror in which modern society in these islands can reflect upon itself.

Foreword

Richard Carstensen

Naturalist, Discovery Southeast, Juneau, Alaska

On the phone, just now (June 2015), Ruth Tittensor and I compared the summer weather outside our homes on far sides of the world. Quite the same, we concluded – cool, grey and moist. No surprise; that's partly why we were talking in the first place.

An American coastal tree, evolved to thrive in my well-watered, fire-free climate, has eagerly galloped over the moistly moderate UK and Republic of Ireland. Roughly contemporaneous with the British invasion of American rock music, Sitka spruce executed a counter-coup. Esteemed and groomed by European silviculturalists, the spruce plantations, and more feral, tree-by-tree advances into moor and sheep pasture, are less welcomed by lovers of open, rural landscapes. Social tension and ecological conquest make a potent mix in the hands of a good story-teller like Ruth Tittensor.

Another way in which I feel connected to Ruth is in our cross-disciplinary approaches to study of natural and cultural history. We've each tried to 'lean back' a bit, to admire the wild and unforeseeable sweep of evolution, succession, and cultural metamorphosis. From that perspective, the entrenched debate over what is 'natural, traditional, indigenous', or 'alien, aggressive, weedy', can seem a bit parochial. Aesthetics aside, what have been the deeper ecological implications of the spread of Sitka spruce into Britain and Ireland or of the massive loss of old-growth spruce forest in my (western) hemisphere? When we learn to 'turn on the projector' – to visualise changes in landform, demography, forest structure, species range – the concepts of 'native' and 'newcomer' relax into relativity.

That said, I agree with ecologist Dan Simberloff that invasive species should be held 'guilty until proven innocent'. Nor should we ever deem 'proof' absolute. Here in Alaska, a few dispersed patches of ornamental Japanese and Bohemian knotweed (*Fallopia japonica/F. japonica* x *F. sachalinensis*) could be found in yards and gardens when I arrived about 40 years ago – innocently minding their own business, I thought. We might then, with enough foresight, have eradicated it. Now, as any British naturalist could have warned us, it will be much harder. Sitka spruce has been present in Britain and Ireland for almost as long as Brazilian pepper-tree (*Schinus terebinthifolius*) has grown in Florida. For most of that time, few in Florida suspected the innocuous pepper-tree would eventually 'bolt', overwhelming and displacing coastal mangrove communities, as well as scrub and pine flatwood habitats.

Is this a meaningful comparison? Southern Florida – a cultural and ecological cul-de-sac until a few centuries ago – is a hotspot of threatened, specialised, endemic species.

Britain and Ireland have been cultural and ecological crossroads for millennia, swept by wave after wave of hardy, world-travelling flora and fauna. Which of the ephemerally 'native' British species might now – or in the future – fall victim to authorised or surreptitious spread of Sitka spruce? What will be the long-term costs and contributions of this conifer to biodiversity in its invited hemisphere? Although surely, nobody knows, who is best prepared to guide us?

For my money, that cultural ecologist, who appraises ignorance, and patiently chips away at it, one fact, one story, one hypothesis at a time. Ruth Tittensor has scaled out to a global view of Sitka spruce. She examines its evolutionary and natural history, as well as the roles it played for cultures as diverse as seventeenth century Alaskan Tlingit – to whom iron adzes salvaged from Asian flotsam were novelties – and mid-twentieth century ecologists, then perceived, Ruth remembers, as 'eccentric bearded men in sandals' (hmmm, that still describes me pretty well).

Reading *Shades of Green*, I realise how much we have to learn from each other. If Sitka spruce eventually comes to define the 'new temperate rainforests' of Britain and Ireland, where better for their foresters to preview them than in an Alaskan Landmark Forest, one of 76 one-acre (0.4 ha) plots we established on Tongass Forest stream sides? In these forests, old-growth spruces rise more than 60 m (200 ft) over bear-chewed salmon carcasses, in thickets of fruiting Devil's club (*Oplopanax horridum*) and Grey currant (*Ribes bracteosum*). And if Alaskan second-growth spruce-hemlock 'plantations' ever replace the old-growth forest currently favoured by our anachronistic timber industry, who better to coach us than Ruth's colleagues in silviculture and forest ecology, who for more than a generation have been iteratively adjusting the balance of timber yield, landscape aesthetics, and habitat values?

I hope every world-changing species someday earns a biographer as capable as Ruth Tittensor, and a tribute as penetrating and open-minded as *Shades of Green*.

Geography

Territorial boundaries shift so that place-names familiar to one group of people may have no later meaning. This section is to help readers follow the historic changes within Europe and North America. Prehistory is discussed using present territorial names though these had no meaning at the time.

United Kingdom and Ireland

Ireland is used here for the whole island of Ireland. Before 1542 Ireland was a flexible grouping of small kingdoms and the Lordship of Ireland under the Normans or Anglo-Normans. The 'Kingdom of Ireland' existed from 1542 (when Henry VIII proclaimed himself King of Ireland) until 1800. It became part of the UK in 1801 under an 1800 *Act of Union.*

Éire is the Irish language name for Ireland.

Irish Free State was that part of Ireland (26 of its 32 counties) formed when Northern Ireland became part of the UK in 1922.

Northern Ireland Those six Irish counties (Antrim, Armagh, Down, Fermanagh, Londonderry and Tyrone) partitioned from Ireland in 1921 and incorporated into the UK in 1922.

Republic of Ireland was formed in 1949 from the Irish Free State.

Ulster was the northern of five ancient provinces of Ireland.

Scotland originated as a kingdom during the tenth century. The Scottish and English crowns were united in 1603, but not the parliaments. In 1707 an *Act of Union* formally united Scotland with England and Wales to form Great Britain.

Scottish Highlands refers to that part of Scotland north and west of the Highland Boundary Fault (from the Isle of Arran in the west to near Aberdeen in the north-east); mainly mountainous.

Scottish Lowlands refers to that part of Scotland south of the Highland Boundary fault; low-lying, less mountainous.

Wales was a loose union of Welsh kingdoms until united by Gruffudd ap Llywelyn in 1055. It returned to two or more kingdoms again after his death. In 1282, the Welsh kingdoms were conquered and annexed by England as the Principality of Wales, formally united with England in 1536 as 'England and Wales'.

England refers to the country formed by union of several Anglo-Saxon kingdoms in 927 AD. It was an entity until 1282, when Wales was annexed.

Great Britain refers to England, Wales and Scotland since 1707 when they were united in crown and parliaments.

United Kingdom was created in 1801 by the union of Great Britain and Ireland.

United Kingdom of Great Britain and Northern Ireland (UK) describes England, Northern Ireland, Scotland and Wales since 1949.

North America

America here describes the early British territories south of French Canada, named after the Spanish explorer Amerigo Vespucci; part of the USA in modern times.

L'Acadie or *Acadia* An area of south-east Canada and north-east USA where descendents of seventeenth-century French colonists lived: now New Brunswick, Nova Scotia, Prince Edward Island, parts of Québec and Maine.

Alaska An area of sparsely-populated Arctic land to the north-west of Canada, acquired by the USA from Russia in 1867 and admitted to the USA in 1959 as the fiftieth state.

Beringian land bridge The land 1609 km (1000 miles) from north to south which joined Siberia and Alaska during times of low sea levels during the latest ice age and which allowed movement of flora, fauna and humans between the two continents.

British Columbia originated as the northern sector of the Oregon Territory and became a British colony in 1846. It joined the Canadian Confederation in 1871 as the sixth province. It is now a western province of Canada adjoining the Pacific Ocean and Alaska.

Canada The name originated from an Iroquoian word meaning 'the village'. French explorers used 'Le Canada' for all the French colonies of eastern North America; also an alternative name for New France. In 1867 it became the Canadian Confederation and then the Dominion of Canada, but is now known as 'Canada'.

Haida Gwaii is the Haida name and official Canadian name for the Queen Charlotte Islands archipelago.

Lower Canada was the name for all the early settlements along the lower St Lawrence River and around the Gulf of St Lawrence.

Maritimes The eastern provinces of Canada: New Brunswick, Nova Scotia and Prince Edward Island.

Newfoundland A large island at the mouth of the St Lawrence River between Nova Scotia and Labrador, the subject of rivalry between Spain, France and Britain for the Grand Banks cod fisheries. It joined Canada as a province along with Labrador in 1949.

New France referred to the French colonies of Acadia, Lower Canada, Upper Canada, part of Newfoundland and Louisiana in the early eighteenth century. It was an alternative name for 'Canada', but became the Province of Québec after 1763.

North America describes modern Canada and USA.

Nova Scotia The most south-easterly province of Canada.

Oregon Territory/Country An area on the Pacific coast of North America centred on the Columbia River and administered by British and American settlers. In 1846, the land north or the Columbia River became British; the southern sector became an 'organised incorporated territory' of the USA in 1848 and in 1859 became the state of Oregon within the union of United States.

Panhandle is an unofficial name for Southeast Alaska.

Québec is a city close to where French explorer Jacques Cartier and his ship's crew spent winter 1535–1536 in the frozen St Lawrence River. Samuel de Champlain built it as a wooden fort in 1608 on the site of a First Nations settlement 'Stadacona'. The modern name Québec is taken from the First Nations 'Kebec' which describes where the River narrows suddenly. It is now the capital city of Québec province.

Québec Province was an area of early French settlements 'Le Canada', later Lower Canada along the St Lawrence River and Gulf of St Lawrence. Now the modern province developed from Upper and Lower Canada in 1841.

Queen Charlotte Islands form an archipelago of about 400 islands across the Hecate Strait from Prince Rupert on the coast of British Columbia. They were incorporated into British Columbia in 1863. They are the heart of the Haida First Nations territory.

Southeast Alaska is a narrow, archipelagic region of Alaska adjoining northern British Columbia.

United States of America was a group of thirteen British colonies on the Atlantic seaboard of North America which declared independence in 1776. Territories were added to form the modern federal republic.

Upper Canada was a province of British Canada between 1791 and 1841, an east–west strip along the northern shores of the Great Lakes. 'Upper Canada' refers to its position higher up the St Lawrence River than 'Lower Canada' (now Québec province).

Acknowledgements

I am immeasurably grateful to have met and benefitted from the wisdom, knowledge, experience, support and help of many people during the preparation of this book. A few words are barely sufficient thanks for their interest and input, but I appreciate it all.

Landowners and their representatives in UK and Republic of Ireland

Andrew Barbour (Atholl), Sir Archibald Grant (Monymusk), Graham Thompson (RSPB, Forsinard), John Sutherland (Corrour), Lord Cawdor (Cawdor), Lord Moray and Gareth Whymant (Darnaway), Marquess of Waterford and Kevin Butler (Curraghmore), Michael Caughlin (Drumlanrig), Sir Robert Clerk (Penicuik), Roderick Leslie-Melville and Thomas Steuart Fothringham (Murthly), Stefan Fellinger (Muirhead), Forestry Commissions: England, Scotland and Wales (Kielder, Galloway, Inverliever, Clocaenog, Gwydyr), Forest Service NI (Knockmany, Castlecaldwell) – who freely gave access to their land and answered queries.

Forestry Agencies in Isle of Man, Northern Ireland, Republic of Ireland and UK

Alan Duncan, Alan Fletcher, Andy Leitch, Bill Mason, Bill Meadows, Chris Quine, Colin Reilly, Dave Williams, David Ellerby, David Jardine, Glenn Brearley, James Duffey, John Walmsley, Jonathan Taylor, Graham Gill, Noel Melanaphy, Rob Soutar, Rhod Watt, Sam Lines, Sam Samuel, Shaun Williams, Steve Lee and Trevor Fenning – who provided specialist information and escorted forest visits.

Organisations and businesses in UK

Alex Blake (Christie-Elite Nursery), Christian ap Iago (Timber Garden Build), Colin Kennedy (Scottish Woodlands Ltd), Elizabeth Roberts (Zacharry's), Ian and David Kingan (James Kingan & Son), Mark Gardner (Royal Botanic Garden, Edinburgh), Patrick Hunter-Blair and William Crawford (Royal Scottish Forestry Society), Rob Mackenna (James Jones Ltd), Sophie Whittaker (Norbord), Susan Anderson (Eikon Design) – who willingly gave their personal expertise.

Organisations and businesses in USA and Canada

Brent Cole (Alaska Specialty Woods), Brian Kleinhenz, Chuck Smythe, Dixie Hutchinson, Ishmael Hope and Zachary Jones (Sealaska Heritage Institute), Deanna Stad (Weyerhaeuser), David McMahon (Atlantic Canada Aviation Museum), Debbie Young-Canaday (Haida weaving), Debby J. Rosenberg (author and poet), Debra Lyons and Matt Hunter (Sitka Trail Works), Jane Lindsey (Juneau-Douglas City Museum), John F. C. Johnson (Chugach Alaska Corporation), Katrina Pearson (Taku Graphics), Luke A'Bear (Sitka Conservation Society), Rick and Suzan Armstrong (Baranof Island Brewing Co.), Patrick Heuer (USDA Forest Service), Steve Henrikson (Alaska State Museum), Wes Tyler (Icy Straits Lumber) – who opened vistas of Sitka spruce and its associated culture past and present.

Specialists

Brian Klinkenberg (E-Flora BC, Canada), Caroline Wickham-Jones (archaeology), Chris Smout (wilderness philosophy) Chrisma Bould and Margaret Mackay (Gaelic names), Douglas Malcolm (silviculture), Elizabeth Roberts (Russian Alaska), Henry Phillips (data for Republic of Ireland), Larissa Glasser (Arnold Arboretum, seeds), Laura Parducci (refuges), Margaret Davidson (Thorne River Basket), Michael Carey (Sitka spruce in Republic of Ireland), Michael Kauffmann (Sitka spruce in California), Paul Haworth (raptors), Richard Carstensen (Alaskan ecology), Richard Dauenhauer (now deceased, Russian Alaska), Richard Oram (Medieval environments), Richard Tipping (peat, Caithness Flows), Richard Unger (spruce beer), Stephen Sillett (Sitka spruce and redwoods), Þröstur Eysteinsson and Edda Sigurdís Oddsdóttir of Skógrækt Ríkisins, Icelandic Forest Research (Sitka spruce in Iceland), Terence Fifield (Thorne River Basket).

De Havilland Mosquito Museum, Hatfield, UK

Bob Glaseby, Gerald Mears, Ian Thirsk and their colleagues – who demonstrated how Sitka spruce was used in Mosquito aircraft.

Musicians

Edwin Beunk (Enschede), John Koster (Vemillion), Norman W. Motion (Edinburgh), Robert Levin (Harvard), Ulrich Gerhartz (London) – who explained the timber composition and functioning of soundboards.

First Nations craftspeople

Haida Elder Delores Churchill and her July 2014 basket-making class (Sitka Campus, University of Alaska), and Tlingit artist Teri Rofkar (Sitka) – who explained how Sitka spruce is used in First Nations' craft work.

Colleagues and family in UK and Canada

Andrea Moe, Bill Linnard, Catriona MacPhee, Derek Tittensor, David Neil, Janet Hendry, Jeannine Moe, Keith Hobley, Kirsty Tittensor, Michael Chalton, Penny Wooding, Ralph Tittensor, Richard Marriott, Richard Roberts, Roger Perry, Rosemary Tittensor – who helped with information sources, personal experiences, contacts, photography, companionship during travels and away-homes while working.

Assistance with photographs

Bob Briggs, George Brown, Jeffrey Jewell, Jeffrey Joeckel, Kayla Boettcher, Marilyn Holmes, Neil Cox, Neill Campbell, and Sophie Whittaker gave unstinting help in obtaining specialist photographs. Norman Davidson of the Forestry Memories website www.forestry-memories.org.uk provided photographs and background information about twentieth century Scottish forestry.

Special thanks

Clare Litt, Julie Gardiner, Mette Bundgaard and Tara Evans put this book through publication; John Walters and Bill Mason spent much time editing drafts; Kirsten Hutchison provided many books and articles from Forestry Commission libraries; Syd House of Forestry Commission Scotland mentored progress. Richard Carstensen and Richard Oram wrote stimulating Forewords. Very special thanks to Andy Tittensor who supported and helped me enormously during four years of work.

I am especially grateful to The Marc Fitch Fund for assistance towards research costs, the Scottish Forestry Trust for grant-aiding a visit to Pacific North America, and the Strathmartine Trust which paid for printing the colour illustrations. Forestry Commission Scotland, Forest Enterprise and Forest Research kindly allowed me use of their facilities.

CHAPTER ONE

The Most Hated Tree?

Sitka spruce is now the most important commercial forest tree species in Ireland.
Padraig Joyce and Niall OCarroll 2002

And yet, it is popularly denigrated, even vilified, and the undiscerning public is blind to its virtues, and is encouraged to be so by legions of misguided conservationist 'green', 'natives only' and anti-forestry bodies.
Alan Mitchell 1996

An astonishingly lush, tall, damp and dank, dense and dark, group of forest ecosystems clothe the Pacific Ocean coast of Canada and the United States of America. Towering trees grow right down to the sea shore, along the banks of inlets and estuaries and on steep, coastal mountainsides typically to about 500 m (1640 ft), but occasionally up to 2700 m (8858 ft) altitude.

These unusual 'temperate rainforest' ecosystems have developed and colonised a narrow coastal belt from northern California to Alaska, a distance of 3600 km (2237 miles). Although it is usually no more than 80 km (50 miles) wide from the ocean to its inland purlieu, it develops further inland along river flood-plains and fjords. In a ribbon of mild, misty, high rainfall, maritime climate, these complex and biologically-productive forests are dominated by a few groups of plants: conifers, lichens and bryophytes (mosses and liverworts).

Conifers are woody shrubs or trees with an ancient geological lineage, appearing in the geological record during the Carboniferous period about 300 million years ago (MYA). Their foliage is usually evergreen, of green scales or needle leaves. The branches are arranged in whorls round the trunk becoming gradually smaller up the tree. This gives most conifers a single trunk and regular tree-shape, often pyramidal or conical. Their shape is one reason UK silviculturalists (people who grow and manage woodlands) today prefer growing conifers to broadleaved trees like oak or ash which have irregular branches. Conifers are easily grown, with straight trunks, while their whorls of branches are easily removed, giving a clean 'bole', suitable for cutting into shorter lengths by machinery. Millions of pit-props to support the tunnels of deep coal mines were produced like this from conifers during the nineteenth and twentieth centuries in Europe. Modern timber, too, has to be long and straight for today's construction spars, planks and paper-making machinery.

In contrast, the asymmetrical, irregular growth pattern of broadleaved trees with their 'kneed' and angled branches was exploited in past centuries and millennia. Their trunks were important for constructing boats, for structural components of medieval buildings like crucks (curved timbers to support a roof) and harbours; their irregular branches were suitable for wooden wheels, implements, tools, clogs and charcoal.

We now rarely travel in wooden boats, or walk in clogs, so modern sawmills and pulp mills buy mainly conifer timber, to manufacture the different products needed by modern society: *us*.

Commercial timber in North America comes from many types of deliberately-planted conifer and broadleaved forests and from indigenous forest ecosystems. Forests are regenerated by deliberate planting of little trees from nurseries, or naturally from seeds produced by the tree canopy which drop or blow onto the ground, germinate and grow. In well-forested parts of Europe, the same methods are used. But in Britain and Ireland, forest cover was reduced time after time during prehistory and history, so that by 1900 AD less than 5% of their land area was woodland. So there is little, if any, indigenous forest extant in these countries. Most commercial timber is produced from tree seedlings germinated and grown in nurseries, then planted out as forests. Sitka spruce is one of the most common conifers in modern British and Irish forests.

The Sitka spruce (*Picea sitchensis*), with its prickly, blue-green needle-leaves and orange-brown cones, is one of the most abundant, tall and large-girthed conifers in the temperate rainforests of western North America. In some localities it is the dominant or only tree species. However, since Sitka spruce is naturally confined to temperate rainforests, which are restricted geographically, it is also rare. It is highly valued in modern North America as an economic tree for its good quality construction timber and paper pulp. (Chapters Two and Three explain the origin and features of spruce trees).

Other conifers contributing to the North American temperate rainforests are Douglas fir (*Pseudotsuga menziesii*), Western red cedar (*Thuja plicata*), Western hemlock (*Tsuga heterophylla*), Mountain hemlock (*T. mertensia*) Pacific silver fir (*Abies amabilis*), Grand fir (*A. grandis*) and Yellow cedar (*Chamaecyparis nootkatensis*). These are important conifers commercially as well as ecologically. A variety of broadleaved trees also contribute to the temperate rainforest, including Bigleaf maple (*Acer macrophyllum*), Red alder (*Alnus rubra*) and Black cottonwood (*Populus trichocarpa).*

There are said to be more Sitka spruce trees in Britain and Ireland than in North America today. It was first introduced into Europe during the early nineteenth century, when exploring was arduous and dangerous. Plant hunters sent seeds of many trees, including Sitka spruce, home to Europe from North America. British scientific and horticultural societies were looking for trees which would grow on the poor soils and thrive in the wet, windy climate of the uplands, as well as for attractive plants to enhance parks and gardens of country mansions. Many other seeds and plants – hundreds of species – arrived in Europe from North America. Other continents were also searched for potentially valuable and beautiful trees and herbaceous plants.

Maps 1 and 2, of the UK and Republic of Ireland and North America respectively, show their geographical features relevant to this book.

The potential of Sitka spruce for Scotland's landscapes was recognised by the first explorer to send seeds back and later by landowners themselves. It grew well and looked beautiful in their gardens and arboreta. During the later nineteenth and twentieth centuries, landowners and foresters in many parts of Britain and Ireland tested its growth in plantation conditions. They discovered that it would establish and grow fast in some of the exposed, wet conditions and difficult soils of the north and west. This ability, as well as its products, endeared it to professional foresters, private landowners and timber merchants.

But for today's public, Sitka spruce is the tree species in their countryside, which is probably the most loathed … even though the public usually loves trees.

For instance, the two common oaks, Pedunculate (*Quercus robur*) and Sessile *(Q. petraea*), represent historic, cultural, aesthetic and user appeal (the Knightwood Oak, the Bloody Oak, King James Oak and Cefnmabli Oak). But the British public cannot find any similar empathy with the Sitka spruce, even though it is now the most common tree species in upland landscapes. People do not see in Sitka spruce the grandeur and rarity of the riparian Black poplar (*Populus nigra* subsp. *betulifolia*) or the religious and mythical appeal of the yew (*Taxus baccata*), nor can it be used to make the traditional longbows of prehistory and medieval times.

- Unlike the stately English elm (*Ulmus minor* var. *vulgaris*), Sitka spruce has not been a familiar hedgerow tree of lowland farms.
- Unlike the hazel (*Corylus avellana*) and limes (*Tilia* species) it does not form coppice woodlands with spectacular assemblages of spring flowers.
- Unlike the Field maple (*Acer campestre*) it does not metamorphose into orange autumn foliage, not does it rustle plaintively in the wind like the aspen (*Populus tremula*).
- Unlike the Wild cherry (*Prunus avium*), Crab apple (*Malus sylvestris*) and Rowan (*Sorbus aucuparia*) it offers neither floriferous spring canopies nor edible fruits.
- British and Irish people denigrate Sitka spruce with 'introduced' and 'conifer' when reflecting upon trees in the countryside.

Britain has merely three indigenous conifer species: yew, juniper and Scots pine. It has no native spruce species. Yew grows singly, or in groups, in churchyards, along important land boundaries and in other woodlands; there are a few pure yew woodlands in southern England and Ireland.

Juniper (*Juniperus communis*) has declined, grazed away by centuries of sheep and cattle ranching in the uplands. A few woodlands of tall juniper grow in Scotland, while patches of the prostrate sub-species scent the air of southern English chalk hills.

The Scots pine (*Pinus sylvestris*) and hazel formed large forests in middle Britain during the warm, dry Boreal period from about 9000 to 7000 years before the present (BP). When the climate subsequently cooled and wetted, Scots pine retreated north. Today, native Scots pine woodlands grow only in some Scottish glens, along some loch shores and islands.

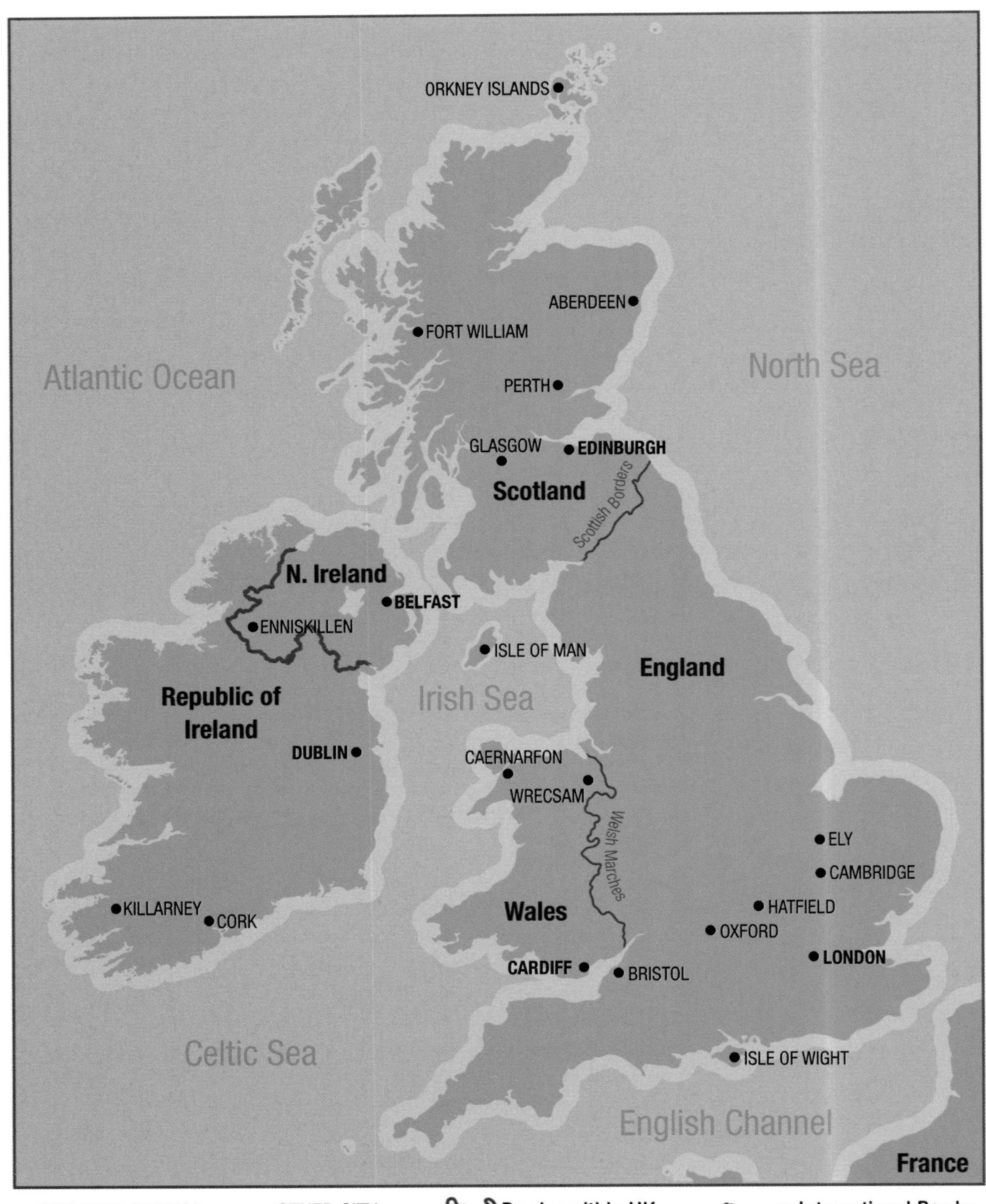

• **COUNTRY CAPITAL** • OTHER CITY Border within UK International Border

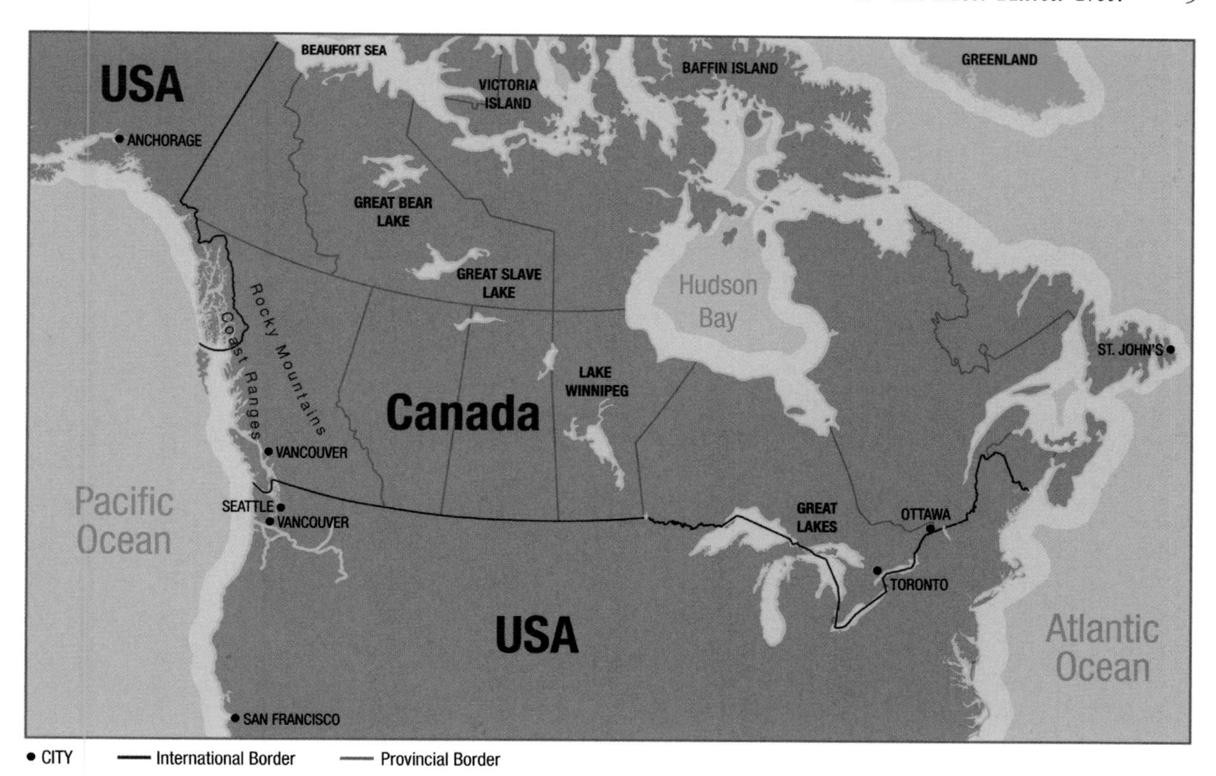

MAP 1 *(opposite)*. UK and Republic of Ireland, prepared by Susan Anderson of Eikon Design, Ayrshire, UK.

SOURCE: ATLASES.

MAP 2 *(above)*. North America, prepared by Susan Anderson.

SOURCE: ATLASES.

After the last Ice Age, broadleaved trees gradually colonised Britain and Ireland, clothing much of the landscape with deciduous forests from about 8000 BP. Without human interference, they might still form the dominant native vegetation. British and Irish people are used to broadleaved trees – not needled conifers – forming the hedges, woodlands, forests and coppices of their beloved countryside. However, not only are there only three indigenous conifer species, there are few broadleaved species either compared with continental Europe.

During the latest glacial episode of the Ice Ages (called the Devensian in Europe) 110,000 to 12,000 BP, what are now Britain and Ireland were part of the European continent. Most plants and animals became extinct in ice-covered northern Europe. But from recent genetic and distribution evidence, we know that some arctic-alpine plants and invertebrates survived in small numbers on ice-free peaks or 'nunataks' amongst the ice fields. And some trees somehow survived in Scandinavia: DNA evidence suggests that Scots pine and Norway spruce (*Picea abies*) lived on through the last glaciation on nunataks in Norway. Palaeolithic (Old Stone Age) *people*, however, had gone south to warmer places – as they do now!

After an exceptionally cold spell at about 18,000 BP, temperatures gradually increased, ice sheets and glaciers started to melt; flora, fauna and people began colonising the bare, wet, rocky landscape of northern Europe. Water from masses of melting ice flowed into the sea and the land lifted because the huge weight of ice dwindled. What are now Ireland and Britain were still joined by low-lying wetlands; Britain was joined to Germany and Denmark by a marshy plain recently christened Doggerland; southern England was separated from Belgium, the Netherlands and France only by the a river we call Fleuve Manche, flowing through a trench to the Atlantic Ocean.

Mesolithic (Middle Stone Age) hunter-gatherers initially migrated into Ireland, Britain and Scandinavia by land when the ice had completely retreated after about 12,000 BP. Those who lived in marshy, riparian and littoral areas led a good existence. There were abundant edible fish and shellfish in fresh, sea and brackish water; abundant large and edible wild animals such as Red deer, Wild boar and seals; abundant edible wildfowl from large cranes to tiny teal; and edible plants.

About 10,000 BP, the marshland link between Britain and Ireland was inundated by the rising sea. This new Irish Sea now separated the island of Ireland from the still-joined Britain and Europe.

During several centuries after 7800 BP, with further changes in relative sea and land levels, the low-lying marshes of Doggerland were gradually overwhelmed by sea-water. People's environments were lost to encroaching seas and at least one tsunami. This catastrophe forced them to walk, swim or travel by home-made boat to drier land. The people who went west or north-west eventually became segregated from humans on the European mainland by the new North Sea.

By 7500 BP, sea levels had risen sufficiently that the Fleuve Manche was flooded by salt water: what we now call La Manche or the English Channel finally separated our fringing, north-west archipelagos from the main continent.

Only about 2000 years elapsed between the last of the ice and the formation of the new archipelago. Some wild plants and animals had migrated from their nunataks and ice-time asyla in south and south-eastern Europe. But once the Irish Sea, North Sea and English Channel had formed, migration and colonisation of wild species into Europe's north-west extremities required sea crossings.

A mere 43 tree species (37 discounting six rare endemics) reached Britain within the joined-up time and fewer – 28 – had reached Ireland (Table 1). And relatively few other wild plants and animals colonised these islands. British and Irish wildlife are marked by their paucity compared with mainland Europe.

Perhaps on account of this, 'native' species are felt as special. People welcome introduced trees into parks, gardens and arboreta but not into their much-loved, rose-tinted countryside! So 'introduced conifers' are anathema to the public whose popular history and culture tells of oak 'Wooden Walls', wands of hazel for dowsing, holly (*Ilex aquifolium*) berries symbolising the winter solstice, ash (*Fraxinus excelsior*) for Irish hurley sticks; and rowan trees to protect Highland dwellings from witchcraft.

In modern times, nature conservation dogma holds that an 'introduced' or 'non-native' species is necessarily of lesser value than a native species. 'Native'

TABLE 1. Native* tree species in Britain and Ireland.

English Name	Latin Name	Britain	Ireland
Alder, Common	*Alnus glutinosa*	+	+
Ash	*Fraxinus excelsior*	+	+
Aspen, Eurasian	*Populus tremula*	+	+
Beech	*Fagus sylvatica*	+	-
Birch, Downy or Hairy	*Betula pubescens*	+	+
Birch, Silver	*B. pendula*	+	+
Cherry, Bird	*Prunus padus*	+	+
Cherry, Wild	*P. avium*	+	+
Crab Apple	*Malus sylvestris*	+	+
Elder	*Sambucus nigra*	+	+
Elm, Wych	*Ulmus glabra*	+	+
Field Maple	*Acer campestre*	+	-
Hawthorn, Common	*Crataegus monogyna*	+	+
Hawthorn, Midland	*C. oxyacanthoides*	+	+
Hazel	*Corylus avellana*	+	+
Holly, European	*Ilex aquifolium*	+	+
Hornbeam	*Carpinus betulus*	+	-
Juniper	*Juniperus communis*	+	+
Lime, Large-leaved	*Tilia platyphyllos*	+	-
Lime, Small-leaved	*T. cordata*	+	-
Oak, Pedunculate	*Quercus robur*	+	+
Oak, Sessile	*Q. petraea*	+	+
Poplar, Black	*Populus nigra*	+	-
Poplar, Grey	*P. canescens*	+	-
Rowan	*Sorbus aucuparia*	+	+
Sallow, Common	*Salix cinerea* ssp. *atrocinerea*	+	+
Sallow, Common	*S. cinerea* ssp. *cinerea*	+	-
Scots Pine	*Pinus sylvestris*	+	?[1]
Strawberry Tree	*Arbutus unedo*	-	+
Whitebeam	*Sorbus aria*	+	+
Whitebeam, Irish	*S. hibernica*	-	+
Whitebeam, Rare Endemics (Scotland)	*S. pseudofennica* *S. arranensis*	+ +	- -
Whitebeam Rare Endemics (England)	*S. vexans* *S. subcuneata*	+ +	- -
Whitebeam Rare Endemics (England)	*S. bristoliensis* *S. devoniensis*	+ +	- -
Wild Service Tree	*S. torminalis*	+	-
Willow, Almond	*Salix triandra*	+	+
Willow, Bay	*S. pentandra*	+	+
Willow, Crack	*S. fragilis*	+	+
Willow, Goat	*S. caprea*	+	+
Willow, White	*S. alba*	+	+
Yew	*Taxus baccata*	+	+
TOTAL		43 (37[2])	28

* Native here means migrating naturally to Britain and Ireland in current interglacial and still present.
[1] Possible survival (Hall 2011).
[2] Excluding rare endemics.

SOURCES: HALL (2011); STACE (2010); FLORA OF NORTHERN IRELAND (2010); NATIVE WOODLAND TRUST HTTP://WWW.NATIVEWOODLANDTRUST.IE/EN/LEARN/IRISH-TREES.

or indigenous species reached and settled in Britain and Ireland without human assistance and before final separation from the European continent.

Ecosystems with introduced species (whose inward migration was facilitated by humans) have been designated as less valuable for nature conservation than 'natural' or 'semi-natural' ecosystems of native species. Wheat or pea fields, Corsican pine (*Pinus nigra*) or Sitka spruce plantations are therefore low in the conservation hierarchy; raised bogs, fens, oak, hazel or lime woodlands are, however, high on the conservation scale.

This conservation emphasis on the native and natural has caused ecological research on conifer plantations to be neglected. It is therefore frequently assumed that modern, planted Sitka spruce forests are much poorer in associated species and ecological dynamics than native woodlands of oak, lime or Scots pine.

Planting with Sitka spruce and other conifers was the British government's answer to the nation's desperate need for home-grown timber and rural renewal during the twentieth century. Experiments had been carried out in Scotland during the nineteenth century to test the growth and suitability of many imported tree species for forestry in the oceanic climate and degraded soils of Europe's Atlantic seaboard. Sitka spruce was the most successful tree in terms of its survival and fast growth as well as its timber structure and functioning. Douglas fir was recommended for more fertile soils, Scots pine for dry, heathery soils and larches for damp sites.

In Germany too, many North American tree species were tested as commercial forest trees in a more continental climate. In 1905, the recommendation was for species from the *east coast* of North America, such as Banks (now Jack) pine (*Pinus banksiana*), Weymouth pine (*P. strobus*) and False acacia (*Robinia pseudoacacia*).

During the middle and later twentieth century, hundreds of millions of little Sitka spruces were planted in Britain and Ireland by government agencies and private landowners. They were needed – in particular – to provide pit props for coal mines, an industry of national importance which powered ships, trains and manufacturing. But timber was also needed for post-war building construction and paper. Other European nations with oceanic coastlines, such as Iceland, Denmark and Norway, also planted Sitka spruce trees.

New tree-plantings were intended to resuscitate existing privately-owned woodlands. British and Irish forests and woodlands had degenerated through many centuries of overuse, over-extraction and over-grazing by stock and deer (heavy grazing reduces growth of seedling trees). They were further devastated by the timber needs of two World Wars. Governments decided that their new plantations should be established on treeless uplands and mountains. The intention was to increase the total area of productive commercial forest and the eventual supply of home-grown timber. Governments also expected the new forests to provide rural employment and to decrease the country's dependence upon imported wood.

During the twentieth century, new plantations appeared in the uplands of Scotland, Wales and northern England. These once-unenclosed, upland landscapes – Britain's classic 'open' landscapes – are appreciated and loved by ramblers, romantics, naturalists, the public – and by farming families who make a living from them! They also happen to be semi-natural ecosystems affected by long-term changes in climate and somewhat degraded by millennia of over-grazing by domestic animals. The hill lands of north-west Britain and of Ireland are unsuited to growing most cereals, so pastoral farming has been their predominant use since prehistory. These 'wild', open, mystic (and misty!) uplands are much-loved by the country's public and diaspora. But they are actually historic living landscapes, farmed and used for the past 6000 years.

Of course Sitka spruce was not the only introduced or native tree species planted. Many other conifers and broadleaved trees had been used in plantations before. European larch (*Larix decidua*), Norway spruce, Corsican pine, and non-native Scots pine were the first, then trees from eastern North America and more recently its compatriots from the Pacific coast temperate rainforest. Native broadleaved trees such as beech (*Fagus sylvatica*) and rowan were also used in plantations.

Sitka spruce became abundant because it particularly suited those types of lands available and produced timber where few other trees would grow. New state forests were planted on the poorest land deemed unfit for farming, mostly in the north and west. It was in these cool, high-rainfall, acid soil, maritime areas that the majority were planted. Private landowners throughout the country grew whatever trees suited their own soils, needs and markets, but again, not on their good farmland. It was not until the 1980s that financial incentives encouraged significant woodland planting by private owners on farmland soils.

'Afforestation' describes the process of planting trees on treeless land. It was necessary to afforest on a large scale in twentieth-century Britain and Ireland because tree-cover was only 5% and 4% of their respective land areas at the start of the century.

To obtain a reserve of timber quickly, hill land was bought by the government's forestry agency. Trees were planted, starting even before the First World War, but quickening in pace from the 1930s to 1980s. New forests of 6000–40,000 ha (15,000–98,840 acres) grew up. The seedling trees were reared in nurseries until 4 years old, planted very close and in large, few-species blocks. So they grew into somewhat even-aged and even-coloured forests which contrasted with the variegated heather, grass and rush moorlands left unplanted around them. At first, there was little emphasis on landscaping, on ecology, on archaeology or natural drainage patterns and hydrology.

The national need for timber took precedence.

Some of the British public were and are convinced that the new state forests devalue landscapes, ecology, and nature conservation. They maintained that the forests were visually inferior to the previous open moorlands, mountains and blanket peat on which the trees were planted. Many have in their mind's eye a picture of landscapes pre-afforestation, and they would prefer them to stay that way, unchanged.

The large twentieth century forests consisted of blocks of deliberately close-spaced, short-lived trees. After 35–50 years' growth and with minimum management, each block was harvested completely. At this age the trees are considered to be mature and commercially most valuable. This contrasts with the typical natural life-cycle of 200–700 years or more for Sitka spruce in temperate rainforests. After cutting down all the trees in a compartment or coup ('clear-felling'), the 'empty' coups were once more planted with a similar set of four-year-old tree seedlings from forest nurseries. The rationale behind these methods is discussed later.

This type of management still takes place, but coups are nowadays smaller and may be planted with several tree species to create a more varied landscape and ecological potential. For about fifteen years of growth, Sitka spruce forests look close-packed, prickly and uninviting. Subsequently, despite some 'brashing' (cutting the lower, dead branches from the trunk), the close-growing trees have a full or 'closed' canopy, with seemingly little flora and fauna. In earlier decades, conifer species were sometimes planted in unsuitable areas, such as on the slopes above Silver Flowe National Nature Reserve in Galloway, southern Scotland. The new forest affected the hydrology of the peat bogs in the valley below.

The results of large-scale afforestation were seen and described negatively by a significant group of the public as – *conifer, introduced, large-scale change, sparse flora, difficult access, dark-green blankets*... People disliked the sudden change away from long-time open landscapes. Deep ploughing had formed corrugations of 'ridge-and-furrow'. But lines of raised soil make for easy planting and tree management, while furrows allow drainage of the very wet ground typical of north-west Britain and much of Ireland.

People felt the amenity value of uplands was being destroyed. Naturalists and ecologists avoided them for study or research, assuming that forests of introduced conifer species could not illuminate major ecological problems. Would attitudes have been different if people had realised that the Sitka spruce trees provided them with everyday and necessary products? Had there been disquiet in past centuries when arable fields were ploughed to produce exactly the same ridge and furrow system on which farm crops (food) were grown?

British and Irish citizens would miss Sitka spruce from their homes whether as roof joists or panels; their kitchens would miss the particle-board units constructed from Sitka spruce. They would certainly miss the unlimited paper available. Even more important, they would miss the bark as mulch for their gardens! Piano players would need new soundboards of another material.

When open spaces in towns and peri-urban Britain were being gobbled up by development in the second half of the twentieth century, people needed somewhere else outdoors to relax. The new state forests were deliberately made available for easy public access via wide tracks, fire-breaks and way-marked routes. Urban people could now visit large forests where they could walk, bicycle or horse ride, set off in their pony and trap, have picnics, watch birds, do orienteering and other activities difficult in car-dominated cities. Despite this, some twenty-first century urban Britons would still be delighted if Sitka spruce were replaced with other trees in the countryside!

First Nations (Native) people of the Pacific Coast of North America regard Sitka spruce through very different eyes. They value it highly for its beauty, spiritual significance and the many items it supplies them. All parts of this tree, from root to crown, are used to make, for instance, canoes, woven bowls and glue. Although to Europeans temperate rainforest is 'wilderness', to First Nations it is where they live their daily lives, make necessary goods, eat and have spiritual attachments.

In contrast to the paucity of conifers in Britain, Sitka spruce in Canada is one of five native species of spruce and one of 31 conifers. And in the USA, Sitka spruce is just one of eight native spruce species and one of over 100 conifers.

Sitka spruce and its rainforest compatriots grow tall and wide, so that one mature tree can provide a huge amount of timber. North American temperate rainforest has been logged (felled) throughout its range during the past century and a half. Tall conifers growing in thick forests are tempting to logging companies, and to politicians who allow or promote logging. So a large proportion of 'virgin' or 'natural' temperate rainforest has been cut-over.

Once a natural rainforest is clear-felled, it is thought to take three centuries for it to re-grow towards something akin to its pre-logging, 'old-growth' ecological status – if some soils remain. Some large stands of temperate rainforest have been selected for protection in a near-virgin status in Canada and in the USA to ensure the continuation of this unique and important living heritage. Examples are the Olympic National Park and the Pacific Rim National Park Reserve.

Researchers in ecology are spellbound by the structure, properties and the ecological processes taking place within temperate rainforests, especially the close relationship between Pacific salmon (*Oncorhynchus* species) and Sitka spruce, and between soil fungi and Sitka spruce. The educational and artistic value of the temperate rainforest and its trees is also well appreciated in North America.

Trees can live long compared with the human life-span. The oldest trees in north-west Europe are a group of Norway spruce, growing at 910 m (2985 ft) on Fulu Mountain, Dalarna province, Sweden, which are 9500 years old. The oldest trees in North America are California's Bristlecone pines (*Pinus aristata*) which are nearly 5000 years old. Nothing quite like these has been found yet in the temperate rainforest, but its conifers, including Sitka spruce, are overwhelming in other ways.

For instance, Sitka spruces are among Canada's tallest trees. One of the biggest is at San Juan on Vancouver Island, 62.5 m (205 ft) high and 3.7 m (12 ft) in diameter at breast height. In fact, Sitka spruce is thought to be the fourth tallest tree in the world. The oldest Sitka spruces we are certain about in North America are about 800 years old, but they are still there after they die – more on that later…

During post-glacial British and European prehistory, timber and wood of a chosen size and shape were *selected* from what was available in the widespread forests. They were cut with axes and shaped with adzes to build boats, for home construction and to provide the uprights for wooden henges (circles) in the landscape or seashore. Methods such as coppicing, shredding and pollarding were used to remove parts of a tree, for small wood, branches and twigs; the parent tree was left to grow. Spear hafts, woven wood pathways, well-linings, and leafy branches for stock to eat were some of the other early products.

Modern silviculture is more *proactive*: whole trees and forests are deliberately grown and harvested with the end-product in mind from the start. Parent trees are not part-used and left to grow. In the UK especially, short-lived, even-aged plantations produce the bulk of our needs. The forestry methods of the eighteenth–twenty-first centuries, where areas are completely cleared and then replanted, have sparked concern and debate in many countries. However, the positive result of this concern and its resulting debate has been progress towards more and different methods of planting, growing and harvesting trees. There are now attempts to reduce the speed and scale of landscape change, take more account of archaeology, hydrology and nature conservation and, especially where afforestation takes place, allow time for complex ecosystems to develop.

The Atlantic seaboards of north-west Europe have their own – tiny remnants of – temperate rainforest in maritime climates. The North Atlantic Current (a branch of the Gulf Stream) ensures mild, humid climates while regular low pressure systems arriving from the Ocean give frequent, high rainfall and strong south-west winds. Average annual rainfall on the south-west coast of Scotland is up to 1800 mm (68 in) and along the north-west coast, up to 4577 mm (180.2 in).

The tree flora is poor compared with North American temperate rainforests of similar latitudes. However, they contain lush growth of many species of lichens, mosses and liverworts growing on the trees, a characteristic feature of temperate rainforests.

Ireland is almost always damp, humid and therefore very green! It receives up to 2000 mm (79 in) of rainfall annually, distributed between about 250 days. Temperatures vary from 6°C in winter to 15°C in summer and frost is rare. Few native woodlands remain in Ireland, but pollen analysis suggests their likely tree species were sessile oak, elm and yew as the main trees, with hazel, birches (*Betula* species) and Common alder (*Alnus glutinosa*); ferns were abundant. Sessile oak is nowadays the dominant tree of the Irish rainforest remnants; the 1200 ha (2965 acres) in Killarney National Park, Kerry, which supports prolific algae, fungi and ferns is the best example.

Along the west coast of Scotland, where the climate is similarly maritime but soils generally less fertile, remnants of temperate rainforests are dominated by hazel, Sessile and Pedunculate oaks, Silver (*Betula pendula)* and Downy birch (*B. pubescens*). These trees support large numbers and amounts of epiphytic lichens, mosses and liverworts. The same groups of plants cover the ground below the trees in a luxuriant lushness of greens.

The indented, Atlantic, coastline of Norway has a long growing season and receives over 2000 mm (79 in) of rainfall annually, falling throughout the year, keeping the coast and fjords constantly humid. From latitude 62° to 67° N the coast supports tiny areas of coniferous temperate rainforest growing stunted Norway spruce with Grey alder (*Alnus incana*), Downy birch and aspen; there is rich lichen, moss and fern flora, especially on old trees. From 58° to 62° N, the rainforest is deciduous: Common alder, ash (*Fraxinus excelsior*), oaks, Small-leaved lime (*Tilia cordata*) and Wych elm (*Ulmus glabra*). There is an extraordinarily rich flora of epiphytic lichens growing on these trees.

Sitka spruce and its compatriot conifers of North American temperate rainforest have joined the vegetation of Europe. I hope this book will encourage wide appreciation of Sitka spruce for its contribution to the forest ecosystems and First Nations culture of Pacific Coast North America, and to the landscapes, ecology and economies of the UK, Republic of Ireland and north-west Europe.

CHAPTER TWO

'The Tree from Sitka'

Some native spruces have been named the provincial tree in three Canadian provinces – the white spruce in Manitoba, the red spruce in Nova Scotia and the black spruce in Newfoundland and Labrador.

Source: www.cwf-fcf.org

Alaska designated Sitka spruce (*Picea sitchensis*) as the official state tree in 1962. Named for Sitka Sound in Alaska, the Sitka spruce is the tallest conifer in the world. Moist ocean air and summer fog are the main factors that account for Sitka spruce's large growth. Sitka spruce trees provide good roosting spots for bald eagles and peregrine falcons. Deer, porcupines, elk, bear, rabbits, and hares browse the foliage.

Source: www.statesymbolsusa.org/Alaska/tree_sitka_spruce.html

Conifers and broadleaves

Conifers such as spruces belong to the *Gymnosperms* which first appeared in the late Carboniferous period about 300 MYA when large horsetails and ferns became fossilised and formed coal. Gymnosperms dominated the Earth's vegetation at times until about 65 MYA when many had become extinct.

Conifers are now the biggest group of gymnosperms on Earth. Spruces (*Picea*) along with Silver firs (*Abies*), cedars (*Cedrus*), larches (*Larix*), hemlocks (*Tsuga*), Douglas firs (*Pseudotsuga*), pines (*Pinus*) and three small genera make up the family *Pinaceae*.

Most broadleaved trees belong to the *Angiosperms* or flowering plants, which are the most recent group of plants in geological time. They originated about 140 MYA in the early Cretaceous period and diversified into many new forms concurrently with insects and dinosaurs.

Nowadays, the angiosperms – which we call 'flowers' – are the most numerous of Earth's flora, with 250,000 to 320,000 species classified into 12,000 genera in about 400 families. In comparison, there are nowadays 850–1000 species of gymnosperms, classified into about 80 genera in fifteen families.

'Trees' are tall plants with one woody stem. 'Wood' is made of lignin, a complex of polymers, which is deposited in the cell walls of land plants; it is supportive and strong so that a tall plant can remain upright. 'Bushes' or 'shrubs' are also woody, but are usually multi-stemmed and smaller than trees.

Conifers are nowadays mostly evergreen and their leaves are hard needles or overlapping green scales. Their seeds develop without any protection, on the scales of female 'cones' which later grow big and woody; smaller, male cones bear pollen.

FIGURE 2.1. Signing of the Act designating Sitka spruce as Alaska State Tree, 1962.

PHOTO: COURTESY ALASKA STATE LIBRARY, DORA M. SWEENEY PHOTO COLLECTION, P421-574.

FIGURE 2.2. Heavily-coning Sitka spruce close to the town of Sitka, Alaska.

PHOTO: ANDY TITTENSOR, 2014.

Broadleaved angiosperm trees produce thin, flat leaves, which are seasonally deciduous in the north-temperate zone of earth. They mostly produce showy flowers and their seeds develop, protected, shut inside a fruit and encircled by the petals.

Origin of its name

Haida First Nation

The First Nations who have lived with Sitka spruce for the last several millennia gave this tree its earliest names (Table 2). Haida called it '*kíid*' plural '*kíidaay*' in their distinct language.

TABLE 2. Vernacular names for Sitka Spruce.

Language/Nation	Names for Sitka spruce
English (Canada and USA)	Airplane spruce, British Columbia sitka-spruce, Coast spruce, Menzies spar, Menzies spruce, Sequoia silver spruce, Sitka spar, Sitka spruce, Silver spruce, Tideland spruce, West Coast spruce, Western spruce, Yellow spruce[1]
First Nations	Kíid, kíidaay[2] (Haida); Shéiyi[2] (Tlingit); Naparpiaq[14] (Chugach, Alutiiq)
French Canadian	Épinette de Sitka, Épicéa de Menzies[3]
Canadian Gaelic	craobh spruiseadh ['spruce tree'][4]
English (UK and Éire)	Sitka spruce
Irish Gaelic	Sprús Sitceach[5]
Manx Gaelic	Jus ny Twoaie[6]
Scots	Spruch, Sprus, Sprush ['spruce tree'][7]
Scottish Gaelic	spruis Sitka[8]
Welsh	sbriwsen Sitca, spriwsen Sitca (s); sbriws Sitca (pl.); pefrwydden Sitca [floras and dictionaries][9]
Austria	Sitka-fichte[10]
Czech	Smrk Sitka[11]
Denmark	Sitka-gran[11]
Estonia	Sitka kuusk[11]
Finland	Sitkankuusi[11]
France	Épicéa de Sitka[12]
Germany	Sitkafichte[11]
Iceland	Sitkagreni[11]
Netherlands	Sitkaspar[11]
Norway	Sitkagran[11]
Poland	Świerk Sitkajski[11]
Portugal	Espruce-marítimo[11]
Russia	Ель ситхинская[13]
Slovakia	Smreka Sitka[11]
Sweden	Sitkagran[11]

Sources

1 Canadian floras and web sites.
2 www.sealaskaheritage.org/Haida%20curriculum/PDFs/SPRUCE%20TREES/Spruce_haida_booklet.pdf and Ishmael Hope, Sealaska Heritage Institute.
3 Jeannine Moe, Halifax, NS, Canada.
4 C. Bould, School of Scottish Studies, Edinburgh University.
5 Noel Melanaphy, Forest Service Northern Ireland.
6 John Walmsley and Colleagues, Isle of Man.
7 Dictionary of the Scots Language, www.dsl.ac.uk.
8 Catriona McPhee, Skye, Scotland.
9 W. Linnard, Cardiff, Wales.
10 Stefan Fellinger, Sandl, Austria.
11 Liber Herbarum II www.liberherbarum.com/Pn3943.HTML
12 www.plantnames.unimelb.edu.au/new/Sorting/Picea.html.
13 Dumbleton, C. W. (1964) *Russian-English Biological Dictionary*, London. Oliver and Boyd.
14 John F. C. Johnson, Chugach Alaska Corporation, Anchorage.

Tlingit First Nation

A *Kwáan* is the group of a First Nation which inhabits a region and uses the surroundings, water and resources. Of the twenty *Kwáan* of the Tlingit First Nation, the *Sheey At'iká Kwáan* or *Sheet'ká Kwáan* described the people 'on the outskirts, the edge or beside Sheey'.

'Sheey' means 'branch' and it was the Tlingit name of an island in Southeast Alaska. Europeans, probably Russians, transliterated the name from *Sheey* or *Sheet'ká* to *'Sitka'* or *'Sitcha'*. The Tlingit derived their name for Sitka spruce from *Sheey* the name of the island: *'Shéiyi'* or 'the tree from the outskirts or edge of the sea'. Similarly, Russians called it 'the spruce from Sitka'.

In 1805 AD, the Russians changed the indigenous name of the island Sitka to 'Baranof Island', although Sitka was still sometimes used. Nowadays, 'Sitka' refers only to the city on Baranof Island and to our special tree. (Thanks to Ishmael Hope and Tlingit Elders for explanation of the name).

English Vernacular

In Middle English (the vernacular of England between 1100 and 1500 AD) *'pruce'* meant something 'from Prussia' (Prussia was the territory now north-east Poland, Lithuania and Belarus). *'Spruce'* probably developed from 'pruce' and was used for the Common or Norway spruce native to north-east Europe, Scandinavia and western Russia. Britain and Ireland have no native spruces, but about 1500 AD they introduced Norway spruce. Another early item 'from Prussia' was 'spruce beer', a jigsaw piece in the Sitka spruce story.

Latin

Two words in Latin provide an internationally agreed name for every known plant and animal. For instance, the name *Picea* describes all spruces, while *Picea sitchensis* describes only Sitka spruce; *Picea abies* is Norway spruce and *P. glauca* is White spruce. The word *Picea* is said to derive from early Indo-European *'pik'* or *'pi'* meaning sap or resin, possibly cognate with Ancient Greek 'pitch.' *Picea* could equally derive from *'pike'*, *'pick'* or *'pic'* describing anything sharp and pointed – from a pickaxe to a woodpecker's beak or spruce needles.

In 1741 AD, German physician-naturalist Georg Steller was taken on the Russian Great Northern Expedition, which set sail from the Kamchatka peninsula to seek north-west America. He stepped onto a small Alaskan island and pushed his way through its thick forests which he described as 'spruce' in his journal.

In 1792 Scotsman and surgeon-naturalist Archibald Menzies was on board Captain Vancouver's ship *Discovery*, checking at stopping-places along the coast of British Columbia for a needle-leaf tree from which to make the sailors an anti-scorbutic drink. He found several conifers for the purpose, including what we now call western hemlock and Sitka spruce. Busy with sailors' health he had no time for the niceties of naming new trees, but sent specimens back to London for others to study.

FIGURE 2.3 *(left)*. Amongst the redwoods of California lives the tallest living Sitka spruce at 96.92 m (318 ft).

PHOTO: STEPHEN SILLETT, COURTESY OF INSTITUTE FOR REDWOOD ECOLOGY.

FIGURE 2.4 *(right)*. Sitka spruce is one of the tallest conifers in the world: these old trees are in Inverliever Forest, Argyll.

PHOTO: RUTH TITTENSOR, 2013.

Early in the following century, an American expedition led by Meriwether Lewis and William Clark set out to establish American claims to the centre of the continent and, particularly, the west coast especially around the Columbia River. In 1806, Meriwether Lewis discovered Sitka spruce in what is now Oregon and gave it a recognisable description. Not being a botanist he called it by the generalised name for conifers: 'fir'.

In 1824, Albert Dietrich, a German professor of botany, started the formal process of naming in Latin, by calling all spruces *Picea*.

Scots gardener David Douglas found Sitka spruce in 1825 around Puget Sound. He described it in detail and called it *Pinus menziesii,* to honour his earlier compatriot. However, the botanical description was not published until 1832 in the UK.

German Karl Mertens was surgeon-botanist aboard the Russian survey ship 'Senyavin' between 1826 and 1829. He collected thousands of plant specimens from the Aleutian islands and coastal Alaska and sent them to his compatriot August von Bongard, working in St Petersburg. In 1827 Herr Bongard received conifers from Baranof Island and named one after the original name of the island: *Pinus sitchensis* (see Bongard, [M] 1831/1833).

His name took precedence over David Douglas's *Pinus menziesii* because it reached publication first. Albert Dietrich, still studying spruces in 1832, confused things with another Latin name: *Picea rubra*.

French botanist Élie-Abel Carrière realised that *Pinus sitchensis* was actually a spruce, not a pine, and in 1855 published his name as *Picea sitkaensis*. Soon after, international botanists (thankfully!) agreed and accepted the Latin name of: *Picea sitchensis* (Bong.) Carr. Messrs Bongard and Carrière, authors of the internationally-accepted names are called the Species Authorities.

The species name *sitchensis* is the Latin derivation from *Shéiyi'* or *Sheet'ka* and Sitka or Sitcha, the tree from the ocean-side of Sheey.

Description

Size and age

In good growing conditions, Sitka spruce is the most imposing and largest, in height and girth, of all spruce species on Earth. It is the third-tallest of all

FIGURE 2.5. This old Sitka spruce was felled because it risked falling in a public area of Sitka National Historical Park.

PHOTO: ANDY TITTENSOR, 2014.

FIGURE 2.6. Sitka spruce trees typically buttress at the base as they grow.

LEFT: PHOTO: ANDY TITTENSOR, TONGASS NATIONAL FOREST, 2014. RIGHT: PHOTO: SYD HOUSE, OLYMPIC NATIONAL PARK, 2010.

FIGURE 2.7. This big Sitka spruce grows in Benmore Botanic Garden, Argyll where an annual average rainfall of 2384 mm (94 in) suits it and the bryophytes living on it.

PHOTO: ANDY TITTENSOR, 2009.

conifers: only Coast redwood (*Sequoia sempervirens*) and Coast Douglas fir (*Pseudotsuga menziesii* var. *menziesii*) can grow taller.

A common height of a mature Sitka spruce tree in North America nowadays is 38–53 m (125–175 ft) high and a common diameter at breast height is 1–2 m (3–6 ft). However, they can often attain a height of 50–73 m (164–229 ft) and 2–4 m (8–12 ft) diameter. *Really* big Sitka spruce reach 76–80 m (250–262 ft) high and trunk diameter 5–7 m (16–23 ft). Many even bigger trees were probably felled for industry during the past two centuries before large sizes were recorded.

A few tall Sitka spruce grow on Vancouver and Queen Charlotte Islands (Haida Gwaii) in British Columbia, on the Olympic Peninsula, Washington, and amongst the coastal redwood forests of California. The tallest Sitka spruce on the Olympic Peninsula is said to be 80 m (262 ft) high. The Carmanagh Giant on Vancouver Island, said to be the tallest tree of any kind in Canada, is 96 m (315 ft) high. In contrast, on the sub-arctic coast of the Gulf of Alaska, Sitka spruce grows much slower and rarely so tall. In the misty, oceanic climate of Southeast Alaska it grows slowly, but eventually becomes tall and large-girthed: a specimen 76.2 m (250 ft) high was recently discovered.

Seven hundred to 800 years old is a good age for Sitka spruce in its southern range, with 300–500 years being more common. In the slower-growth conditions of Alaska it lives several centuries longer, but exact dating is difficult as big trees often have heart rot. For comparison, Douglas fir typically reaches 750 years, and 1200 years occasionally, while Western red cedar grows for 1500 years.

Shape

Young Sitka spruce trees growing alone outside forests have a broad conical shape with spreading lower branches near the ground. Because branches form regular whorls round the main trunk each year, a young tree can be aged by counting the whorls. As they mature, trees often become a tall spire.

On close-grown plantation trees, the early whorls die leaving the lower trunk with dead branches and a compact green crown.

Foliage and bark

The young shoots are soft and pale pinky-brown but the older twigs turn orange-brown, grooved and knobbly. The leaf buds are 5–10 mm (0.2–0.4 in) long, pale brown to start with and then purplish.

You can recognise Sitka spuce by its very hard, stiff, needles which grow from pegs on the twigs and have a keel below. Needles are 1.5–3 cm (0.6–1.2 in) long, with an extremely sharp point, making the foliage very prickly to handle: grab hold of a spray and find out! The needles are bright green on their upturned sides but whitish-blue with a glaucous sheen underneath. This gives the Sitka spruce a characteristic blue-green colour. The pegged needles seem to be in two rows making the shoot flat, mimicking silver firs.

FIGURE 2.8 *(below left)*. The bark of a young Sitka spruce, Gwydyr Forest, Caernarfonshire.

PHOTO: ANDY TITTENSOR, 2013.

FIGURE 2.9 *(below right)*. The bark of an old Sitka spruce, Benmore Botanic Garden.

PHOTO: ANDY TITTENSOR, 2009.

FIGURE 2.10. Cross-section of a trunk showing the annual growth rings.

PHOTO: ANDY TITTENSOR, 2014.

The bark of young trees is thin, smooth and grey-brown. On older trees it flakes into purple- or silvery-brown, roundish plates. They lie loosely on the surface of the trunk until a tree gets even older when they lift and make the trunk rough.

In its natural forest, the lower trunk of a mature tree is clear of branches and the base often enlarges into huge buttresses. The trunk and branches provide living places for a mass of 'epiphytes' – plants using the tree surface as a foundation on which to live. These can be mosses and liverworts, lichens or ferns, which all take in moisture through their surfaces and benefit from the high humidity of the rainforest environment.

Roots

The roots of Sitka spruce are very important organs. Bigger, structural roots hold the tree in the ground, while fine roots take in water and nutrients for the tree to grow. Roots growing 12.2 m (40 ft) downwards into caves have been observed in the karst landscape of Prince of Wales Island! Plantation trees have roots going down only 1–2 m (3.3–6.6 ft) but they spread horizontally below the soil surface as a large, shallow root-plate.

Its roots are also important as the organs which host over 100 species of fungi within or on their surfaces, fungi known as 'mycorrhizae'. They colonise the roots of Sitka spruce and other tree seedlings. The mutual relationship between mycorrhizae and trees assists trees and their seedlings by enhancing water and nutrient uptake; it assists the fungi by providing nutrients made by the tree. Mycorrhizae on Sitka spruce roots are most numerous when the tree canopies close with one another and when the litter on the forest floor is deepest.

The unseen by us, underground portion and processes of Sitka spruce trees are as important as what we can see above the ground.

Wood

The wood of Sitka spruce is soft, light-weight and strong compared with its weight. It is straight-grained and non-porous. The heart-wood is pale pinkish-brown or yellow-brown, while the sapwood is creamy-white. It is easily worked, producing a fine, smooth surface. The trunk wood or 'timber' is an important product to modern societies, used for paper and construction.

Cones and seeds

Once the tree has reached 20–40 years old, female cones grow at the ends of main branches in the top one-third of a tree. They are small and red, turn yellow-green then brown when ripe. The male cones of Sitka Spruce develop on branches lower in the tree; they are blunt and ovoid, yellow with pollen, then turn orange-brown. To see the young cones closely, you need to climb the prickly tree or else to ask a kind forester to cut it down!

FIGURE 2.11. As a Sitka spruce is felled in Inverliever Forest, Argyll, the male cones shed their pollen in a cloud.

PHOTO: RUTH TITTENSOR, MAY 2013.

FIGURE 2.12. Close view of ripe male cones.

PHOTO: ANDY TITTENSOR, 2013.

FIGURE 2.13. Young female cones.

PHOTO: RUTH TITTENSOR, 2014.

Pollen is shed in spring and if lucky enough to be blown onto a female cone, fertilises an ovule: cells within a pollen grain and ovule join to form an embryo. By summer, the embryos have become seeds which ripen in the autumn five to seven months after pollination.

The cone scales are rounded, thin and papery, with ragged margins and small bracts. In North America, the cones open in late fall, releasing their seeds; cones drop onto the ground from then until early winter. In Britain and Europe the cones open earlier, release their seeds and drop, complete, throughout the winter and following spring, though some may remain on the trees for several years. The mature cones are cylindrical, 5–10 cm (2–4 in) long and 2 cm (0.78 in) broad, but when they open they reach 3 cm (1.2 in) across. The seeds are dark red-brown, 2–3 mm (0.08–0.12 in) long and have a wing 7–9 mm (0.28–0.34 in) long to help them blow away.

In its native habitat, Sitka spruce produces about 20 seeds per female cone and a tree may have about 1000 cones, so about 20,000 seeds per tree per year. In a lifetime of, say, 300 years, this adds up to 6 million seeds. Of course, in a tree's lifetime, only *one* seed has to germinate in a suitable place, grow and reach maturity for the parent tree to have reproduced successfully. In good conditions, a seed, attached to a delicate wing, can travel up to 1 km (0.6 miles) from the parent tree; the species travels slowly.

Distribution

Sitka spruce is a magnificent tree yet geographically the most restricted of all seven American spruce species. It grows naturally along 3600 km (2237 miles) of the Pacific coast of North America, from California to Alaska. It is abundant upwards to about 500 m (1640 ft) altitude although it can grow up to 1158 m (3800 ft) on mountains and further inland where conditions are suitable. It is particularly tall and abundant on Haida Gwaii and the west side of Vancouver Island.

Average annual rainfall over 1000 mm (39.37 in) is really the minimum needed for it to grow well. Where it flourishes in British Columbia, average annual rainfall is 2378 mm (94 in) and there is precipitation on a mean of 217 days annually. On western Vancouver Island rainfall is 3016 mm (119 in)

FIGURE 2.14. A felled Sitka spruce in Inverliever Forest showing the developing green or pink seed-cones and smaller brown male cones.

PHOTO: RUTH TITTENSOR, 2013.

FIGURE 2.15. Developing seed cones of Sitka spruce.

PHOTO: RUTH TITTENSOR, 2013.

FIGURE 2.16. Superficial root-plate of a plantation Sitka spruce after windblow; Whitelee Forest, Ayrshire.

PHOTO: RUTH TITTENSOR, 2008.

annually. These ocean-side places have a 'maritime' climate with over 200 days frost-free. In lower rainfall areas, such as California, coastal fog supplies the necessary high humidity in summer. Although Sitka spruce can grow in temperatures down to -20°C (-4° F) in Alaska, the young shoots may be killed by spring frosts so a mild, damp maritime climate is ideal.

Sitka spruce grows very well along shorelines in the salt spray zone, along inlets and estuaries, and in river flood plains, even where flooding may continue for several weeks. It is considered an important riparian species because it contributes considerable organic material to water systems, while tolerating a high water-table. In mountain ranges further inland, Sitka spruce grows best in damp valleys.

In deep, moist, soil, with a good supply of phosphate, potassium and nitrogen it grows well. A soil of alluvial gravel seems to produce the best trees. An acidity of pH 4.0 is typical in British Columbia. However, it does grow on a wide range of soils throughout its natural range. In the maritime climates of north-west Europe it is grown commercially on many soils including deep, acid, peat, but exposure to high winds affects it stability.

Sitka spruce seedlings grow slowly for the first four or five years, but can then put on one or one-and-a-half metres of growth a year. The main growth period is from May to August and there may be a second period of growth during September, especially in damp soils.

Genetics

Sitka spruce, like all spruces, indeed all the pine family, has 24 chromosomes per cell, and sometimes extra, floating, 'B' chromosomes which affect the date a tree starts to produce cones in spring. It is an out-breeding species with obvious differences between trees from different parts of its long latitudinal range. Individuals differ, for example, in height and growth form, susceptibility to low temperatures, dates when they start to grow annually, foliage colour and anatomy.

Hybridisation

Growing mainly between mountains and ocean, Sitka spruce overlaps geographically with only a few other species of spruce in western North America.

Along the coasts of South-Central Alaska and on the Kenai Peninsula near Anchorage, genetically-pure stands of White spruce and Sitka spruce grow naturally. Hybrid swarms – in which individual trees contain genes of both species and show a range of traits between the two parent species – have been found nearby by alert researchers.

Along some inlets of the British Columbia coastline, the ranges of Sitka spruce and White spruce overlap again. Here, the two species hybridise especially along the Bulkley, Nass and Skeena rivers. The hybrids have been named *Picea* x *lutzii* Littl.

Engelmann spruce (*P. engelmannii*) is a more inland species, so localities available for hybridisation with coastal spruces are few. However, from about 50° to 60° N and on the eastern margin of its range, Sitka spruce populations just about overlap with both White and Engelmann spruce There has been 'introgression' of genetic material between all three species. This involves mating between individuals of each species to produce hybrid offspring. The hybrids later 'back-cross' with individuals of any of the three parents to produce a wide variety of types which may show features of any or all the parent species.

Chapter Three teases something of the shadowy past of Sitka spruce from the incomplete story of spruce-tree evolution and colonisation on several continents.

FIGURE 2.17. In a gap in Inverliever Forest, Argyll, this comely Sitka spruce is surrounded by youngsters.

PHOTO: RUTH TITTENSOR, 2013.

CHAPTER THREE

Origin, Migration and Survival on the Edge

> With huge volumes of water locked up in ice sheets, global sea level was about 120 m lower than present at the Last Glacial Maximum (LGM), exposing large expanses of continental shelf and creating land bridges that allowed humans, animals, and plants to move between continents. Migration from eastern Russia to Alaska, for example, was possible via the Bering land bridge.
>
> Jamie Woodward 2014

Origins

Fossil leaves, stems, wood, cones, seeds and pollen can be identified to genus or even species by paleobotanists as clues to the geological history of trees.

Early gymnosperms, which appeared in the Carboniferous period about 300 MYA, included cycads, gingkos, and conifers resembling modern Monkey-puzzle trees (*Araucaria araucana*). Many new gymnosperm species evolved during the next 50 million years and, despite a mass extinction of life on Earth about 251 MYA (when 70% of land life disappeared), many species, including conifers, appeared in the following 52 MYA of the Triassic period. Becoming abundant, they dominated Earth's vegetation during the later, warm Cretaceous period (145–65 MYA) when herbivorous dinosaurs munched on the Monkey-puzzles.

Meantime, angiosperms had appeared and by the end of the Cretaceous period, reached par with gymnosperm species. About this time there was mass extinction of all dinosaurs, most marine life and many monkey-puzzle type conifers.

So by the early Tertiary period (about 65–35 MYA), angiosperms, with their huge variety of species, advanced, insect-pollinated flowers and mainly broad leaves reached numerical and ecological supremacy on Earth.

Fossils related to modern trees were not common in North America until the beginning of the Oligocene epoch, about 35 MYA. *Pinus*, *Tsuga*, *Sequoia*, *Alnus*, *Quercus* and *Ulmus* (pine, hemlock, redwood, alder, oak and elm) and – *Picea* (spruce of some kind) – grew at latitude 69° N in what is now the Mackenzie Delta in the far north west of Canada. It is uncertain whether spruces originated in North America or Asia and how they spread between those continents and Europe.

History of Spruces in North America

Between 35 and 10 MYA mixed coniferous forests at high latitudes and altitudes expanded south and downhill where cooling conditions suited them. They consisted of a great variety of conifers including pines, firs, cypresses, spruces plus angiosperm trees such as Snowy mespil (*Amelanchier lamarckii*) and chestnut (*Castanea* species). Spruce fossils and microfossils at many studied sites showed that spruces extended over a big expanse of North America and in varied conditions at this period.

Starting about 2.8 MYA (the Pleistocene epoch), the Earth's northern hemisphere cooled and plant diversity decreased. Deciduous angiosperm forests contracted especially from higher latitudes, to be replaced by tundra and expanding coniferous forests. Grasslands increased. Deciduous gymnosperms related to *Ginkgo* and *Metasequoia* and even some angiosperms disappeared. Ice started to form in the Arctic region.

Nevertheless, species typical of temperate climates survived, especially in western North America. These included the coniferous evergreen forests containing *Picea* and *Pinus* (the most abundant), *Abies*, *Tsuga* and either *Larix* or *Pseudostuga* species. The vegetation of Alaska was now mainly coniferous forest, although in the Bering (Sea) locality there were low willows, shrubs of the rose family, grasses and sedges.

Broadleaved deciduous forests gradually became restricted to southern and eastern North America where descendants still survive. In the north and west they were replaced by forests of many conifers. In Middle America they were replaced by prairie vegetation.

This was the evolutionary stage and geographical distribution that the vegetation of North America had reached when ice sheets spread out to lower latitudes from the Arctic. Today's genera of gymnosperms and angiosperms had evolved. We would recognise many of the plant species, but their geographical distribution was different from today. Sitka and other spruces were widespread in evergreen, multi-species conifer forests in the north and west of the continent.

The Pleistocene epoch lasted until about 12,000 BP; there were big, fast, geological and climatic changes. Four severe ice ages were each separated by a short, warmer 'interglacial' time. The rate of evolution reduced during ice times. Populations of flora and fauna rearranged and redistributed themselves as Earth's climate altered significantly and northern lands disappeared under ice or became polar-desert with little vegetation. Many gymnosperms became extinct, leaving few on Earth today. Nevertheless, some still live here, including spruces.

Sitka spruce during the Ice Ages

There was climate change in the northern hemisphere as a whole, and not just on and around the ice sheets. The North American jet stream split into two, fierce winds and storms affected the southern USA, and the seasonal patterns of rain and storms altered. Sea level went up after ice melted, and went down when the land re-bounded or ice formed again. Temperatures in the oceans as well as on land dropped, and patterns of water vapour in the atmosphere changed.

The landforms of North America and northern Europe were re-moulded by extensive and deep ice sheets in the past two million years. Thick, moving ice scraped away much evidence of the plants and animals which were alive before that.

Ice age processes left a legacy: the current species and their distributions across northern continents. Plant geographers, palaeo-biologists and archaeologists have worked at piecing together the post-glacial story of temperate rainforests and their species including Sitka spruce.

The fourth and latest, main glacial period is called the 'Wisconsin' in North America and the 'Devensian' or 'Baltic-Scandinavian' in Europe.

The Wisconsin alias Devensian ice lasted from 110,000–10,000 BP: 100,000 years of ice cover! At their maximum, the ice sheets came as far south as

FIGURE 3.1. Nunatak pokes up from the Greenland ice sheet, 2009.

PHOTO: NICOLAJ K. LARSEN, AARHUS UNIVERSITY.

FIGURE 3.2. Sitka spruce rainforest between ice and ocean: below Mendenhall Glacier and around Auke Bay, seen from Saginaw Channel, Alaska.

PHOTO: ANDY TITTENSOR 2014.

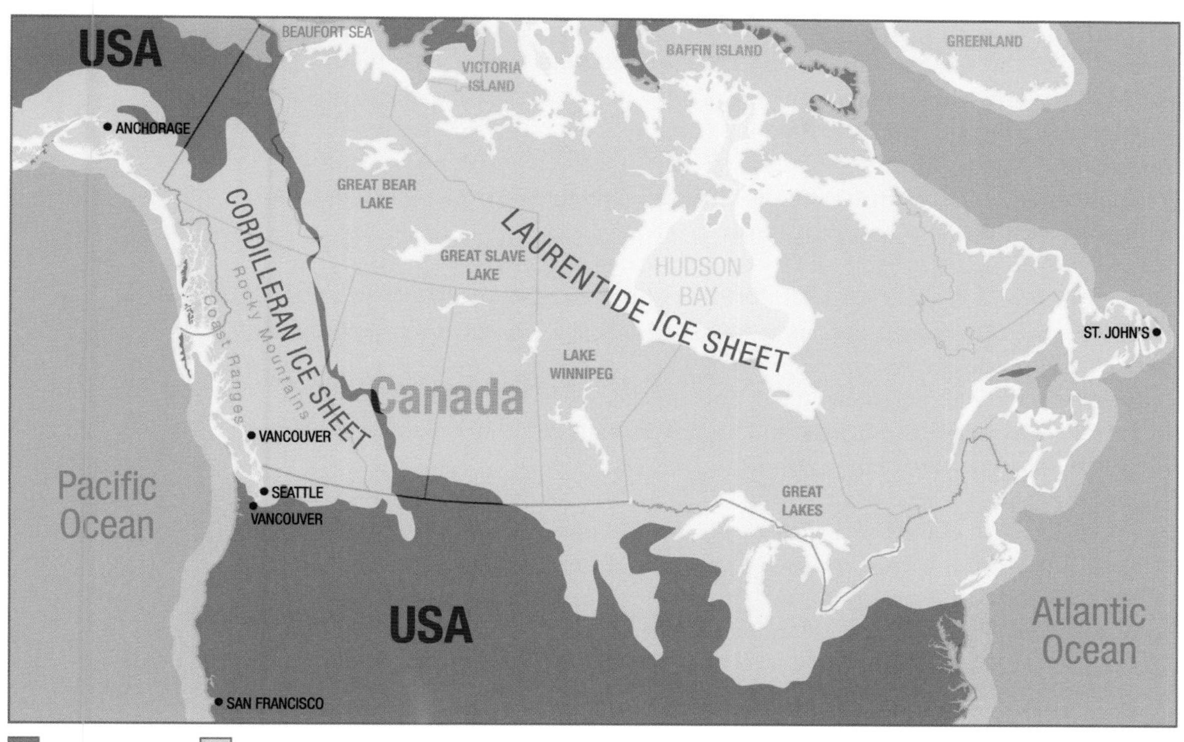

MAP 3. Wisconsin ice sheets. Prepared by Susan Anderson.

AFTER RITCHIE (2003); WYNN (2007); DIGITAL GEOLOGY OF IDAHO: HTTP://GEOLOGY.ISU.EDU/DIGITAL_GEOLOGY_IDAHO/MODULE12/MOD12.HTM AND HTTP://GEOLOGY.ISU.EDU/DIGITAL_GEOLOGY_IDAHO/MODULE13/MOD13.HTM.

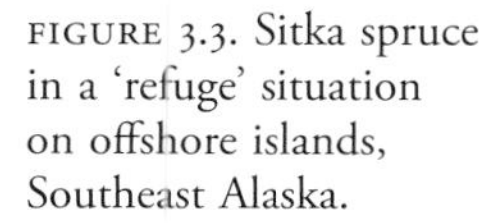

FIGURE 3.3. Sitka spruce in a 'refuge' situation on offshore islands, Southeast Alaska.

PHOTO: ANDY TITTENSOR, 2014.

Seattle-Portland in the north-west USA and to south of the Great Lakes in the east (Map 3). In Britain and Ireland ice extended, at its maximum, from the Northern Isles southwards leaving only the south and south-east ice-free. Ice also covered the Irish, Celtic and North Seas, all of Scandinavia, the Baltic States, Baltic Sea and western Siberia. Much of North America and northern Europe lay beneath 2–3.5 km (1.5–2 miles) of ice.

The 'Laurentide' ice sheet covered most of continental Canada and some of the USA. The smaller 'Cordilleran' ice sheet covered the western tenth of Canada, the very north-west tip of the USA and coastal Alaska. It extended tongues of glaciers out over the Pacific Ocean, an important piece of the Sitka spruce story. Alaska, apart from its coast, was – amazingly – free of ice because the climate there was arid (no water to form ice). Small plants typical of tundra managed to survive in places in its polar desert.

Sea levels around Europe and North America were 100–120 m (328–394 ft) lower than now. So a Beringian land bridge connected Siberia and Alaska. The North American continent had an exposed coastal shelf along what is now the Pacific Ocean's eastern margin.

Fungi, bacteria and possibly algae and lichens may have lived in and on the frozen snow and ice, but most plants and animals could not survive in the extreme cold, ice-covered environment.

But there were ice-free places within and close to the main ice sheets: Sitka spruce and other conifers managed to persist in some of these life-sustaining refuges. *Where* Sitka spruce survived during the fourth glaciation and where and *how it colonised* since the ice melted at the start of this interglacial period were *important to its present distribution pattern.*

Survival on the edge

When water was locked up into ice, sea level was lower than now, leaving a persistent corridor of ice-free land along the Pacific coast down to 100 metres (328 ft) below present sea-level: there are still remnants of underwater forests there.

The following coastlines gave ice-free habitation to some plants and animals:

- *Alaska*: the outer coast of Glacier Bay of Dall Island; Kodiak Island and part of Prince of Wales Island; the seaward edges of Baranof Island; 100 km (*c.*62 miles) length of shoreline along the southern Yakutat Foreland.
- *British Columbia*: Brooks peninsula on northern Vancouver Island and the seaward edge of Haida Gwaii.
- *Washington*: south and west of the Olympic Mountains.

Sitka spruce survived just beyond the south-west corner of the Cordilleran ice sheet in several locations near the coast. Towards the end, but during of the coldest time of the Wisconsin glaciation between about 20,000 and 18,000 BP, only tundra, steppe and arctic-alpine vegetation survived in valleys between ice-covered mountains on the southern edge of the Cordilleran ice sheet in the west of Washington. But on slopes leading down to the Pacific Ocean, six genera of conifers, including spruce and Douglas fir, survived between advancing tongues of ice.

Between 20,000 and 16,800 BP Sitka spruce survived on the coastal plain west of the Olympic Mountains along with Western white pine (*Pinus monticola*), Lodgepole pine (*Pinus contorta*), Mountain and Western hemlock. But in the rain-shadow to the east of the mountains, only tundra and a few small Engelmann spruce and Lodgepole pine could survive.

Later, between 16,000 and 15,000 BP, Sitka spruce, Engelmann spruce and Mountain hemlock were growing close to the ice in the southern Puget Trough. By 14,000 BP, alders, spruce (probably Sitka spruce) and Western hemlock still grew between tongues of ice on the west side of the Olympic Peninsula.

Much further north, on the low-lying coastal plain of Haida Gwaii, Sitka spruce persisted during full-glacial times with Lodgepole pine, willows and herbaceous plants. As species of maritime habitats, they could survive because the ocean brought cool (not cold), moist conditions to the most western margins of the continent.

Not all spruces survived the ice ages: a species which, with oak trees, formed upland woodlands south of the ice sheets in eastern USA from 25,000 BP, became extinct by 12,000 BP, during the period of ice melt.

Ice melt: early ecosystems

After about 14,000 BP temperatures gradually rose and ice started to melt in both North America and Europe. The Laurentide and Cordilleran ice sheets segregated and ice-free land appeared between them. Although *coastal* North America was becoming ice free, *continental* ice sheets still covered parts of inland Canada and Alaska.

12,000 BP is taken as the start of the *Holocene* or *post-glacial* period; the weather became warmer with seasonal differences between summer and winter.

Sea level changes relative to land – new islands, new waterways or changing courses, changed landforms scraped and worn by glacier melt, slope, aspect or altitude, the type of debris dropped and left – helped decide the area of land available for colonisation and which plants and animals colonised a locality.

Arctic tundra vegetation developed on newly appearing ground, which was sometimes frozen a couple of feet under the surface. Tundra life at high latitudes endured a very short growing season with very long day-lengths. A few species of low plants such as bryophytes (mosses and liverworts), lichens and small shrubby heathers (*Ericaceae* family) existed. Summer temperatures were too low for most tree seeds to germinate and grow, so tundra normally lacked trees. Animal life was also species-poor and sparse.

Moving ice and melt-water made hollows which became lakes. Frozen ground prevented summer melt-water from percolating far down, so when water, dead plants and animals collected, bogs developed; they sustained sedges (*Cyperaceae* family), bog mosses (*Sphagnum* species) and invertebrates.

Along the coast below the Chugach Mountains in the Gulf of Alaska, ice-melt started about 14,000 BP, but did not reach the inland, glacier-filled valleys until 9000 BP. On the coast, tundra vegetation of ferns, sedges, alder, birch and willow grew for about eight millennia. Eventually, by 6000 BP, alder and birch formed open woodland. However, on colder sites, boreal conifers such

FIGURE 3.4. After ice melts from the land, tundra then shrubs and trees appear: Sitka spruce colonising a side valley of Tracy Arm Fjord, Alaska.

PHOTO: ANDY TITTENSOR, 2014.

as Black (*Picea mariana*) and White spruce were successful. It was not until after 3000 BP that temperate conifer forests containing Sitka spruce developed in that part of lowland Alaska.

In Southeast Alaska, the islands of the Alexander Archipelago began losing their ice cover between 14,000 and 12,500 BP, depending upon altitude and distance from the coast. Although tundra developed first, Lodgepole pine trees reached parts of Southeast Alaska like Heceta Island only 600–1000 years after the glaciers melted from the coast. They colonised, established and flowered on infertile debris left by melting ice in the tundra. By 12,000 BP, with a still-warming climate, Alaskan forests of Lodgepole pine with undergrowth of ferns and sedges had developed.

These pine woodlands lasted until a wetter climate produced conditions suitable for alder woodlands to develop and sustain themselves for several centuries. Sitka spruce followed tundra, pine and alder woodland from coasts, inland along river valleys and up mountain slopes. It first colonised some parts of Southeast Alaska only a few centuries after ice melt and tundra.

Sitka spruce had colonised Heceta Island by 9200 BP, and by 8500 BP was joined by Western hemlock and Mountain hemlock, followed by Nootka cypress and Western red cedar. As vegetation and soils developed, what we would recognise as a version of rainforest – home of Sitka spruce – was beginning to form.

FIGURE 3.5. Sitka spruce flourishes on cliffs close to the melt water of the South Sawyer Glacier, Tracy Arm Fjord, Alaska.
PHOTO: ANDY TITTENSOR, 2014.

Pollen analysis data from a high altitude lake on the archipelago's Prince of Wales Island showed forests growing there even earlier. Lodgepole pine woodland was growing by 13,715 BP, while woodlands of alder and ferns started to overtake the pine by 12,875 BP. Mountain pine (*Pinus mugo*) and Sitka spruce began to colonise about 11,929 BP, Western hemlock soon after.

After 14,000 BP in north-west Washington, Lodgepole pine trees grew on ice-free ground which had previously been the Puget glacier lobes. They grew abundantly on wet, stony, debris-filled ground where there was little competition from conifers which needed warmer, humid conditions. As early as 12,000 BP, Sitka spruce, Douglas fir and Western hemlock had joined the Lodgepole pine and by 10,000 BP had formed open woodland.

On unglaciated ground beyond the southern margin of the ice, typical lowland trees of temperate climates – such as Sitka spruce, Western red cedar, Douglas fir, Balsam poplar (*Populus balsamifera*) and Red alder – had colonised the Puget trough locality by 11,200 BP. They interpolated themselves amongst typical 'montane' conifers, such as Mountain hemlock and Grand fir which had survived close to the ice. By 10,000 BP, as warmth increased, the 'temperate' conifers increased and the 'montane' conifers became restricted to higher altitudes.

The ice-free refuges and increasingly ice-free coast of British Columbia and northern Washington were a main route for Sitka spruce (and other plants) to colonise north to Alaska during the post-glacial centuries 10,500–8500 BP. Sitka spruce and Mountain hemlock forests reached their most westerly limit in south central Alaska from 4000 BP onwards. They are still colonising westwards.

Formation of temperate rainforests

When ice melted during the early Holocene period, huge areas of coastal, and then inland, landscapes were opened up for colonisation by plants and animals.

But it took several thousand years for the early rainforests to develop. Tree species could migrate only at the pace at which they grew, matured and produced seeds, the distance their seeds could travel, whether seeds reached appropriate ground in suitable climate – and so on. And there was a time-lag of 500–1000 years needed for plants to respond to shifts in climate: they cannot get up and walk away as can animals! It was not until 8000–3000 BP that temperate rainforests we might recognise had formed.

Sitka spruce migrated long distances during the Holocene and is still travelling slowly. Where it gets to, and how soon it reaches there, depend upon several factors, not least its genetic make-up. Historical and environmental factors – such as location of its ice-age havens and potential barriers such as the Hecate Strait at 48–140 km (30–89 miles) wide, or the Coast Range average 300 km (186 miles) wide and generally 2000–4000 m (6562–13,123 ft) altitude with ice fields across the summits – also determine its distribution at any time.

Although their species have lived on earth for millions of years, today's North American temperate rainforests are ecosystems formed by geologically recent re-combinations of tree species. Even now, new associations are likely to be assembling as environments change.

Native distribution of spruces

As we near the end of this current interglacial period, spruce species are widely distributed in the temperate (mid-latitude) and boreal (high latitude) zones of the northern hemisphere: Northern and Central Europe, Asia Minor, Caucasus, China, Himalayas, Japan and North America. They are often the dominant species of large, dense forests; spruces also form mixed forests with other types of conifer or with broadleaved angiosperm trees.

Americas

There are about 36 recognised spruce species worldwide (Table 3). The three Central American species *Picea chihuahuana*, *P. mexicana* and *P. martinezii* grow south of the limit of the most recent ice sheets.

Two of the seven North American spruces, *P. breweriana* and *P. pungens*, also grow south of the ice limits. The remaining five, *P. engelmannii*, *P. glauca*, *P. mariana*, *P. rubens*, and *P. sitchensis*, grow in regions which experienced the Pleistocene glaciations. This influenced where they now grow and the ecosystems in which they participate.

TABLE 3. World spruce species.

Latin name	English name	Continent
Europe and America		
Picea abies	Norway spruce	Europe
Picea obovata	Siberian spruce	Europe, North-East
Picea omorika	Serbian spruce	Europe, Balkans
Picea orientalis	Caucasian spruce	Europe, Caucasus
Picea breweriana	Brewer's spruce	North America
Picea chihuahuana	Chihuahua spruce	Central America
Picea engelmannii	Engelmann spruce	North America
Picea glauca	White spruce	North America
Picea mariana	Black spruce	North America
Picea martinezii	Martinez spruce	Central America
Picea mexicana	Mexican spruce	Central America
Picea pungens	Blue/Colorado spruce	North America
Picea rubens	Red spruce	North America
Picea sitchensis	Sitka spruce	North America
Asia		
Picea alcoquiana	Alcock's spruce	Asia, Japan
Picea asperata	Dragon spruce	Asia, China
Picea aurantiaca	Orange spruce	Asia, China
Picea brachytyla	Sargent's spruce	Asia, China
Picea crassifolia	Qinghai spruce	Asia, China
Picea farreri	Burmese spruce	Asia, China
Picea glehnii	Glehn's spruce	Asia, Japan, Russia
Picea jezoensis	Jezo spruce	Asia, China, Japan
Picea koraiensis	Korean spruce	Asia, Korea
Picea koyamae	Koyama's spruce	Asia, Japan
Picea likiangensis	Likiang spruce	Asia, China
Picea maximowiczii	Japanese Bush spruce	Asia, Japan
Picea meyeri	Meyer's spruce	Asia, China
Picea neoveitchii	Veitch's spruce	Asia, China
Picea obovata	Siberian spruce	Asia, Urals-Okhotsk
Picea orientalis	Caucasian spruce	Asia, NE Turkey, N Iran
Picea purpurea	Purple spruce	Asia, China
Picea retroflexa	Green Dragon spruce	Asia, China
Picea schrenkiana	Schrenk's spruce	Asia, Kazakhstan
Picea smithiana	Morinda spruce	Asia, Himalayas
Picea spinulosa	Sikkim spruce	Asia, India, Bhutan
Picea torano	Tiger-tail spruce	Asia, Japan
Picea wilsonii	Wilson's spruce	Asia, China

Hybrids not listed. Data courtesy Martin Gardner, Royal Botanic Garden, Edinburgh.

Europe

All three European spruces grow in regions which were under ice during the Pleistocene period. Common or Norway spruce, is abundant and widespread from Scandinavia, across east Europe to the Ural Mountains, with a small

FIGURE 3.6. Pioneers – trees and people – migrated along the west coast of North America after ice melt; coastal forest of Sitka spruce, Cape Meares, Oregon.

PHOTO: SYD HOUSE, 2010.

area in central Europe. *P. omorika* a relict species from the Tertiary, survives in about 30 small populations in the Balkans, while *P. obovata* grows across Siberia eastwards to the Pacific coast.

Picea orientalis grows on mountains at the boundary of Europe and Asia in the Caucasus and north-east Turkey only.

Asia

There are 23 spruce species native to Asia. Those in China, Japan and parts of Siberia have been lucky enough to exist through a stable, fairly warm climate for 65 million years or so. Why? Because these regions were not glaciated during the Pleistocene period.

Spruces are now widely distributed in China where there are cool, moist, temperate conditions. During the Pleistocene, spruce forests were found over a much larger area of China and Taiwan than today, because cool temperate conditions were more widespread. Many Chinese species are endemic, which means they are restricted geographically to one region or site and occur nowhere else in the world. Similarly, most Japanese species are endemics living only on a few mountain ranges or islands

FIGURE 3.7. Driftwood and living organisms move long distances with ocean currents and wind, collecting as here on the Washington coast.

PHOTO: SYD HOUSE, 2010.

Contentious origin

It is accepted that spruces have one evolutionary origin, but there is disagreement as to whether this was Asia or North America. Fossil spruces from the Oligocene epoch (35–23 MYA) appeared at similar periods in Japan, Europe and North America, so these fossils give no clue to spruces' nativity. But some researchers suggest spruces appeared much later in Asia and Europe, implying they migrated to Asia across the Beringian land bridge from an origin in North America. The land bridge allowed colonisation of spruces between Asia and North America – but how often and did it take place in both directions?

In North America there are few spruce species but each is widely distributed; in Asia there are many species but each is very restricted geographically. Geological and climatic changes during the ice ages precipitated repeated expansion and retraction in the range of spruce species, in both Asia and North America. This caused repeated isolation, rendezvous and hybridisation, resulting in the complex pattern of spruce species on Earth today.

CHAPTER FOUR

At Home in North American Rainforests

> Coastal temperate rain forests, as areas where the land meets the sea, are part of some of the most complex and most dynamic systems on Earth, comprising terrestrial, freshwater, estuarine and marine ecosystems.
>
> Ecotrust and Conservation International 2012

> The undertaking of trees (and other plants) is so commonplace in nature that it is easy to take for granted... The death of a tree neither spills blood nor smells bad. Instead, trees may be nibbled on for years by insects, and after they die, through the agency of beetles, fungi, and bacteria, they slowly and unobtrusively disintegrate into the soil, a recycling that makes the rest of life possible and is life as well.
>
> Bernd Heinrich 2013

Geographical extent

During our lifetimes, Sitka spruce is a canopy tree of the Pacific Rainforest Bio-region, which stretches from San Francisco in the south to beyond Cook Inlet in the north. This bio-region is restricted to a narrow band between the coast and parallel mountain ranges inland to the east; it has very high rainfall and humidity. Natural forest fires are infrequent.

The southern limit of Sitka spruce is by Doyle Creek, Mendocino County, California at latitude 39° 36' N. Its current northern limit is the southern shore of the Knik Arm of Cook Inlet, Alaska at latitude 61.4° N. Kodiak Island and the Katmai National Park form the current western limit of native trees at longitude 155° W, latitude 58.34° N. During the past 60 years, offspring have been colonising westwards around the Gulf of Alaska at a rate of 1.6 km (1 mile) per century.

Map 4 shows the current native distribution of Sitka spruce and those Aleutian Islands where it was planted during the nineteenth century.

Archaeologists working on Haida Gwaii found tools in a cave used by ancestral people about 12,800 years ago. They hunted bear on tundra where Lodgepole pine trees were growing. Humans continued living along the coast even as Sitka spruce replaced shoreline Lodgepole pine and alder woodlands about 10,000 BP. The progress of First Nations during the succeeding ten millennia coincided with rainforest development. People discovered and used the changing resources of rainforests, including Sitka spruce, Western hemlock and Western red cedar.

Continual change

If you return to your favourite woodland or forest 50 years after you first visited you'll probably recognise most of the species, yet the forest has altered. It takes time for an ecosystem to assemble from individual species. Ecosystems vary from one place to another and from one time to another. Their species may spread, decline or evolve; climates and soils change, animals might eat one species more than another, altering their proportions.

Lodgepole pine and willows had colonised the post-glacial tundra of northwest America by 12,000 BP. For two millennia the weather was drier but warmer than now and a combination of trees unusual to us grew together on Haida Gwaii and south to the Oregon Coast Range: Mountain hemlock, Western hemlock, Douglas fir, Sitka spruce and Grand fir. Today, these tree species occupy different niches and localities within the rainforest zones.

Drier times continued after 10,000 BP and yet other combinations of tree species grew along the Pacific coast. Douglas fir and alders formed southern forests. Douglas fir and Sitka spruce grew together in moist places on Vancouver Island and the west of the Olympic Peninsula.

It was not until 7000–4000 BP that decreasing the summer sunshine led to a much moister atmosphere along the Pacific coast from California to Alaska. The contemporary distributions of tree species started to change again and tree species re-assembled.

In the Gulf of Alaska, Sitka spruce colonised and formed forests on coastal land, spreading as far north and west as Icy Cape on the Chukchi Sea. From Southeast Alaska south to Vancouver Island and the Olympic Peninsula, Sitka spruce and Western hemlock became exceedingly common and formed rainforests as joint dominants. Alder, however, declined. Inland, Sitka spruce was a minor component of the developing forests. In northern British Columbia, Western hemlock dominated the forests, accompanied first by Lodgepole pine and then Western red cedar. But further south, Western red cedar and Pacific silver fir accompanied Lodgepole pine. In what is now Washington, Douglas fir became an important constituent of drier rainforests.

Between 5000 and 4000 BP Western red cedar expanded from the south, particularly at lower altitudes. Yellow cedar expanded at higher altitudes and latitudes. Western hemlock and Sitka spruce became abundant along the coast and at lower altitudes. Douglas fir, Western red cedar, Western hemlock and Grand fir replaced what had been oak woodland in southern Washington.

An early, 10,700 BP woodworking site on Haida Gwaii showed Sitka spruce and Western hemlock were already used to make tools and objects; wooden tools and objects became common after 5000 BP. As Western red cedar and, to a lesser extent, Sitka spruce, proliferated, they became valuable to human communities, while houses of Western red cedar dating from after 3000 BP are common at archaeological sites and recently-abandoned villages. Early European explorers saw these two species in use by contemporary communities.

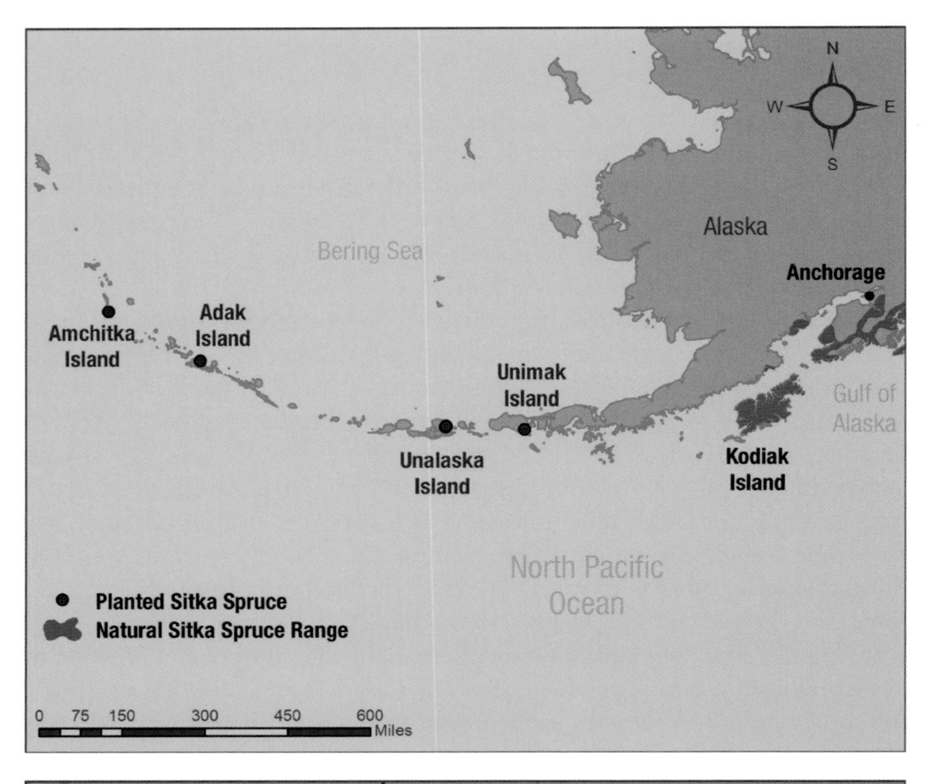

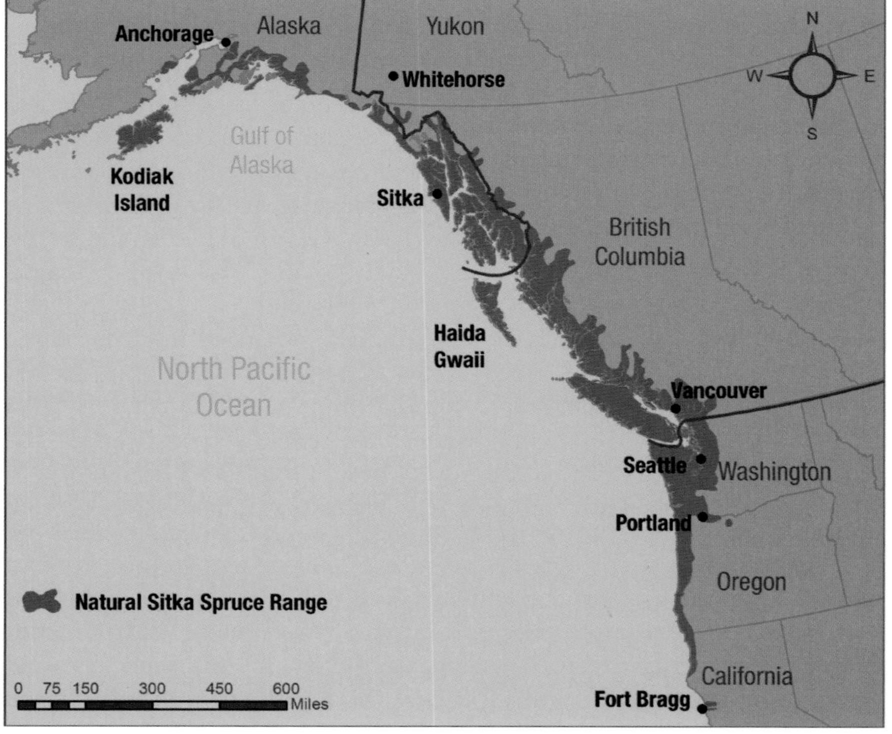

MAP 4. Distribution of Sitka spruce, prepared by Luke A'Bear, Sitka Conservation Society. Above: Plantations on Aleutian Islands; below: Present native distribution.

PREPARED WITH DATA BY KIND PERMISSION OF SAMUEL (2007); KLINKENBERG (2014); KAUFFMANN (2013); CARSTENSEN (2014).

FIGURE 4.1. Pacific Spruce Corporation stand of Sitka Spruce near Lincoln City, Oregon.

PHOTO: JOHN D. CRESS, SEATTLE, 1920S. © WEYERHAEUSER USA.

FIGURE 4.2. Sitka spruce in misty conditions amongst Californian redwoods.

PHOTO: STEPHEN SILLETT, COURTESY INSTITUTE OF REDWOOD ECOLOGY.

Between 4000 and 2000 BP, combinations of trees which we recognise nowadays as temperate rainforests had assembled. They are dynamic systems which do, and will, change if there is windblow, soil erosion, volcanic activity or altered snow cover. Individual trees are not as old as 2000 or 4000 years: several generations of trees have been produced in that time depending on their individual longevity. Although Sitka spruce was growing in coastal Alaska between 12,000 and 9000 BP and in north-west Washington by 12,000 BP, it was not until 3000–2000 BP that it became the abundant constituent of the rainforests with which we are familiar.

From northern California, temperate rainforest can be found along the Oregon and Washington coasts. It has developed its greatest extent and magnificence on the archipelagos and islands of British Columbia. Vancouver Island and Haida Gwaii are well-known for both their rainforest and Sitka spruce.

In Oregon and Washington, Sitka spruce grows well inland, for instance to the edge of the Cascade Mountains, about 170 km (106 miles) from the coast. But in British Columbia, where the climate is more suitable, the temperate rain forest with its Sitka spruce is found from the shoreline eastwards more than 200 km (124 miles) inland. From the west coast of Haida Gwaii east to Smithers on the edge of the Skeena Mountains, its range extends 400 km (248 miles) inland. At the very south of its range in California, Sitka spruce grows only 60 km (37 miles) inland.

The altitudinal range of temperate rainforest and Sitka spruce vary with the environmental conditions. Forests reach down to the sea along both exposed coasts and sheltered estuaries. Sitka spruce grows where the air is saline from salt spray and in saline soil. Upwards, it lives on the lower slopes or 'sub-montane' vegetation zone. It is usual for it to reach up to about 300 m (984 ft) altitude. On mountain slopes near the ocean, it grows to about 900 m (2953 ft). But on the Coast Mountains north of Vancouver, it grows to 1500 m (4921 ft) altitude; in sheltered places on Vancouver Island it reaches 1000 m (3281 ft). In sub-arctic Alaska, however, Sitka spruce and Mountain hemlock are limited to below 200 m (656 ft) altitude.

Landscapes

The landscapes occupied by temperate rainforests are extreme and magnificent: the Ocean coastline runs far inland as fjords, inlets and estuaries. Islands are large, small and very numerous. Glacial lakes are numerous and there are ice fields over higher mountains and glaciers calving into the Ocean.

The jagged coastline is paralleled closely to the east by chains of rugged mountains. These reach 3048–4267 m (10,000–14,000 ft) high and form one section of the Ring of Fire.

Rainforests occupy floodplains of rivers and estuaries, mountain-sides and seashore. The ocean, freshwater and land are closely intertwined physically and biologically.

Climate

The 3600 km (2237 miles) of temperate rainforests grow in variations of a maritime climate, usually characterised by persistently moist conditions, coastal fog and high rainfall, with at least 10% of rainfall in the summer months. There is usually 60–80% cloud cover monthly. The air is cool with mean temperature of 4°–12°C (39°–54°F). Seasonal change is minimal, with few temperature extremes. The maritime climate regimes have developed as a result of two main features: the adjoining ocean and the mountain ranges to the east.

The Pacific Ocean moderates temperatures and produces very high rainfall. The mountains determine where the rainfall drops. Frequent storm systems roll in from the Pacific Ocean and the mountains to the east catch the rain clouds as they move inland, causing high precipitation which falls as rainfall down their slopes via streams and rivers – back to the sea. The rain shadow area just to the east of the mountain ranges has a much lower rainfall. It also has a hotter climate than the coastal fringe where the rainforests grow.

High rainfall is received in all months of the year, but most is between November and March. The rain shadow receives less than 381 mm (15 in) a year, while the coastal regions of the Olympic National Park (for instance) receive 2921 mm (115 in). But annual rainfall of 3048 mm (120 in) or even 5300 mm (209 in) is not uncommon. The southerly coasts of lower-rainfall California receive summer coastal fog which ensures suitable conditions for both Sitka spruce and redwoods.

The more maritime the climate, the smaller the range of mean monthly temperatures and the more indistinct are seasons. Even in winter, low temperatures and frost are rare due to the ameliorating influence of the nearby ocean. Species intolerant of frosts can therefore thrive.

Latitude, longitude and the altitude from sea level to mountain top affect the climate and therefore the species content of rainforest. As you travel north you find that some trees, such as Western red cedar, reach their climatic limits. And from west to east the climate becomes less oceanic and more continental.

The maritime climate supports an abundance of some of the most magnificent conifer trees on the planet; they need considerable water to grow so large and old. They are tall, shady and support a relatively high biodiversity. Many species, such as the Spotted owl (*Strix occidentalis*) and Marbled murrelet (*Brachyramphus marmoratus*), are dependent on these forests and cannot live elsewhere.

Soils

The ice ages, different geological outcrops and a long latitudinal range mean great variety in soils. At the start of the Holocene, the melting ice left wet, rocky, stony landscapes almost bare of vegetation.

Now, most temperate rainforest soils are rich in organic matter and plant nutrients formed by decomposition of dead organisms – from dead grizzly bears

FIGURE 4.3. Sitka spruce forest on the coast of Oregon north of Brookings.

PHOTO: ALAN FLETCHER, 1978.

FIGURE 4.4 *(left)*. Sitka spruce in gorge rainforest of many tree species, Vancouver Island.

PHOTO: GEORGE BROWN, 2012.

FIGURE 4.5 *(below)*. Riparian Sitka spruce flourishing on Vancouver Island.

PHOTO: GEORGE BROWN, 2014.

FIGURE 4.6. Sitka spruce in its typical misty conditions forming rainforest around Salmon Lake, Silver Bay, Alaska.

PHOTO: ANDY TITTENSOR, 2014.

FIGURE 4.7. Sitka spruce at its western limit, Hallo Bay, Katmai, Alaska.

PHOTO: MICHAEL FITZ, KATMAI NATIONAL PARK AND PRESERVE, 2013.

and tree trunks to slugs and leaves. Slowly-decaying standing trees, trunks on the ground, decaying leaves and dead animals eventually add to the soil's humus. These forests' high biomass provides considerable organic material to the soil surface. However, in the cool temperatures, it decomposes slowly causing thick, soft humus to accumulate. Temperate rainforests store immense amounts of carbon in the dead organic matter and large trees: about 1000 tonnes per ha (446 tons per acre).

The abundance of moisture throughout the year affects soil processes. Wetlands of standing water produce swamp, marsh and bog forests in northerly latitudes; Sitka spruce is the usual dominant tree in swamps. Moisture explains why there are few catastrophic fires in temperate rainforests compared with many North American forests where fires are natural regenerative processes.

Sitka spruce grows in several other soil types. On the better-drained, alluvial soils, it is dominant along with Western hemlock. On the wetter soils it is accompanied by Western red cedar. On alluvial flood plains and where there have been landslides, Sitka spruce colonises along with Black cottonwood and alders. It is the only conifer to colonise the saline and splash zones by the shore.

On the newly-forming soils of Glacier Bay, Alaska, Sitka spruce contributes to vegetation succession, behaviour of special interest. Most of the glaciers are retreating, leaving rocky moraines along the coasts. On this newly exposed land, mosses of the genus *Racomitrium* colonise the stony ground first, followed by dwarf shrubs, such as Yellow mountain avens (*Dryas drumondii*) which forms large mats. Forty years on, shrub willows and alders develop continuous vegetation cover on the once-glaciated landscape. Sitka spruce seeds itself into the dense alder cover, germinates, grows and after 60 years has developed a continuous-canopy forest, the alder gone.

At each stage, dead plants and animals leave organic matter, which builds up and, with fragments of stone, forms a young soil. The alder bushes contain nitrogen-fixing bacteria in their roots: nitrogen in the soil increases when the roots die and decompose. The acidity of the soil increases (pH drops from 8.0 to 5.4 and then to 4.2) because calcium carbonate is leached out. After one to two centuries, a Sitka spruce-Western hemlock forest has formed. Modern glacier melt started about two centuries ago, so all these stages can be seen, starting by the shore and travelling inland to the oldest forests.

Glacier Bay is approximately the same latitude as Scotland (Glacier Bay 58.5° N; Inverness, Scotland 57.5° N) where Sitka spruce grows so well nowadays as a plantation tree.

In its native range, Sitka spruce does best in rich, organic soils with high levels of inorganic compounds of potassium, calcium and phosphorus. Wetter soils, particularly where there is water seepage, also promote growth. Without seepage, it needs a constant supply of groundwater within 100 cm (39 in) of the surface.

FIGURE 4.8.
Temperate rainforest is overwhelming; Baranof Island, Alaska.

PHOTO: ANDY TITTENSOR, 2014.

Latitudinal forest zones

Although 'maritime' describes the overall climate of the rainforest bio-region, there are variations within the long coastline, so the proportions of the main conifer species fluctuate. Four main forest zones are recognised.

The *Coast Redwood Forest Zone* is developed along the warm coast of California and southern Oregon. Although the dominant trees are Coast redwood and Pacific silver fir there are many tree species in the canopy. Sitka spruce grows sparsely in these forests where mist and fog provide enough humidity. In these foggy places, epiphytic ferns, bryophytes and lichens are abundant. The tree-line can reach 2744 m (9003 ft). Winters are mild and wet, summers are cool and dry, snow is uncommon along the coast except on mountain tops. This is the least maritime climate of the rainforest coast.

It merges into the *Seasonal Rainforest Zone* in Oregon, through Washington to mid-Vancouver Island and the southern British Columbia mainland. In wet places Western red cedar dominates the rainforest, while Sitka spruce and Western hemlock form a lush forest adjoining the coastline and in foggy valleys. Douglas fir and Western hemlock dominate a forest type which gets catastrophically burned about once a century. Summers are dry and cool, winters mild or cool and snow is rare.

The *Perhumid Rainforest Zone* develops from the north of Vancouver Island to Southeast Alaska via Haida Gwaii and the mainland of British Columbia. The rainforests grow amongst peat bogs and meadows. They are dominated by Sitka spruce and Western hemlock with Pacific silver fir and Yellow cedar. Western red cedar reaches its northern limit in southern Alaska. There are no redwoods and little or no Douglas fir. High rainfall in all months and a small range in monthly mean temperatures make conditions suitable for a rich understorey of shrubs and herbaceous plants. The climate of north-west Britain and of Ireland, where Sitka spruce is planted, is most similar to this zone.

In the *Subpolar Rainforest Zone* round the Panhandle and Gulf of Alaska the forests grow in a landscape of muskeg, alpine meadows and glaciers. The climate is very cool and wet and the soils are poor compared with further south. Sitka spruce and Western hemlock are abundant, with Mountain hemlock usually at a higher altitude. Yellow cedar is sparse in the south of this zone. The nearby mountains hold long-term snow-cover. Away from the coast, there is tundra vegetation and the tree line is only 200 m (656 ft) altitude.

Sitka spruce grows in all four zones, but is dominant in the Perhumid and Subpolar Zones, where rainfall is highest.

Productivity of temperate rainforest

We tend to think of the more widespread and publicised tropical rainforest as the most diverse ecosystem on Earth. However, temperate rainforest surpasses it in biological productivity and biomass. One reason is that fire, a natural event in many forests, is not common in temperate rainforest. It is too wet!

Energy flow is slower in the cooler temperate rainforests, so dead organisms decay more slowly and humus builds up. There is also a huge weight of living organisms, including trees, shrubs and epiphytic ferns, mosses, liverworts and lichens growing on or hanging from the tree trunks and branches; many animals such as chipmunks, squirrels, crossbills and woodlice eat these plants, and other animals eat them. Biomass (weight) of living and dead organic matter is thus very high in temperate rainforests. Although Biodiversity (number of species) is less than tropical rainforests, it is much higher than in other types of North American forests.

Structure and composition

For Europeans, whose forests are almost all managed, a temperate rainforest is a revelation. It is drenched in wetness, has soggy ground, pools of water and damp tree trunks. Branches have their own soil and epiphytic vegetation, while mist and raindrops pervade the air. Whereas we could (just!) climb to the top of a European tree, the enormous height of rainforest canopy dwarfs us. Seedlings, saplings and young trees are abundant. Temperate conditions allow lush growth all year round, producing a painter's dream of green colours.

North Americans also find rainforests places of wonder. Lawrence Millman went on a trip to the Gwaii Haanas National Park Reserve for the National Geographic Society in 1998. In his article he wrote:

> A cold rain was slashing away at sea and shore, mountains and forest. Likewise a fog had crept in… Rain and fog – the meteorological two-step of the Pacific northwest. The former exalts flora to majestic proportions and swells up rivers so that salmon can swim upstream and spawn. As for the latter, it can draw a curtain of invisibility over whatever the former exalts… However time-honored the weather, I was determined not to let it spoil my encounter with a place Haida artist Bill Reid once compared to the Peaceable Kingdom. So I donned my anorak and, pocketed my compass, and went ashore. Soon I found myself walking through, if not the Peaceable Kingdom, at least a temperate variation on the theme of Amazonian lushness. Sitka spruces and western red cedars soared 200 feet or more into the sky, having cultivated rich mineral subsoils for thousands of years. Their roots now hunched up like gargantuan crab legs or sprawled out like gargantuan serpents. On the ground was a seemingly endless cushion of moss, squishing at my every footfall. Not content to cover only the ground, this moss crawled up the trunks of trees too.
>
> From the branches of these same trees dangled wispy green lichens called old-man's beard – a fitting complement, I thought, to an old-growth forest…
>
> Millman 1998

You could visualise unmanaged, old-growth forests as multi-storey car parks, each storey with its own living components.

At the top storey, the main canopy of mature conifers can be over 61 m (200 ft) high and is usually 250–1000 years old. This is occupied by insects such as bees and aphids, birds such as Townsend's warbler (*Setophaga townsendi*) and Chestnut-backed chickadee (*Poecile rufescens*), mammals such as the Douglas' Squirrel (*Tamiasciurus douglasii*), Northern flying squirrel (*Glaucomys sabrinus*) and Townsend's chipmunk (*Neotamias townsendii*). Sitka spruce trees make good roosting places for Bald eagles (*Haliaeetus leucocephalus*) and Peregrine falcons

FIGURE 4.9. Rich undergrowth, big trees and and dead timber are typical of old-growth Alaskan rainforest.

PHOTO: ANDY TITTENSOR, 2014.

FIGURE 4.10. Well-vegetated fallen tree boles are a common feature in the damp atmosphere.

PHOTO: ANDY TITTENSOR, 2014.

FIGURE 4.11. Fallen trunks provide 'nurseries' for Sitka spruce seedlings to start their lives.

PHOTO: ANDY TITTENSOR, 2014.

FIGURE 4.12. Epiphytes (mosses, liverworts lichens and ferns) form distinct ecosystems on high branches of a 78.9 m (259 ft) tall Sitka spruce.

PHOTO: STEPHEN SILLETT, COURTESY INSTITUTE OF REDWOOD ECOLOGY.

FIGURE 4.13. Mosses enclose Sitka spruce branches below the canopy top and contribute considerably to the biomass of the rainforest.

PHOTO: STEPHEN SILLETT, COURTESY INSTITUTE OF REDWOOD ECOLOGY.

(*Falco peregrinus*). Spruce grouse (*Falcipennis canadensis*) forage on conifer needles in the winter. The canopy of an individual conifer is huge and structurally complex, forming a three-dimensional habitat where several generations of a species may live.

Fourteen or more conifer species contribute to the temperate rain forest as a whole, including: Grand, Silver and White (*Abies concolor*) firs, Pacific yew (*Taxus brevifolia*), Western larch (*Larix occidentalis*), Incense cedars (*Calocedrus* species) and Port Orford cedar (*Chamaecyparis lawsoniana*). Broadleaved trees contribute too but are much less abundant. Bigleaf maple and Vine maple (*Acer circinatum*) are common; Red alder may become dominant near water or where ground is disturbed by windblow or landslide. Quaking aspen (*Populus tremuloides*) and Douglas maple (*Acer glabrum*) are some of the less common broadleaved trees of the temperate rainforest.

There is an understorey, dense in places, of deciduous, woody shrubs, young conifer and broadleaved trees. Shrubs include the Salmonberry (*Rubus spectabilis*), Red elderberry (*Sambucus racemosa*), Pacific rhododendron (*Rhododenrdon macrophyllum*) and wild currants (*Ribes* species). Mammals like the Black bear (*Ursus americanus*) and Black-tailed deer (*Odocoileus hemionus*) feed in this storey. Many animals eat the prickly leaves of young Sitka spruce, for instance, deer and elk, bears, porcupines, rabbits and hares.

Herbaceous plants, many of them with relatives throughout the northern hemisphere, live on the ground storey. Examples are: Twinflower (*Linnaea borealis*), Trail plant (*Adenocaulon bicolor*), Bunchberry (*Cornus canadensis*) and Red columbine (*Aquilegia canadensis*). Evergreen ferns are also abundant, including the Deer fern (*Blechnum spicant*) and Licorice fern (*Polypodium glycyrrhiza*). Many species of salamanders and a few snakes live amongst the herbs, while snails and slugs rasp their away across the forest floor. Dark-eyed juncos (*Junco hyemalis*) look for seeds and Long-tailed weasels (*Mustela frenata*) hunt for mice and shrews.

There is a thick mat of lichens, clubmosses and bryophytes on the soil surface. There are said to be over 1100 species of lichen in old-growth forests, their biomass and diversity reaching a peak when the forest is 350 years old. Evergreen ferns form abundant patches. But sometimes the ground grows only algae, mosses and clubmosses or is covered with dead needles and broadleaves.

In the basement beneath the ground, the soil is a busy place with miles of fungal filaments (more about them later). There are an estimated 3000 species of fungi in rainforest soils of British Columbia, together with thousands of invertebrates such as beetles and nematodes, millions of bacteria in a cubic metre – and plant roots.

Standing dying, dead and decomposing trees are called 'snags'. They are abundant and important. For instance, at least nineteen species of mammals and 37 bird species nest in their cavities. The Pileated woodpeckers (*Dryocopus pileatus*) and Spotted owls (*Strix occidentalis*) are famous cavity-nesters. Ospreys (*Pandion haliaetus*) make big stick nests on the tips of snags. Wood- and bark-boring beetles colonise snags and provide food for birds such as woodpeckers.

The dead, decaying and decomposing tree trunks are vital to the continuation of the temperate rainforest. Having been alive, their death and decay contribute nutrients to future life. Hundreds of species of decomposers feed on a dead trunk to convert it, during more than a century, to humus. For instance, bark beetles, slugs, termites, mites, fungi and bacteria.

Fallen trunks provide a nursery bed for young tree seedlings and saplings which grow in lines (obviously!) on them. These 'nurse logs' eventually become integrated within the soils humus while the saplings may grow into trees. When close rows of different conifer seedlings take root on them, they can grow so close together that their trunks become joined. Sitka spruce may be grafted to Western hemlock! A nurse log which is 15 m (50 ft) long can grow several thousand tree seedlings along its length. Its role is vital to the continuation of the rainforest both alive and dead. Seeds germinate when and where they fall, are blown or taken, so that trees live in populations containing many ages and sizes.

Some old-growth forest has a closed tree canopy casting deep shade. Western hemlock is one of the few conifers whose seedlings can survive low light at soil level. Other conifer seedlings, including Sitka spruce, need more light.

Glades, avalanche tracks and landslips are a vital ingredient of rainforest processes. They contribute to ecological patchiness, regeneration of trees and feeding places for forest-edge species. Seedlings and saplings grow in the light gaps. Shrubs grow thickly in glades and along open riverbanks and floodplains.

Temperate rainforest has developed recently on thousands of offshore islands as well as on the continent edge. As a result, more plant and animal species than usual have evolved into distinct, island, subspecies. For instance, the Alexander Archipelago wolf (*Canis lupus ligoni*), Haida ermine (*Mustela erminea haidarum*) and Prince of Wales flying squirrel (*Glaucomys sabrinus griseifrons*).

Processes

Large-scale forest structure is determined by natural disturbances such as tree fall or insect outbreak (creating a gap), flooding (causing inundation and swampy ground), landslides, avalanches or windblow (producing big openings). Fire affects rainforests only once every millennium or so, a reason their trees can live to a great age.

After any of these happenings, plants colonise the open areas – and do so in a particular sequence according to the rock type, soil, angle of slope and wetness. This sequence is called succession. After a landslip, fast-growing, open-ground herbaceous species come first, then alder and conifer seedlings, then canopy tree species. Sitka spruce has a prominent role in landslide succession overtaking other tree species after about half a century. Along with less abundant and smaller Western hemlock, it eventually becomes the dominant tree. The whole succession sequence takes more than two centuries, after which old-growth forest gets established again.

As well as landslips, Sitka spruce grows abundantly on other dynamic surfaces, such as avalanche slopes, alluvial flood-plains, river terraces and the salt-spray zone by the ocean.

Ocean, brackish estuaries, freshwater rivers and temperate rainforests are not only close physically, but are interrelated via ecological processes. *Salmon, bears* and *Sitka spruce* – apparently quite separate species with completely different life-styles – illustrate this interrelationship and interdependence.

The higher-latitude temperate rainforests support the greatest salmon runs on Earth along rivers such as the Fraser and Skeena, Chilkat and Kenai. From the rivers' origins at high altitudes, they transport glacial meltwater, which carries fine sediments, which are nutrients used by aquatic plants or dropped lower downstream to form muddy wetlands. Landslides send sediments into waterways too, adding nutrients to the waters, promoting plant growth and productivity. Floods return organic matter to the land, where it is used by Sitka spruce and other plants.

Sitka spruce trees which fall into rivers and streams provide cool, shady, spawning pools, places for young fish to grow and hiding places amongst their branches. Decomposing trees also produce nutrients for aquatic life.

When the salmon return from the Pacific Ocean in huge numbers to their home rivers to spawn, they bring with them oceanic nutrients within their large bodies. They die after spawning and when their bodies break-up, the nutrients are unlocked and added to the riparian and aquatic ecosystems.

Grizzly (Brown) bears (*Ursus arctos horribilis*) and Black bears lay down fat for the winter by feasting on spawning salmon. Part-eaten fish carcases discarded in the forest decompose and add nutrients to the soil, available to the roots of Sitka spruce and other plants. Bears' faeces deposited on land also decompose into nutrients available to plants. Wolves and Bald eagles feast on salmon, taking yet more oceanic nutrients into the forest.

Salmon are dependent on old-growth rainforests. Where there has been extensive logging, their numbers decline: inappropriate logging causes rapid and heavy water run-off which erodes stream banks, changes the water temperature and clogs the clear spawning pools with heavy sediment. Thus, although salmon are freshwater and marine animals, they depend upon the land forests for their survival.

Ian and Karen McAllister wrote:

> The ancient temperate rainforest of British Columbia is a moss-draped, mist-shrouded forest built on ecological foundations that took 10,000 to 14,000 years to evolve – a unique combination of plants and animals that in turn migrated here from ecosystems as old as 70 million years. Growing slowly but virtually year-round in the wet, moderate climate, the ancient trees can endure for a thousand years or more, contributing to a total forest mass (biomass) of some 1,000 tonnes per hectare – exceeding that of tropical rainforests.
>
> McAllister and McAllister 1997

FIGURE 4.14. Black bear (*Ursus americanus*) feeding on salmon near the estuary of the Conuma River, Vancouver Island, British Columbia.

PHOTO: GEORGE BROWN, 2014.

FIGURE 4.15. George Brown holds a King salmon (*Oncorhynchus tshawytscha*) caught from the Big Qualicum River, Vancouver Island, 2014: it was returned to the water.

PHOTO: WILLIAM FINDLAY, 2014.

FIGURE 4.16. George Brown watched by William Findlay holds a Coho salmon (*O. kisutch*) from the Gold River, Vancouver Island, 2014: it was returned to the water.

PHOTO: DAN ANDERSON, 2014.

CHAPTER FIVE

Sitka Spruce in the Lives of First Nations

For thousands of years the coastal temperate rainforests of North America supported one of the highest densities of nonagricultural human settlements on the continent.

Peter Schoonmaker *et al.* 1997

Sitka Spruce wood is light and strong. First Peoples often used it to make pegs for cedar boxes. The Kwakwaka̲'wa̲kw and their neighbours sometimes used it to make digging sticks, herring rakes, arrows, bark peelers and slat armour.

Nancy J. Turner 2007

Sources of information

The relationships of coastal First Nations with Sitka spruce and rainforest are spiritual, practical and necessary for survival. There are several ways of discovering more about them. Archaeological studies and finds have given information about both trees and people in post-glacial and early-modern times.

Strong oral traditions, memories and knowledge are relevant to environments anywhere in the world. A willingness to listen and learn from them is important because modern enquirers are likely to have quite different world-views and mind-sets even from elderly members of their own communities – and certainly from First Nations people.

The earliest contacts between Europeans and west coast First Nations took place during the fifteenth century. European explorers' diaries and journals from then and for several centuries are pertinent to late pre-European civilisations of the Pacific Coast.

Participating in hunting, fishing and foraging activities of modern groups is also revealing and educational.

The sum of evidence has given the world an instructive and humbling picture of people and nature along the Pacific coast during the past 15,000 years.

Early colonisation

At least two distinct human populations in East Asia contributed to the early inhabitants of North America (DNA studies, blood group distribution and languages provided evidence). One group lived at the eastern margin of Asia while the second group originated further west and south, contributing up to 38% of First Nations' ancestry.

Even as the Wisconsin ice sheets were at their greatest extent, some humans reached North America. These Palaeo-Americans are thought to have been the first humans to enter and inhabit the American continent. They were probably hunters of large animals, and may have travelled from Siberia across the land bridge of Beringia. They started out about 47,000 BP and (probably) walked as far as the dry interior of Alaska where they settled for a 20,000 year period. Some people may have gone onwards, east and south along an ice-free corridor opened between the Cordilleran and Laurentide ice sheets.

It was not until after the very cold period between 20,000 and 18,000 BP that people migrated in larger numbers from Siberia. When the ice sheets started to melt, they travelled east and north-east across the extant land bridge of Beringia – and much further. An ice-free land corridor formed along the Pacific coast and east into the valleys of the continent. Between 16,000 and 13,000 BP people coming from Siberia colonised the coastal margins, islands and valleys.

As ice continued to melt, the coastal land-bridge of Beringia was flooded. People used boats to paddle southwards along the coast, onto islands and into inlets. The same lengths of coastline and inlets simultaneously opened up for colonisation by plants and animals. So people could obtain their food from inshore fish, sea mammals and seaweeds, as well as littoral plants and shellfish.

Resources would have been somewhat different from place to place, depending on, for instance, how sheltered a bay was, the rocks a stream flowed through, or the angle of the seashore.

DNA evidence showed that humans lived in the Paisley Caves region of southern Oregon 14,340 years ago. More recent DNA evidence showed that about 14,000 BP, early humans on the north coast of the Olympic Peninsula hunted large mammals – and left a spearhead in the body of a Mastodon (*Mammut* species).

The people who came to live in North America between then and 5500 BP had to be flexible and cope with an ever-changing environment: changing temperatures; changes in amount and patterns of rainfall; submergence of the once-helpful, ice-free corridor and its resources when sea level rose; a quickly-changing landscape; continually-changing vegetation and its dependent fauna.

A young man who lived in Southeast Alaska about 10,300 BP died at twenty-something and left his skeleton in a cave on Prince of Wales Island. It was found in 1996 and analysis of bones and teeth showed he had eaten marine fish, mammals and molluscs.

Coastal archaeological sites in British Columbia, dated to 8000 BP have yielded remains of food resources which show people ate wapiti or elk (*Cervus canadensis*), the estuary fish smelt (*Osmerus eparlanus*), shellfish from fresh and saltwater, and seals.

As the coastal glaciers melted, Lodgepole pine and Sitka spruce seeds might be blown by the wind in all directions from their coastal and southern refuges. Those which went east, north-east and northwards towards and onto newly ice-free coasts and islands, could germinate and grow. They eventually produced cones whose seeds would be carried far away: and so on…

Trees, forests and people

As soils developed depth and fertility, more trees could colonise and spread slowly over the landscape. By the time assortments of conifers, angiosperm trees and other plants colonised the coastal corridor to the nearest mountain ranges, indigenous humans had to learn how to survive, not only from coasts, estuaries and rivers, but also in and with extensive forests.

There is less archaeological evidence about their relationships with forests than with coastal ecosystems because archaeology under woodlands and forests is rather difficult!

However, First Nations specialists have shown us how to interpret evidence of human life in rainforests from marks and cuts on the trees themselves. Such 'culturally modified' trees are being located and listed so they can be perpetuated if logging takes place around them.

A wetter, cooler climate and a fluctuating sea level caused more change for people, flora and fauna. Human populations sorted themselves into hierarchies with an élite social class. Rainforest habitats continued changing. Western red cedar expanded northwards: trees grew big enough to build houses and large canoes, making big settlements and long ocean voyages possible. Wooden objects and tools constructed from cedar are found frequently at archaeological sites.

Coastal people made increasing use of rainforest trees particularly Western red cedar, Sitka spruce, Western yew, Western hemlock and Lodgepole pine.

It is now agreed that there was not any 'pristine' or 'natural' temperate rainforest in North America after the Wisconsin glaciation – natural in the sense of being unaffected or unused by humans. Sitka spruce, other trees, plants and animals colonised and reached their current distributions concurrent with human populations. Migratory humans colonised or used all localities and ecosystems of the coast and temperate rainforest during the past ten millennia.

Temperate rainforests and humans have contributed to ecological connections between ocean, estuary, freshwater margin and land. The places where ecosystems meet – ecotones – are especially rich in species.

A sort of 'joint tenancy' of land and water was held by plants, animals and humans together.

Useful products for people came from plant roots, leaves, stems, fibres and fruits, as well as from multitudes of associated animals – from below the ground to the three-dimensional living-spaces of the high canopy – from grubs to squirrels.

Different parts of trees could be used for foods and drinks, tools, utensils, thread and homes. Herbaceous plants and shrubs grew berries and leaves. Mammals gave meat, fat, sinew, bone and skins. Eggs and meat came from birds

FIGURE 5.1. Tlingit houses and spruce canoes on Sitka waterfront, 1886–1887.

PHOTO: COURTESY ALASKA STATE LIBRARY, WILLIAM PARTRIDGE PHOTO COLLECTION, P88-051.

and fish, feathers from birds. Forested rivers and their estuaries supported huge populations of fish, birds and mammals. Coastal waters grew shellfish, squid, herring spawn and edible seaweed.

Pacific Coast First Nations and their environments

Indigenous Americans are known to themselves and to us as the many ethnic groups of the First Nations. They each have their own language, plant and animal names. Some groups still live, at least partly, using their traditional methods of gathering resources.

But their ancestors depended upon the resources of the surrounding environment *every day for all their needs* (so do we, but we designate resource collection to a few specialists). They were initially nomadic people, making use of the seasonal resources from different ecosystems: marine, littoral, fluvial, riparian, streams, flood-plains, grasslands, forests and arctic-alpine meadows above the tree-line.

Many groups did not settle into resident communities until the past five centuries or so. They share an attitude to, and concept of, their environment which is quite different from the current, western view. First Nations have a strong feeling of custodianship towards the land and its life. Resources are usually managed collectively, based on communal decisions. 'Environment' is an accepted part of their whole world-view, not segregated as a different policy and ethic from social, institutional and philosophical views and policies.

Their world view and ecological knowledge are of the utmost importance, because survival requires it. Spirituality is imbued in all their views of the living and mineral environment. So respect for other life is an accepted and essential part of their world view. Foraging, collecting and hunting are activities which have spiritual meaning as well as providing for their daily needs.

People understand that over-harvesting or wasting a resource can cause numbers or amounts to dwindle, followed by less food for them and their families in the future. Land, rocks, water and life are all treated with the same respect. After hunting and killing animals, the carcasses receive admiration and esteem. Unused parts of an animal are returned to the environment, available to other species from bacteria to Common ravens (*Corvus corax principalis*). Children are taught not to waste food and to kill only with a subsequent use in mind. People of First Nations would harvest a resource according to its prolificity.

Occupying lands long-term, a strong oral and community memory, remembering where plants grow and animals have territories, where fresh water and appropriate soils, shells or wood occur – means that First Nations, as individuals and as groups, identify closely with the environment. Knowledge is still passed down through stories and by young people participating in activities.

Increasing forest complexity

Once sufficient trees reached and colonised the north-west Pacific coast and when biomass had increased, then ecosystems developed greater complexity in structure and processes. A greater variety of ecosystems than just after the end of the glaciation had developed. Plants and animals formed ecological-altitudinal zones from ocean to mountain-top. By that time, these indigenous people were living in resource-rich forest environments. They used a huge number of species not only for food but for their other needs such as homes, boats, tools and utensils.

In recorded times, people of the north-west coast ate at least 50 species of berries, 25 sorts of roots, and green parts of 30 plants. They ate food from 20 species of mammal, 20 birds including waterfowl, 35 sorts of fish and 35 sorts of sea and littoral (intertidal) invertebrates such as clams, mussels and sea cucumbers.

Although Sitka spruce was only one of many plant resources, the root, bark, inner bark, wood, timber, foliage, resin, sap, shoot tips and cones were all used.

What made Sitka spruce so useful?

It is the fortune or mis-fortune of Sitka spruce that it has evolved anatomical features and chemical composition which make it useful to humans.

Those of relevance to North American First Nations were and are, first, its roots, which have thin, stringy, wood fibres. Second, in closed forests it grows straight and tall, losing its branches, so grows into one very long, trunk which can produce many planks. Overall, its timber is strong yet comparatively light. Its bark can be peeled off, the outer and inner bark of the roll separated for different uses.

Tubes of cells in the wood and bark produce and carry a thick, aromatic liquid called *resin* (in North America *resin* is called *pitch*). Resin and *sap* are the basis for *medicines*, *foods* and *glues*. *Branch wood* and *foliage* are good for *fuel*, *bedding* and *tools*. *Young leaves* and *cones* are rich in Vitamin C and edible.

The features which make Sitka spruce so useful to modern western civilisation are somewhat different and are discussed in Chapter Fourteen.

Indigenous peoples and Sitka spruce

Many First Nations and their clans lived and live within and adjoining the Pacific temperate rainforest region (Map 5). Umpqua and Suislaw people live in southern Oregon from the forested Coastal Range to the Pacific Ocean; the Kwakwaka̲'wa̲kw, Nuu-Chah-Nulth and Coast Salish live in the area of Vancouver and its island, Seattle and Puget Sound; the Haida, Tsimshian and Tlinglit live in Alaska and northern British Columbia; the Chugach Alutiiq (Sugpiaq) and Tanaina round the Gulf of Alaska.

The Haida

The land where the *Haida* First Nation people settled, included Haida Gwaii and part of the Alexander Archipelago. Their lands reach as far north as 54° along the Alaskan coast. Haida Gwaii consists of two large and about 150 small islands off the northern coast of British Columbia.

The islands, fjords, estuaries and inlets occupied by Haida are ecologically diverse in temperate rainforest and other habitats. There is variation in rainfall, plants and animals, landscapes and soil. Haida homeland is one of the areas where Sitka spruce grows remarkably abundant, tall and thick. Many other conifers also grow vast in this landscape, for instance Yellow cedar, Lodgepole pine and Mountain hemlock.

10,700 years ago, early Haida lived at Kilgii Gwaay, then a mosaic of sedge and fern meadows amongst hemlock, pine, spruce and alder trees. They ate fish, crabs, periwinkles and mussels from the coast which was submerging beneath the sea. They used Sitka spruce to make wooden wedges, a plank, bundles of roots and braided strands of split roots, as is still done today.

In later times Haida people lived by fishing, gathering shellfish, hunting animals (particularly marine mammals) and gathering plant foods. Plant foods included berries, stalks, tree fibres, seaweeds and roots. These were either eaten raw or preserved by drying for the winter.

With an abundant supply of wild food they eventually settled in semi-permanent villages and developed a rich culture. They travelled long distances in their big Western red cedar canoes, and sometimes acquired slaves from far-off places.

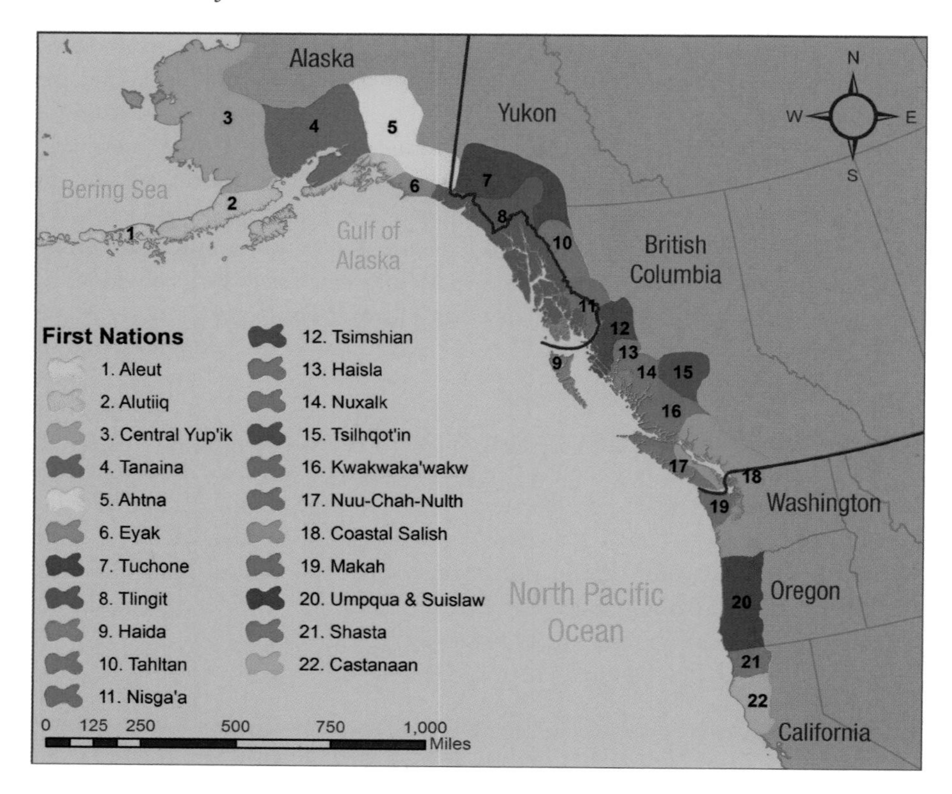

MAP 5. First Nations of the rainforest coast. Prepared by Luke A'Bear.

DATA AFTER: CANADA'S FIRST PEOPLES: HTTP://FIRSTPEOPLESOFCANADA.COM/FP_GROUPS/FP_NWC5.HTML; BRITISH COLUMBIA MINISTRY OF EDUCATION: HTTPS://WWW.BCED.GOV.BC.CA/ABED/MAP.HTM; MAP OF WEST COAST FIRST NATIONS: HTTP://WWW.JOHOMAPS.COM/NA/CANADA/BC/MAPS.HTML.

Roots of Sitka spruce were used by Haida people to weave *cooking utensils*, *baskets* and beautiful *hats*. The baskets are so tightly woven that they are watertight and can be used for liquids and for cooking. The hats are woven, painted or embroidered with traditional patterns. Thread from roots also made *fishing lines* and *twine*, particularly for binding handles to tools.

From the *trunk* of the Sitka spruce the Haida made *canoes*, *paddles*, *totem poles* and *temporary shelters*. In recent centuries, trunks were split to make planks for large houses.

The Haida used *branch wood* for *fuel*. But *knotty branches* were steamed so that they could be easily shaped into *fish hooks* to catch massive cod and halibut.

They made *antiseptic for cuts and burns* by melting down the *resin*. Haida and groups further south made *chewing gum* from *resin*. They put it in cold water first to harden it. The *Haida* had another use for *resin* as *medicinal ointment* and as *glue* to fasten a *head to the shaft of harpoons*.

Boats or homes built of planks needed Sitka spruce *resin* for *caulking*: plant material like moss or fibre was squeezed into the long gaps between planks, and covered with *strips of resin* to make them watertight.

Haida, Tsimshian, Tanaina and Tlingit liked to eat the sweet *inner bark* layer of trees. They garnered it when the sap was starting to rise in May. First they peeled off long, parallel-sided, strips of bark, lay them on the ground, then scraped off pieces of inner bark until it was all removed. They either *ate the pieces* straightaway or dried them and made *sweet cakes* for the following winter.

When the inner bark had been removed Haida used the remaining long strip of *outer bark* to make *roofing* for *their houses, seasonal shelters* and for *fuel.* It is the scars left by removing long strips of bark that can still be seen today as evidence of past woodland management. People did not kill the tree by taking bark this way, because they pulled off only one or two vertical strips from each tree.

Sprays of prickly *foliage* were used during ceremonial *winter dances* to show its protective powers against evil. Young *shoot tips* were eaten raw and made into *jellies, cordials* and '*tea*'.

The great advantage of Sitka spruce compared with other conifers is the light weight of its wood for its considerable strength, and the many uses for all parts of the tree. Most other temperate rainforest conifers had fewer uses. But Sitka spruce is second to Western red cedar as a resource where both occur.

The Tlingit

Tlingit people use and used at least 53 plant species. Their lives were governed by the seasons in which a plant or animal was available and the place or places where it could be found. Tlingit seasonal resources included at least nineteen sea fish, four sea mammals, 23 land mammals, five birds and twenty littoral invertebrates (information from Thornton, 2008).

Unlike First Nations to the south, Tlingit lack Western red cedar in much of their territory. Its northern limit in Alaska is about 55° N, but Tlingit territory continues beyond to the Yukon border. Here, Sitka spruce is a very abundant tree, co-existing with Yellow cedar, Western hemlock, Mountain hemlock and Red alder along the coastline.

For Tlingit people, Sitka spruce is *the* most important tree.

Tlingit use it in ways similar to the Haida and also for items which would be made with Western red cedar further south. For instance, big, ocean-going, *Sitka spruce log canoes* were important and are still made occasionally. They were up to 7 m (24 ft) long with bow and stern prows slanted in line high above the water. Six people paddled many others out from the land to fish, hunt or travel. August von Bongard described how one big tree on Mount Verstovia near Sitka could be hollowed out into a canoe:

> Elle est jusqu'au sommet couverte de forêts épaisses, où les pins et sapins, qui y regnent seuls, atteignent par fois la hauteur prodigieuse de 160 pieds, avec un diamètre de 7 à 10 pieds. Un seul trone de ces arbres suffit aux indigènes pour creuser un canot qui peut contenir jusqu'à 30 hommes avec tous leurs utensils.
>
> Bongard, [M.] 1831/1833

Tlingit, Haida and Tsimshian are famous for their '*formline*' *artwork* using Sitka spruce wood as its base. It is an ancient method originating more than two thousand years ago and many people have taken up painting this way recently.

The basic design elements are simple but can be built up into complex pictures; colours are rich, pictures are stylised and full of cultural meaning to a clan.

Clan stories, events and animals are painted or shallow-carved onto objects of *Sitka spruce wood*, such as *screens*, *boxes*, *ornaments* and *totem poles*. Designs to be woven into cloth are first carved onto a pattern board which a weaver uses as a template.

Tlingit and Haida are famous for their beautiful but functional *baskets and hats, woven from roots* of Sitka spruce. Roots were gathered by women in late spring from young trees. Delores Churchill, a renowned basket-maker kindly invited us to visit her basket-making class in Sitka. She told us that Sitka spruce trees growing on sandy shorelines have the most suitable long, flexible roots near the surface: a length of 1.5–2.4 m (5–8 ft) is best:

> Delores Churchill remembered how her mother, after collecting them, scorched each root in the remains of a fire and then pulled or scraped off the bark. A bear's canine tooth was used. The smoothed root was then coiled, tied lightly and dried. It would be used during the next winter.
>
> She took a coiled root brought in by Corinne Parker, a young basket maker whose great-grandmother had prepared the root about a century ago, coiled and dried it, but never used it. Delores and Corinne had put the coil to soak in water a couple of days previously.
>
> She sat down, took one end of the wet root and with a very sharp knife cut it downwards into half. Then she put the nearer half between her teeth and sliced the far piece downwards into two: a length with an outer, rounded and coloured surface and one inner, flat surface; and a second length with both surfaces flat and white.
>
> Delores then took this second, flat length and sliced it downwards again into two finer, white flat-surface pieces. She continued slicing down several times, producing finer and finer white threads several feet long. This was repeated with the outer length with the one rounded surface. The half root was still held in her mouth to ensure even pressure while she sliced.
>
> The (fewer) darker, threads with one rounded outer surface become warp lengths; the (many) white, flat, inner threads are used for the weft. In the 'old days' women used a sharpened mussel shell not a knife.
>
> From Ruth Tittensor's hand-written notes, 18 July 2014

Spruce-root baskets, made using wet threads and hands, are woven in twining stitch, with double weft threads wrapped around each warp. As the strands are so fine, this produces tightly woven, watertight containers of varied shapes for different purposes. Coloured patterns are sewn on or added with plant dyes.

Basket-making with Sitka spruce roots is a First Nations tradition with an ancient heritage; the oldest basket yet discovered, a food-collecting container, was found in 1994 in the muddy banks of the Thorne River, Prince of Wales Island, Alaska by Dave Putnam:

> As Dave Putnam walked the Thorne River's muddy banks, he held in his mind concepts of a geologically dynamic landscape. He envisioned changes in sea level since the end of the last Ice Age, perhaps 10,000 years ago, the gradual accumulation of fine sediments released by the river into the calm waters at the tidal fringe, and the relentlessly migrating river channel cutting 'windows' into the depths of the

FIGURE 5.2. Tlingit woman weaving baskets and hats from Sitka spruce roots, *c.*1905.

PHOTO: COURTESY ALASKA STATE LIBRARY, VINCENT SOBOLEFF PHOTO COLLECTION, P1-051.

> mud flat. It was in one of these windows that Putnam noted the fringe of tattered fabric peeking from the mud. This fringe turned out to be the Thorne River basket.
>
> To traditional basketmakers of Southeastern Alaska the 1994 discovery of the nearly 6000-year-old spruce root basket is a dramatic symbol of their ancestral connection to the land. They see in the intricacies of the weave an expression of skills passed from mother to daughter for countless generations. To archaeologists this is the oldest example of basketry yet to emerge from the region. In this basket, and in the muds which held it, rests a unique opportunity to make comparisons across time and to speculate about the makers, their lifeways, and the changing world they lived in nearly sixty centuries ago.
>
> Fifield 1998

Margaret Davidson of Seattle recorded the basket with a detailed drawing while it was both refrigerated and submerged in alcohol-water to prevent it turning to dust.

To conserve it, the basket was treated with polyethylene glycol to ensure the plant's cell walls did not degrade and cause the basket to disintegrate; then it was gradually freeze-dried so that it could be exposed to air and displayed.

FIGURE 5.3. Fine roots of Sitka spruce with bark removed.

PHOTO: ANDY TITTENSOR, 2014, COURTESY DELORES CHURCHILL.

FIGURE 5.4. Delores Churchill makes a clean cut at the end of the root.

PHOTO: ANDY TITTENSOR, 2014, COURTESY DELORES CHURCHILL.

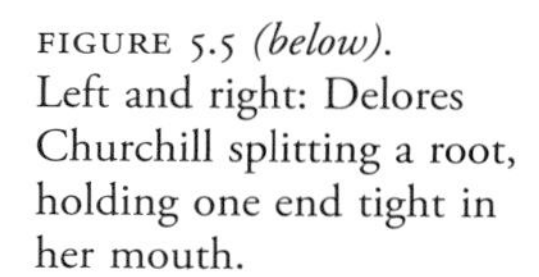

FIGURE 5.5 *(below)*. Left and right: Delores Churchill splitting a root, holding one end tight in her mouth.

PHOTO: ANDY TITTENSOR, 2014, COURTESY DELORES CHURCHILL.

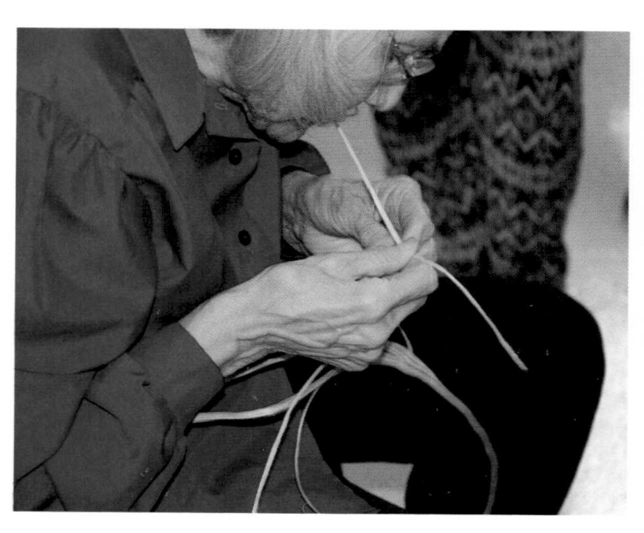

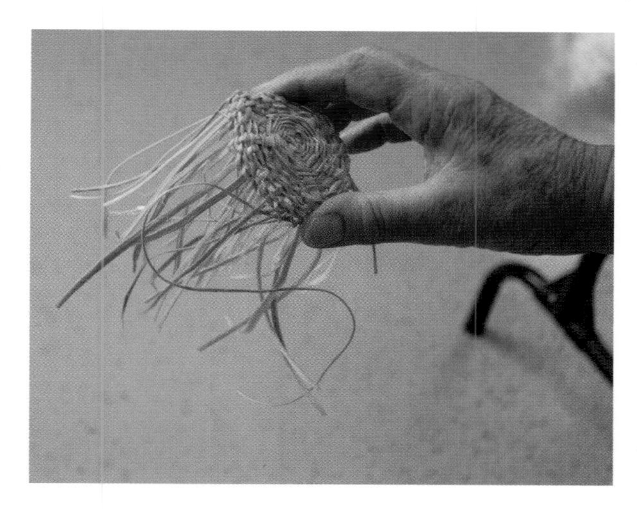

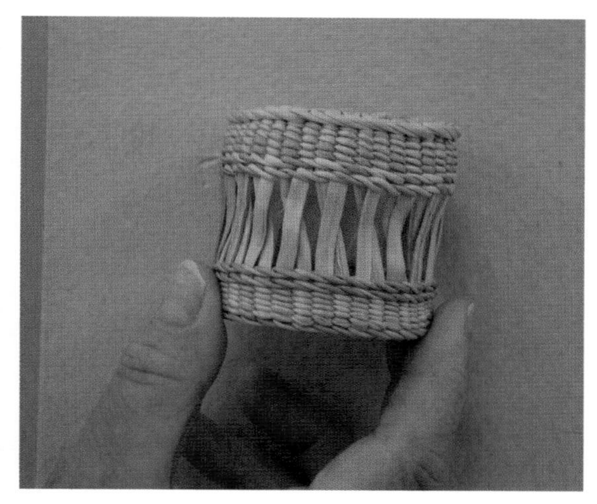

FIGURE 5.6 *(above left)*. Using finely-split roots to weave a small basket.

PHOTO: ANDY TITTENSOR, 2014, COURTESY DELORES CHURCHILL.

FIGURE 5.7 *(above right)*. Finished basket.

PHOTO: ANDY TITTENSOR, 2014, COURTESY DELORES CHURCHILL.

FIGURE 5.8. Large Tlingit basket woven from Sitka spruce roots.

PHOTO: ANDY TITTENSOR, COURTESY SHELDON JACKSON MUSEUM.

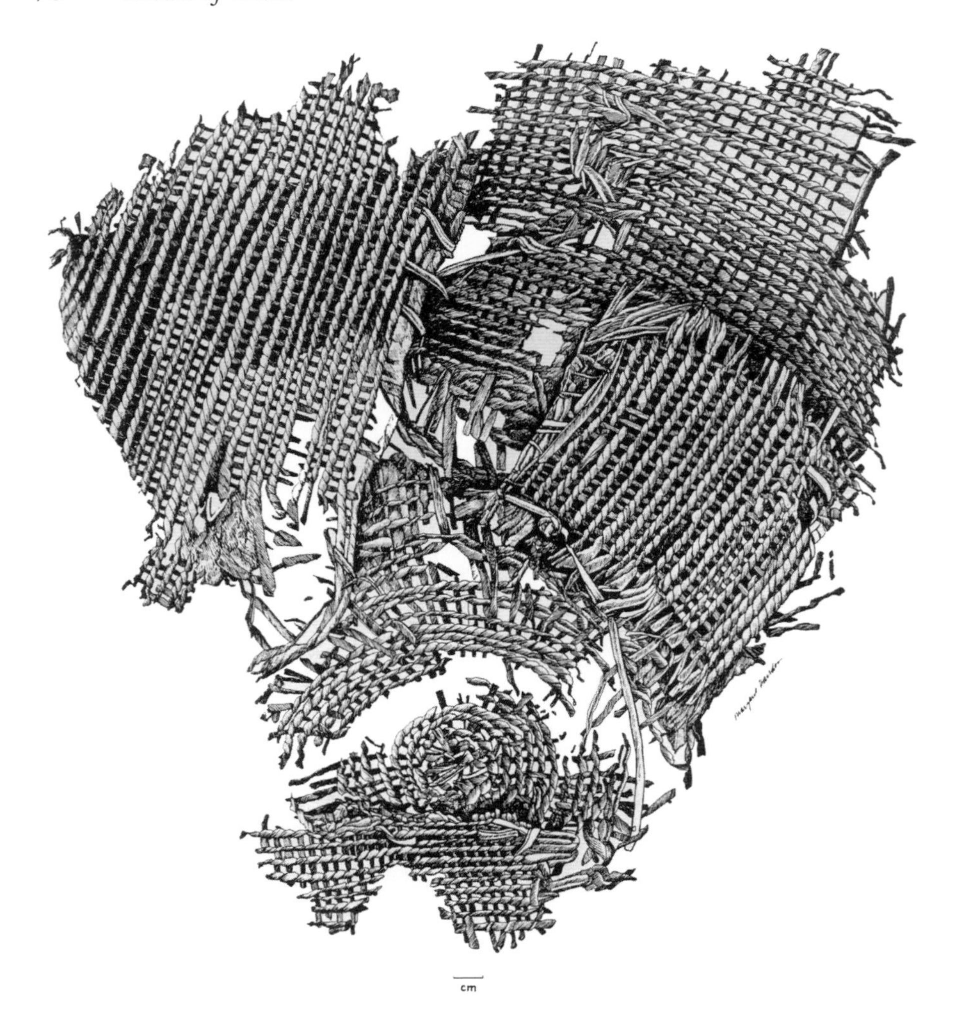

FIGURE 5.9.
Archaeological rendering of one side of the 6000-year-old Thorne River Basket as it looks today.

The basket was dated to 5450±50 years BP by radiocarbon dating: what a great age for it to survive! Six millennia ago, climate became moister and temperate rainforests were accumulating a different complement of trees than today. It would be several millennia before Sitka spruce would reach the ecological status we recognise today.

Margaret Davidson reflected on the Thorne River Basket:

> What I find most thrilling about this find is how old it is – older than the pyramids, for example – and also how nearly complete it is. Furthermore it is woven using a material that requires complex preparation in a weaving technique that is still used today. This basket connects all traditional basketmakers of Southeastern Alaska to the deep past, to the land, and to each other.
>
> Email message from Margaret Davidson, February 2015

The surviving portion of this ancient basket, 28 × 33 cm (11 × 13 in) in size is displayed in Juneau-Douglas City Museum, Alaska.

FIGURE 5.10. Tlingit fishing camp in Southeast Alaska; note the spruce trees and canoes at the shore.

PHOTO: COURTESY ALASKA STATE LIBRARY, WILLIAM NORTON PHOTO COLLECTION, P226-427.

FIGURE 5.11. Salmon shoaling in the River Eve, Vancouver Island: important food past and present.

PHOTO: GEORGE BROWN, 2012.

Fish, particularly salmon, trout and eulachon (*Thaleichthys pacificus*), were a main food resource of the Tlingit people and they caught them in small numbers with spears and hand-held hooks. In living memory, they gathered larger quantities of fish during the spring and summer spawning runs up-river from the ocean, using basket-like cage traps. These were labour-intensive to

FIGURE 5.12. 500-year-old fish trap from the Juneau-Douglas City Museum collection, Alaska.

PHOTO: STEVE HENRIKSON, ALASKA STATE MUSEUM, JUNEAU.

make and were owned by a whole family which camped each year closed to the creeks and rivers in which the traps were placed.

About 200 old weirs of stone, wooden mesh or wood palisades have been found in tidal Alaska waterways, some with gaps in the stonework, which archaeologists think might have contained fish traps which were taken away or got broken and were washed away.

A *basket-like fish trap* was eventually discovered in 1989, buried in the silts along Montana Creek, a shallow tributary of the tidal Mendenhall River, Alaska and only 1.5 km (*c.*1 mile) from the open ocean of the Gastineau Channel.

The Montana Creek fish trap had been made by weaving ten pieces of pliable, green Sitka spruce branch wood equidistant through 44 straight slats of (probable) Western hemlock; they were then curled into hoops so that the whole item formed a big cylinder. Each hoop was lashed by Sitka spruce root ties to each slat to hold the trap in shape. The two ends of each individual spruce hoop were also held together by spruce root ties. A basket-work funnel 60 cm (24 in) long, narrowing to a cone by using alternating lengths of slats woven with five spruce hoops, was inserted into the entrance to guide fish into the 'cage'. Dolly Varden trout were probably the main fish to be caught in the Montana Creek.

This large trap is 2.74 m (108 in) long and 80 cm (31.5 in) diameter; the hemlock slats are 3.5 cm (1.4 in) apart and the Sitka spruce hoops are about 30 cm (12 in) apart. Short cords of spruce root held a hinged door into the trap for extracting fish. This very rare fish trap has been dated to 1300–1500 CE.

FIGURE 5.13. Replica of fish trap in the Juneau-Douglas City Museum collection, Alaska (Ref. JDCM 2006.46.001).

PHOTO: STEVE HENRIKSON.

Along with a modern replica made by local people and the Alaska State Museum, it is on display at Juneau-Douglas City Museum.

Other First Nations and Sitka spruce

The Alaskan Dena'ina and Chugach Alutiiq people used *wood* of Sitka and White spruce to make tools, for instance, *digging-sticks*, *spear-shafts*, *shovels*, *tongs*, *hammers* (mauls and clubs) and *bows, sleds* and *canoes*. They used *bark* for the *roofs*, *floors and sides* of *seasonal, summer shelters*; bark was also a source of *colouring to dye* their *fishing-nets*.

Chugach people used swathes of outer bark to construct the roofs and sides of *smokehouses* and *steam bath-houses*, and to make *containers* waterproofed with *resin*. Chugach not only chewed young male and female *cones* raw or cooked for diarrhoea and bad colds but for *toilet paper*.

Nuu-chah-nulth people tied *branches* onto *submerged wooden fences* in the sea. *Herring spawn* (female eggs or roe) got caught onto the branches which were hauled out and dried, when the spawn was removed and eaten.

Kwakwa̱ka̱'wakw groups of the Quatsino First Nation lived in the north-west of Vancouver Island. They hunted whales and other marine life for food, but used Sitka spruce for *digging sticks*, *big rakes to catch* and *hold herring-spawn* and *rope*; from its *sap*, they made *medicines*. From their *bark stripping* and *plank cutting* activities, Quatsino left culturally modified Sitka spruce in Vancouver Island forests.

Other groups ate young *male* and *female cones* raw or cooked. Some used Sitka spruce *wood* for the *handles* of *fish-hooks* (*gaffs*). *Thread* from *roots* was used for *nets*, for *sewing* together *wooden boxes*, to *tie handles* to *fish-hooks* and *harpoons*, to make *fishing* and *harpoon lines*, *to tie the planks* of *houses*. More information can be found in the Bibliography.

And then…

In the late eighteenth century Sitka spruce was discovered by European explorers as a tree unknown to their science. Some seeds were despatched to Britain in 1831, but for quite different reasons than First Nations' needs. The intended use of Sitka spruce in Britain and Ireland would have been puzzling to contemporary First Nations who lived in such a well-forested and resource-full environment compared with the British and Irish.

The urgency of its despatch to Europe will be discussed in Chapters Eight and Nine.

CHAPTER SIX

Prehistoric Lives and Woodlands in Britain and Ireland

> The biggest revolution of the past 50,000 years was not the advent of the internet, the growth of the industrial age out of the seeds of the Enlightenment, or the development of modern methods of long-distance navigation. Rather, it was a seemingly trivial event that happened rather quickly around 10,000 years ago – the dawn of the age of agriculture.
>
> Spencer Wells 2011

> It is easy to think of the Mesolithic as a long period of relative stability… However, the environment was certainly not stable; the changing temperature and vegetation, along with marked changes in sea-level, must have required ingenuity and adaptation of practices and habits for successive generations of Mesolithic people.
>
> John Bunting 2015

Introduction

Before I unfold my biography of Sitka spruce any further, I will consider the dramatic background to its import and welcome into Britain and Ireland.

As in North America, Europe was ice-covered for long periods. However, the setting of Europe, the position of Britain and Ireland at the continent's Atlantic extremity, their segregation into islands, and the countrywide incorporation of farming into hunter-gatherer cultures five thousand years ago – have guaranteed that the environment, flora and fauna are quite different from North America.

This chapter therefore discusses the changing prehistoric landscape, particularly trees and woodlands. Chapter Seven will outline the interactions between people and their environments after the time of surviving written records under Roman rule in the first to fifth centuries AD.

People, but no spruces

Four main glacial periods in northern Europe and North America spanned the geological Pleistocene epoch, which lasted from 2.6 million to 11,700 years ago. The glacial periods were separated by warmer, interglacial episodes. The latest glaciation, called the Devensian in Britain and the Midlandian in Ireland, started about 110,000 BP and came to an end about 11,700 BP when the ice sheets had melted.

Palaeolithic or Old Stone Age describes the human cultures of north-west Europe during the last one million years until 11,700 BP. As there were long

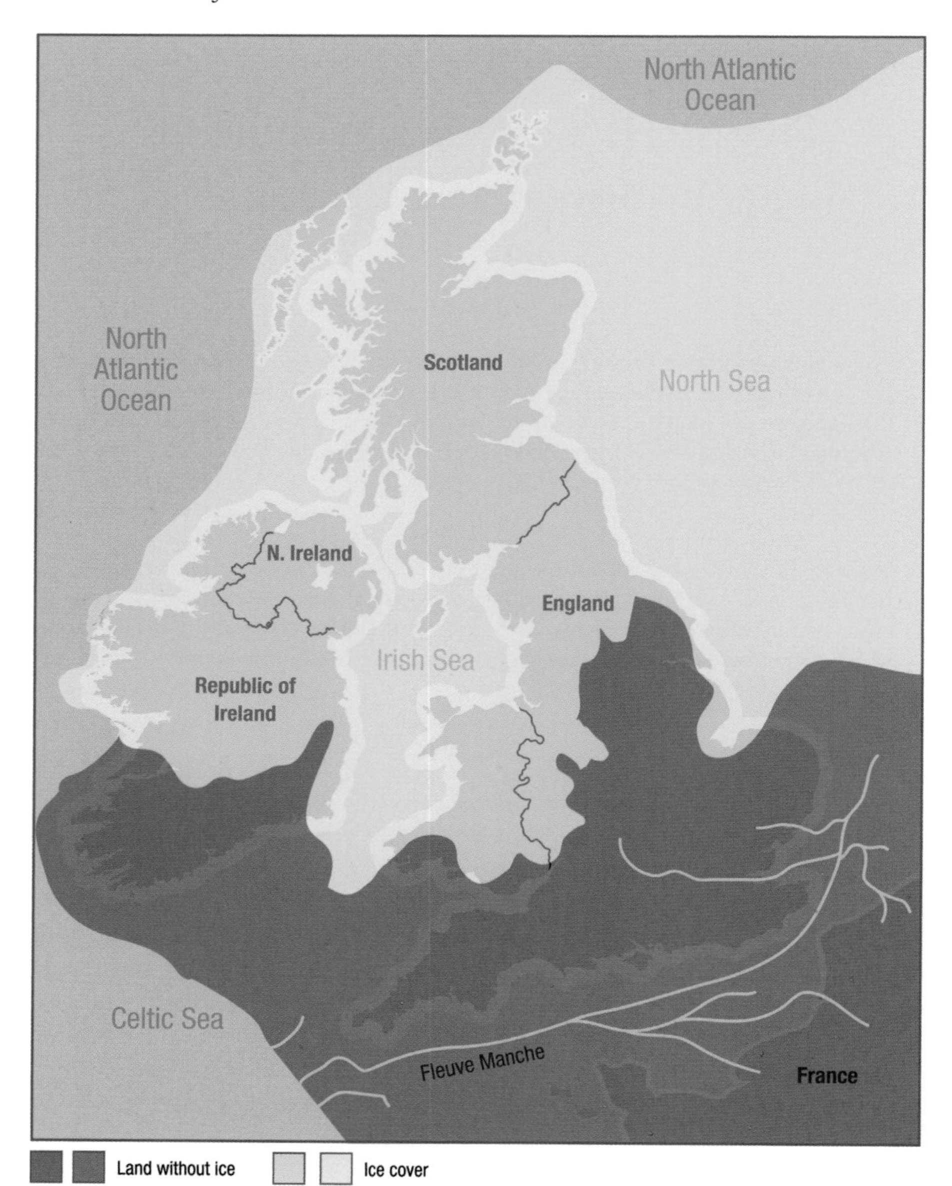

MAP 6. Devensian ice sheet. Prepared by Susan Anderson.

DATA AFTER BRITISH GEOLOGICAL SURVEY: HTTP://WWW.BGS.AC.UK/DISCOVERINGGEOLOGY/GEOLOGYOFBRITAIN/ICEAGE/HOME.HTML?SRC=TOPNAV; FREE MAPS OF IRELAND, ICE AGE IRELAND: HTTP://WWW.WESLEYJOHNSTON.COM/USERS/IRELAND/MAPS/HISTORICAL/ICE_AGE.GIF; QUATERNARY PALAEOENVIRONMENTS GROUP, UNIVERSITY OF CAMBRIDGE: HTTP://WWW.QPG.GEOG.CAM.AC.UK/LGMEXTENT.HTML.

periods of very severe climate inimical to humans during the Pleistocene epoch, Britain and Ireland were occupied only intermittently.

Devensian ice cover fluctuated in extent, with glaciers advancing or melting, but it reached its last maximum between about 26,500 BP and 19,000 BP. The British, Irish and Scandinavian ice sheets were joined for lengthy periods but not constantly. The maximum extent of ice over Ireland and Britain at 22,000 BP is shown on Map 6.

Two evolutionarily advanced human species, Neanderthal and Modern people (*Homo neanderthalensis* and *H. sapiens*) with Old Stone Age culture lived

MAP 7 *(opposite)*. Reshaping of Britain and Ireland after 18,000 BP. Prepared by Susan Anderson.

AFTER COLES (1998); REPRODUCED BY PERMISSION OF VISTA, UNIVERSITY OF BIRMINGHAM AND THE COUNCIL FOR BRITISH ARCHAEOLOGY.

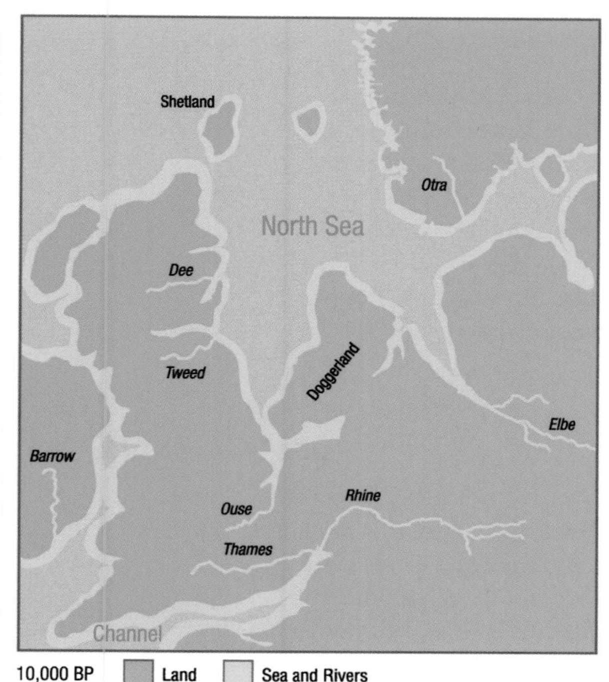

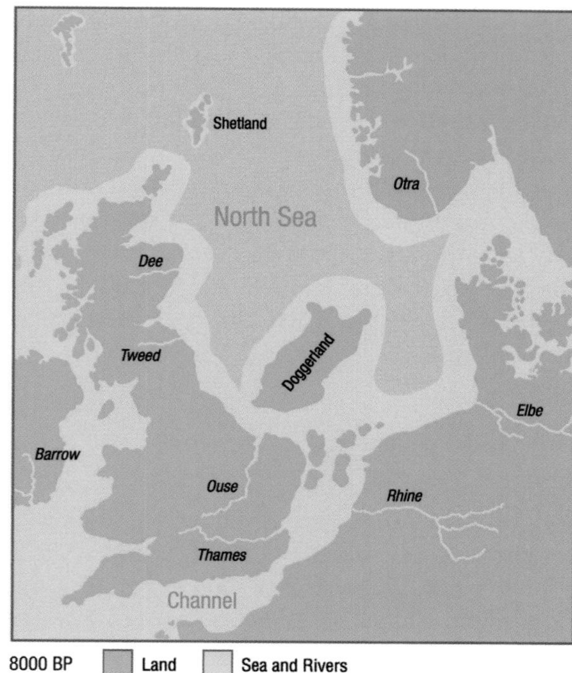

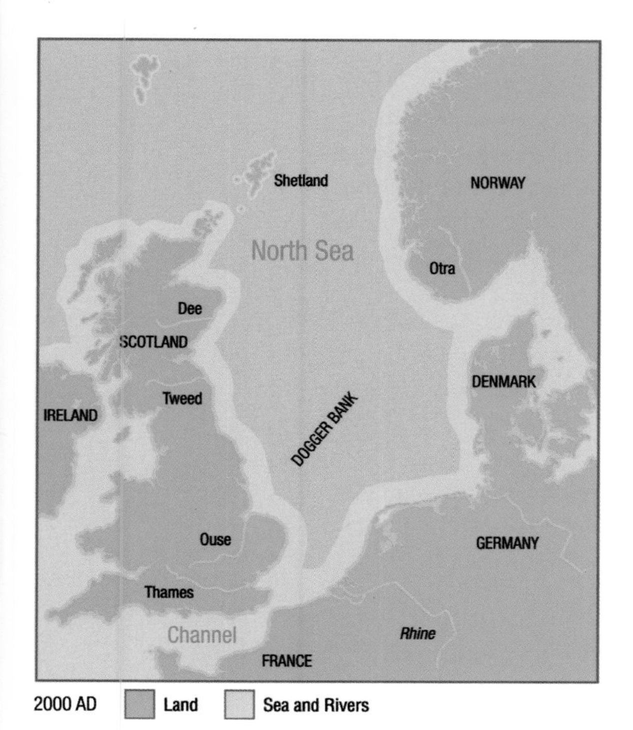

in challenging conditions on tundra landscapes not far from the edge of the ice sheet. These peoples were hunters, fishers and foragers, and they left tools and food remains which survived in certain caves. Neanderthal people had become extinct in western Europe by 40,000 BP but more advanced humans (*Homo sapiens sapiens*) were already established – and we are still here now!

In the early Pleistocene, before ice started to expand and thicken, trees we can identify as spruces formed temperate, mixed conifer-deciduous forests with pine, hemlock, Bald cypress (*Taxodium* species), oak, alder and hickory (*Carya* species) in northern Europe.

During two interglacial periods of the Pleistocene and during a warmer period within the Devensian ice age about 60,000 BP, the Common or Norway spruce grew in what is now Britain.

But unlike North America, no spruce trees of any kind survived the last glaciation in refuges either within, or close to, the ice sheets over what are now Britain and Ireland. Even after all the ice cover had disappeared during post-glacial times, and even when other tree species were migrating and colonising these islands, *no spruce species arrived there of their own accord* – from refuges elsewhere or from the main tree population centres in southern Europe. Common spruce did colonise central and eastern Europe, and reached eastern Scandinavia from a nunatak in Russia. But Britain and Ireland have been spruce-less since before the end of the Devensian glaciation.

The nearest native populations of spruces today are Common spruce, some 1125 km (700 miles) away in Norway and south-east Germany. Indeed, there are only three native conifers which grow in these oceanic islands: a juniper, a pine and a yew (see Chapter One). They are infrequent as woodlands in the modern landscape.

Post-glacial people

Nobody survived on the tundra of north-west Europe during a particularly cold spell between 25,000 and 18,000 BP, but humans returned again after this when temperatures rose slightly. About 12,100 BP, Late Glacial reindeer hunters made a long coastal journey from central Europe around the northern coastline of Doggerland – a low, marshy area between Germany, Scandinavia and Britain – and then round the northern coast of Scotland to the island of Islay.

But Palaeolithic culture in Britain was coming to an end, superseded by cultures of Mesolithic or Middle Stone Age people who colonised new ice-free land and coastlines. They travelled across Doggerland; they negotiated the Channel River (alias Fleuve Manche) from the south; or progressed north along the Atlantic seaboard from Iberia to Ireland and Scotland, reaching Islay about 9300 BP.

Doggerland was rich in edible mammals and birds, marsh and wetland plants. It had been a focus of hunting for Palaeolithic people and continued to be important to some immigrant and resident Mesolithic cultures: people expanded their food intake to include more marine and littoral resources on offer as sea level rose from large volumes of melting ice.

It took several centuries after 7800 BP for sea water to rise sufficiently to inundate Doggerland completely. This new North Sea then broke through the land barrier ar its south-east corner and flooded the Channel River and the wide swath of land either side to form the English Channel – alias La Manche (see Map 7).

The final land connection to continental Europe had been breached. After that, wild animals, plants and humans had to negotiate expanses of water to reach and colonise the islands of Britain and Ireland.

Plant survival amongst the ice

Two kinds of evidence show that the ice sheet over Scandinavia contained some ice-free nunataks during the last glaciation. First, and amazingly, Common spruce trees still alive on mountains at 910 m (2985 ft) are considered to be

9500-year-old relics of nunatak populations: ice did not retreat from around these havens until 9000 BP, well after these now-ancient spruces started to grow:

> Our results demonstrate that not all Scandinavian conifer trees have the same recent ancestors, as we once believed. There were groups of spruce and pine that survived the harsh climate in small, ice-free pockets, or in refuges, as we call them, for tens of thousands of years, and then were able to spread once the ice retreated. Other spruce and pine trees have their origins in the southern and eastern ice-free areas of Europe.
>
> Willerslev 2012

Second, some Common spruce trees which nowadays grow on the known glacial haven of Andøya Island, north-west Norway, are examples of a rare mitochondrial DNA 'haplotype' unique to Scandinavia. Sub-fossils with the same haplotype occur in lake sediments at Trøndelag in central Norway, dating between 22,000 and 17,700 BP, well before the time of ice melt. This haplotype of Common spruce has thus descended from ancestors living on nunataks during the Devensian glaciation.

Geologists have shown that many mountain tops between 700 and 960 m (2300–3150 ft) altitude in north-west Scotland were free of ice during the Devensian glaciation, but it is uncertain what vegetation these nunataks supported. As sea-level was much lower during the Devensian, there were probably some ice-free havens for plants and animals along Atlantic coasts – similar to refuges in coastal North America occupied by Sitka spruce.

Re-colonisation by plants and animals

By about 14,000 BP some of north-west Europe was ice-free: plants and animals could colonise the bared ground and form tundra habitats, followed, in greater warmth, by grassland and low shrubs. However, about 12,700 BP, an event in North America had profound repercussions in Britain and Ireland. A huge quantity of freshwater from the melting Laurentide ice sheet discharged into the Atlantic Ocean, causing the warm waters of the 'Atlantic Conveyor' current to cool and slow *before* they reached the Arctic Ocean. So Ireland and Britain got much colder and tundra returned; giant deer and reindeer became extinct in Ireland.

But about 11,500 BP, the Conveyor returned to its normal path, warmth resumed, so more plants and animals colonised the tundra. For three millennia until 8500 BP, the climate continued to warm, increasing from an average July temperature of 9.5° to 17° C (49°–63° F).

Then, for another 3500 years a warm and wet climate provided 'optimum' conditions of the Holocene in northern Europe. From 3000 BP to the present the climate has gradually cooled again. However, there have been smaller-scale climate fluctuations such as a Mediaeval Warm Time between 1000 and 1250 AD and a Little Ice Age between 1500 and 1850 AD.

The flora of European tundra paralleled those in North America. There were sedges (for instance *Carex bigelowii*, *Eriophorum vaginatum*), grasses (for instance *Trisetum spicatum* and *Deschampsia cespitosa*), mosses (for instance

Polytrichum alpinum) and lichens such as Deer 'moss' (*Cladonia rangifera*). With increasing warmth, shrubs of the heather family (for instance *Ledum palustre, Vaccinium uliginosum* and *V. vitis-idaea*) and cushion plants such as Tufted saxifrage (*Saxigraga cespitosa*) spread. Dwarf willows, such as *Salix reticulata, S. herbacea* and *S. lanata* were also common.

In Britain, larger animals of tundra included Red deer, Wild horse (*Equus ferus*) and Blue hare (*Lepus timidus*). Small species included lemmings (*Lemmus lemmus*) and Red grouse (*Lagopus lagopus scotica*), many duck and raptors. Their remains have been found at caves such as Cresswell Crags, Nottinghamshire and Gough's Cave in the Mendip Hills, Somerset along with human remains.

People's lives on the tundra

Mesolithic people occupied Britain and Ireland for more than half of the Post-glacial (Holocene) period, from about 12,000–6000 BP. They needed fresh water as well as food, and shelter from the weather and predatory animals. In northern Europe and Britain, Mesolithic tundra environments grew plants such as those of the heather and rose family with edible berries, while mammals and birds such as wild horses and grouse could be culled for animal protein. Rivers, streams and lakes contained edible fish and invertebrates such as freshwater mussels and crayfish. Oysters, limpets, whelks and edible seaweeds grew on rocky shores, but their availability varied according to tidal cycles. People also fed from the year-round resources of shallow seas.

A dynamic environment

There were substantial environmental changes in the earlier Holocene period. Changes in sea level occurred when melting ice released a heavy weight from the land – which uplifted. Other land was inundated when ice melt-water raised sea levels. Earth tremors were frequent because the pressure of ice on land lessened but more water in the seas increased pressure on the ocean floor. A tsunami travelling from Norway hit eastern Scotland about 8000 BP and probably travelled inland, affecting coastal populations of people and animals.

With variable weather, waterways changed course, springs dried up or appeared, and lakes developed into marshes and raised bogs. Flora and fauna came, went, increased or decreased according to climate, weather – and how people used them. For instance, bird migrations and salmon runs would take time to develop stable patterns when the European ice sheet retracted, and a continent and islands re-appeared after 100,000 years!

The landscape of Britain and Ireland was a geological and ecological mosaic. Ecosystems assembled but changed composition when their species declined or others colonised. Freshwater could metamorphose into dry land, while rivers in spate spread over their flood-plains. Mesolithic people had to adapt when their food migrated or disappeared, and new foods came or grew in their place.

When tundra gave way to trees, human residents had no option but to form relationships with spreading woodlands. As trees matured, the landscape became multi-dimensional with new habitats: the tree-canopy, trunks, branches, dead trees and wood, saplings, herbaceous flora, and soils with tree-leaf humus.

Colonisation and occupation by trees

As adjoining landscapes became suitable, shrub and tree seeds could start the slow emigration out of their nunataks and away from their ice age sites in the Balkans or Iberia. How fast, far and in what direction a species travelled depended upon its refuge sites (as with Sitka spruce), its fertility, direction of seed dispersal and obstacles like mountains and inhospitable habitat.

The Pyrenees, Alps and Carpathians, aligned east to west, were more of a barrier to tree migration than the north-south alignment of the Rockies and Coastal Range in North America. There were also many smaller barriers, such as southern Scotland's mountains and big estuaries, which affected tree colonisation northwards. Oak took two millennia to reach mid-Scotland from southern Europe but another millennium to reach its north coast.

By 10,000 BP, juniper, three species of birch (*Betula nana*, *B. pendula* and *B. pubescens*), hazel and tall Goat willow (*Salix caprea*) were colonising the tundra of southern Britain. Starting about 9500 BP from south-west Ireland, hazel colonised Britain via the west coasts of Wales, northern England and Scotland, reaching round to north-east Scotland and England 500 years later. Woody nuts which float in fresh or sea water facilitated its travel!

During the dry Boreal period, between 9000 and 7500 BP, Scots pine arrived in England from several directions. With hazel, it overtook the birches and willows, forming a wide zone across middle England. Deciduous species moved along behind the Scots pine from the south and east. Bits of tundra remained in parts of Ireland, North Wales and Scotland. And about 8500 BP, Scots pine expanded rapidly from a haven in north-west Scotland to form a large central forest zone. It reached the northern Orkney Islands where, with birches and Hazel, it formed marshy woodland.

Deciduous, mainly oak, woodland already occupied southern England, but when, about 7500 BP, temperature and rainfall increased, Wych elm immigrated into humid gullies, and limes (*Tilia cordata* and *T. platyphyllos*) colonised calcareous soils.

After the Conveyor-induced cold spell, Ireland was a landscape of ice-scoured hollows and lakes, drumlins and dunes growing tundra species. Coasts suited Sea buckthorn (*Hippophae rhamnoides*), Sea pink (*Armeria maritima*) and Scurvy grasses (*Cochlearia* species). Marsh and reed bed flora filled the hollows and riversides. On unstable soils, heathland of crowberry (*Empetrum nigrum*) developed. On richer soils, grasslands developed, along with low flowering plants, particularly Meadow rue (*Thalictrum* species). But in Ireland there were no aurochs, horses, red or giant deer to feed on grasslands and shrubs.

Hazel formed extensive woodlands on rich, damp soils after 9400 BP. Mature individuals produced huge numbers of nuts (estimated at 0.5 tonnes per ha (0.2 tons per acre)) which were important Mesolithic food.

On drier land, early Irish woodlands were mainly juniper, followed by birches and Eared willow (*Salix aurita*) which formed a sparse canopy. In the maritime south-west, woodlands of Eurasian aspen (*Populus tremula*) and birches developed. Riparian woodlands of willows, poplars and Common alder colonised reed beds and the margins of waterways.

Scots pine and yew migrated into southern Ireland initially. Pine later spread throughout the island, especially on drier sites in the uplands; yew formed pure woodlands in south and west Ireland and colonised peat bogs in the centre of the island. Oak and Wych elm were present before 8000 BP, but later spread, especially on drier ground in the west.

By about 7000 BP deciduous woodlands became the most common feature of lowland Britain and Ireland, forming part of the European Temperate Deciduous Forest biome, recognised by species' adaptations to its extreme seasonality. Location, geology and soils, altitude and frequency of major perturbations also determined the composition and processes of these woodlands.

Woodland fauna

Animal life also changed when trees colonised tundra. For instance, wild horses vanished, while ptarmigan, red grouse and mountain hares became confined to tree-less mountain-tops and upland heaths.

Characteristic species of pine woodlands, for instance capercaillie (*Tetrao urogallus urogallus*) and tiny Crested tits (*Lophophanes cristata*) survive today in pinewood remnants. European red squirrels (*Sciurus vulgaris*), which is adapted to feeding on pine seeds, survived in the subsequent deciduous forests (which offer a wider variety of tree seeds).

Large mammals which colonised along with deciduous trees included: aurochs (*Bos primigenius*), Brown bear (*Ursus arctos*), Red, Roe and Elk deer (*Cervus elaphus*, *Capreolus capreolus* and *Alces alces*), European beaver (*Castor fiber*) and Wild boar (*Sus scrofa*). Passerines such as chaffinch (*Fringilla coelebs*), Wood warbler (*Phylloscopus sibilatrix*) and Pied flycatcher (*Ficedula hypoleuca*) were the main birds of deciduous woodland.

Some species which immigrated into Britain did not reach Ireland (except when introduced later by humans). These include Roe deer, Brown hares (*Lepus europaeus*), polecats and weasels (*Mustela putorius* and *M. nivalis*) and moles (*Talpa europaea*). The modern DNA signature of mammals which did reach Ireland, for instance Pine martens (*Martes martes*), Pygmy shrews (*Sorex minutus*) and Blue hares, suggests that their ancestors travelled there from Europe not Britain. Brown bears were probably the largest mammal colonisers but there are none left in Ireland to provide DNA.

Mesolithic people and woodlands

Pine, birch and dwarf willows grew in the vicinity of Star Carr, Yorkshire, when Mesolithic people visited it after 11,000 BP. A reed-fringed lake was edged by fen with hazel and birches growing close by. For two periods of a century each, people visited Star Carr every spring and burned the reeds and birch trees. Following burning, patches of newly-sprouting shoots tempted deer to congregate, allowing them to be culled.

Star Carr visitors ate Red and Roe deer, Wild boar, elk and aurochs, as well as duck and grebes. However, after a couple of centuries, Willows and Eurasian aspens encroached upon the lake, which gradually dried and disappeared. People never returned to hunt at Star Carr which is now a peat bog amongst farmland.

When hazel was abundant on the Scottish islands of Rum and Colonsay, people roasted and ate large quantities of hazel nuts collected in one season about 9000 BP; the kernels supplied them with large amounts of oil and protein but they left the shells behind.

Mesolithic occupation sites on the coast could be overwhelmed by rising seas, as at Bouldnor Cliff, Isle of Wight. About 8000 BP, people built a residential platform of clay and timbers on a twiggy base by a river. Common alder grew on the river bank, while elm, hazel and oak trees infiltrated the pine trees further from the river. But peat started to grow, trees decayed and died, replaced by reeds and fen species. After seven centuries, everything had been submerged beneath the new English Channel. People abandoned their home, now 12 m (40 ft) below current sea level, and left charred hazel nuts, flints, and pieces of charcoal from their lives by the River Solent.

Archaeologists have found remains of a few Mesolithic timber dwelling-structures as well as seaworthy log boats and coracles. Five hazel-wattle fish traps abandoned in the River Liffey, Dublin were recently discovered in the estuary mud 6.3 m (21 ft) below modern sea level. These caught fish on the moving tide for their owners. For wattle, the hazel had been managed by coppicing on an 8-year cycle between 8100 and 7700 BP, to give flexible wands. Bones of European eels (*Anguilla anguilla*) are abundant in Mesolithic sites in Britain, Ireland and Europe, while elvers (young eels) have been found at a Mesolithic site at Loch Boora, Ireland. Numerous untamed waterways, pools and swamps made it possible to collect edible roots of water plants such as bulrush (*Typha latifolia*), bogbean (*Menyanthes trifoliata*) and water-lilies (*Nuphar lutea* and *Nymphaea alba*).

Peat formation

Peat grows in cool climates with high rainfall, poor drainage and low evapotranspiration. Continuous watery conditions favour Bog mosses (*Sphagnum species*), sedges and carnivorous plants. But when they die, they do not get decomposed in the waterlogged environment. Beneath the green, living surface, dead bodies of plants and animals pile up to make brown peat, which is 90% water! As the surface plants grow, so the dead pile beneath increases in height; the greater weight and lack of air turns the bottom layers black.

Peat formed naturally whenever and wherever conditions were suitable. In the valley of the River Severn peat grew naturally in hollows between sand and gravel ridges left by melting ice. When sea levels rose after 8850 BP, the rivers of East Anglia backed up and brackish water spread out over a dry plain. Alkaline fen peat grew there for about two millennia and hid the evidence of Mesolithic people who had previously lived on the plain.

After 9000 BP there were several episodes when 'blanket peat' grew and replaced the existing soil and vegetation over convex sloping surfaces of the north and west uplands of Britain and the glacial hollows of central Ireland. Blanket peat became so widespread that it produced a significant transformation of the landscape.

In Ireland, Scots pine, which can grow on nutrient-poor soils, spread onto peat bogs there: 'bog pine woodlands' assembled whenever climate and peat became a bit drier, especially for the eight centuries after 8361 BP. Oak trees also grew on Ireland's peat in drier periods between 7200 and 2200 BP, forming 'bog oak woodlands'.

Pine and oaks exist but do not flourish, on peat, so thousands of bog oaks and bog pines which fell into wetting peat in prehistory, were part-fossilised to be dug out of Irish peat bogs centuries later.

Ireland, with an extremely oceanic climate and a landscape of hollows, lakes and long rivers, became the peatiest, boggiest country of Europe! Britain (mainly Scotland) is a close second!

Prehistoric people's activities may have triggered large-scale peat growth. For instance, burning and clearing trees and bushes to encourage new herbaceous growth for wild deer, edible-berried plants for themselves, or to grow crops, was sometimes contemporary with peat growth. However, detailed study of peat stratigraphy and past climates suggest that blanket bog also expanded without human influence, especially when the climate cooled.

Whereas Black spruce (*Picea mariana*) flourishes on acid peat bogs in eastern Canada, there are no native trees which really thrive in these conditions in Britain and Ireland. However, Britain and Ireland's almost tree-less peat landscapes, whether anthropogenic, natural, or both, are pertinent to the story of Sitka spruce thousands of years later.

End of an era

Mesolithic people obtained food for six millennia by hunting, fishing and foraging, 4000 of them in and around woodlands.

However, within a period of five centuries between about 6000 and 5500 BP there was a major shift in the way humans obtained their foods and in the types of foods eaten. A transition to farming set in motion long-term and profound ecological changes which had consequences for people, trees and many other species – still reverberating today and relevant to Britain's modern need for Sitka spruce.

Essence of farming

Farming as a way of life and food production was first carried out by people with new cultures called Neolithic or New Stone Age. There were new styles of living and technology; for instance, big timber and stone monuments were erected throughout the country. Between 5200 and 4200 BP, the stone settlement of Skara Brae on the Orkney Islands raised sheep and cattle, grew barley and wheat, and fished inshore. The Orkney community farming at the Braes of Ha'Breck, lived in and around one timber and several large, stone buildings between 5300 and 5100 BP. They left a large quantity of cereal grains underneath a covering of organic waste on the floor of a stone building.

Farming was pro-active:

- Farmers introduced non-native crop plants such as barley and emmer wheat.
- Barley and wheat were domesticated from wild ancestors, natives of open habitats in warm, dry environments of the Levant.
- Crops replaced areas of complex, many-species ecosystems of native species.
- Farmers introduced cattle, pigs, sheep and goats also domesticated in environments of the Near and Middle East.
- Cattle and pigs, being descendants of woodland fauna, could feed in native woodland; goats and sheep, descendents of open-habitat species, needed grass and herbs to feed; goats ate shrubs also.

Hunting, fishing and foraging continued in parallel with farming. Analyses of prehistoric bones and teeth have shown that people's diets changed: wild, native meat, fish and plants were reduced, but grain 'porridge', meat and milk from domestic animals increased.

About 6000 BP, people in Ireland farmed in sites cleared of trees, leaving behind cereal pollen and bones of domestic animals. They moved on, and their sites were re-colonised. Tree species which had once grown on-site did not necessarily re-grow, in which case heathland or peat habitats developed on the abandoned farmland.

On the coast of County Mayo, north-west Ireland, there were five centuries of intensive cattle ranching within a 1000 ha (2470 acre) field-system, soon after 5500 BP. The fields were cultivated by clearing scrubby Scots pine woodland on poor mineral soil and building walls of rubble and soil capped by stones. Cattle-grazing modified the vegetation, which changed to grassland with clovers (*Trifolium* species), buttercups (*Ranunculus* species) and docks (*Rumex* species).

Signs of Mesolithic culture vanished from Britain and Ireland within five centuries by about 5500 BP.

Woodlands in later prehistory

Ireland's upland pine woodlands went into decline from about 4000 BP, possibly caused by wetter climate and soil, with volcanic ash and acids in the atmosphere affecting their growth and fertility. A few, small individuals struggled on until about 3500 BP but pine then became extinct there.

In Scotland, pine woodland became local and confined. Remnants survive in some glens, along loch shores and on their islands, as beautiful red-barked, dark-green, canopied trees. Scots pine failed to survive to modern times in Wales and England.

Beech, which reached Britain about 3000 BP and hornbeam (*Carpinus betulus*) which came later, were the last two trees to colonise Britain unaided. They both reached no further than southern England and south-east Wales.

Common spruce colonised Scandinavia slowly and with Scots pine formed the dominant species of the taiga forest zone. Mixed-deciduous woodland with oaks and beech assembled in southern Sweden and Denmark; deciduous woodlands akin to temperate rainforest assembled on the oceanic coasts of Ireland, Britain and Norway.

The two most common west European conifers, European larch and Common spruce, did not colonise Britain and Ireland. And no ecological equivalent to Sitka spruce as a major species of the slopes of ocean-facing mountains, of sea-shore and riparian habitats – arrived in Britain. There were no conifer equivalents to hemlocks, Douglas fir, Red cedars and true firs in British and Irish post-glacial woodlands.

Prehistoric people and woodlands

Between 5500 BP and 43 AD, people used trees on a much larger scale than their Mesolithic predecessors. Why and what for?

To increase the scale of cereal-growing, it was sometimes necessary to clear trees with fire, by ring-barking, and by cutting down with effective stone axes hafted onto wooden shafts. This also encouraged woodland-edge plants which could be eaten by people, cattle and goats.

Useful products were obtained from specific species or sizes of trees. For example, they selected twigs and foliage of Wych elm for stock fodder; timber of oak and alder for walkways; yew wood for long-bows; and small whole trees for building. People chose big timbers or whole trees for landscape-sized monuments such as the 4100 BP wooden henge at Sarn y Bryn Caled in the Welsh Marches. The outer circle consisted of twenty big, worked, oak timbers, the inner circle of six more. Seahenge, in a saltmarsh on the Norfolk coast, is another spectacular circle monument, built of 55 small, split oak trunks; a huge upturned tree-base was placed in the centre. The oaks were all felled in spring or early summer 4049 BP, the early Bronze Age.

Several species were selected for the Sweet Track across the low-lying

Somerset Levels. Nearly 2 km (1.2 miles) long, it was built across a reed swamp in either 5807 or 5806 BP, very early in the Neolithic period. Oak planks were attached end to end on a rail of alder, ash and hazel along the swamp's surface, and held in place by pairs of pegs each forming a V on the rail. Below the Sweet Track, and 30 years earlier, the Post Track had been constructed of long ash planks between hazel and lime posts 3 m (9.8 ft) apart

People coppiced trees and shrubs and wove the resulting flexible shoots into wattle for fences, hurdles, fish traps and trackways. They lopped tree branches to collect foliage and twigs for their animals to eat. They hunted native herbivores like deer for food and to reduce competition with farm stock for herbage. They killed carnivores like lynx and wolves which ate domestic stock.

As by-products of farming, new 'anthropogenic' ecosystems assembled. These consisted of native species in new groupings – or of native and introduced species growing together. Prehistoric examples are: stock-grazed grassland, cereal plantations with weeds, hedges and stone walls.

Examples

Neolithic people were the first to remove deciduous woodland from the 625 km^2 (241 miles^2) granite upland of Dartmoor, Devon probably using fire. By 3500 BP (in the Bronze Age) there were at least 5000 timber-based round dwellings on Dartmoor! An amazing 10,000 ha (24,710 acres) was enclosed by stone walls called 'reaves' to make fields. Dartmoor has not been wooded since then: soils over the granite rock deteriorated when tree cover was removed and continuous stock grazing prevented tree regeneration. So you can see reaves on the still-treeless Dartmoor.

Until the Bronze Age, the long River Severn and its tributaries supported vast riparian woodlands dominated by alder (*Alnus glutinosa*). Along the River Avon in Worcestershire, extensive alder woods grew with Pedunculate oak, Wych elm, hornbeam, hazel, hawthorn and elder (*Sambucus nigra*). Red deer and aurochs frequented the woodlands and were hunted and eaten by people – who left the inedible bits. Thick aquatic vegetation grew in the clear waters. However, when people cut away the riparian trees, aurochs and Red deer disappeared from the locality, the ground was inundated by water and evidence of them was covered by alluvium until modern times.

Flag Fen in eastern England was a wetland basin during the Bronze Age. It yielded a basketwork eel trap made of coppiced willow. It differed from modern eel traps in the same locality only in that the prehistoric willow wands were whole, whereas modern traps are made of *split* willow. The pith of split willow can absorb water so the trap sinks. Flag Fen also preserved, in an old river channel, wooden weirs for catching fish.

'Log boats', made of a single tree trunk, were important prehistoric transport. 150 have been found in Scotland. A superb example, the Bronze Age 'Carpow Logboat' rested in the tidal muds of the River Tay until 2001 AD. At 9 m (30 ft)

long, it was built between 3260 and 2910 BP. Within it, 3000 hazel nuts were stashed – but dormice had got at them before the humans!

Iron Age people (*c.*2700 BP on) were experienced farmers and woodsmen. They built sophisticated homes on artificial islands called 'crannogs'. There are hundreds in Scottish and Irish lochs. Oakbank Crannog in Loch Tay was made of alder, oak and elm piles driven into the loch bed to form an aerial platform. On the platform people built a house with a three-layered floor of small alder trees with ferns on top, using over 2000 large and 1000 smaller timbers. Hurdles of coppiced hazel formed house walls and a perimeter fence. The causeway linking crannog and shore needed 40 oak and elm timbers.

At the Butser Experimental Iron Age Farm, Hampshire, a reconstructed Iron Age round house was built in the 1970s, based on archaeological evidence of a 6 m (20 ft) diameter house at Maiden Castle, Dorset. The timber frame needed 50 straight oaks about 40 years old for the ring of wall posts. Six elm trees each 9 m (29.5 ft) long, and twenty straight ash trunks formed roof rafters; wands from 80 hazel coppice stools made the wattle walls. All these materials must

FIGURE 6.1 *(above)*. The 3815-year-old Eclipse Track preserved in peat of the Somerset Levels is one of many which demonstrate woodland management during prehistory.

PHOTO: COURTESY SOMERSET LEVELS PROJECT, 2004.

FIGURE 6.2 *(left)*. The 3000-year-old Carpow Log Boat in 2006 after excavation from the River Tay silts, Perthshire.

PHOTO: DAVID STRACHAN, COURTESY PERTH AND KINROSS HERITAGE TRUST.

have come from carefully-managed woodland. Iron Age people also used wood for ploughs, well linings, fences, tools and fuel for smelting metals, for example.

There is less archaeological evidence of the Iron Age in Ireland. Considerable cultivated land was abandoned and woodland increased in area: hazel, holly and birch on clay soils of the Midland Plain; alder, birch, willow, hazel and holly in the south-west; yew woodlands in the west. By the time Roman armies reached southern Britain in the late Iron Age, Ireland was a pastoral country of nomads with cattle.

Being peaty countries, hundreds of prehistoric wooden tracks in Britain and Ireland have been preserved. Several were discovered in the Somerset and Gwent

FIGURE 6.3. A reconstructed, Early Iron Age lake-dwelling or crannog in Loch Tay, Perthshire, with Sitka spruce plantations on the far side.

PHOTO: ANDY TITTENSOR, 2013.

FIGURE 6.4. Agents of woodland demise for five millennia: Blackface sheep on the Distinkhorn, Ayrshire.

PHOTO: KEITH HOBLEY, 2007.

Levels, while the Corlea trackway in Co. Longford is the longest in Europe at 1 km (0.62 miles). In the estuary of the River Humber after about 4000 BP (Bronze Age) people made wattle bridges and tracks, mainly of hazel, so they and their stock could cross the tidal channels on the saltmarsh. Saltmarshes still make excellent grazing for cattle and sheep around coastal Britain and Ireland today, producing exceptionally tasty meat.

Effects of prehistoric farming on woodlands

Decline

In the 5800s BP, the natural woodlands on the sandy hills above the Somerset Levels peat had a canopy of oak, elm, lime and ash, a lower growth of alder, hazel and holly, with alder, willow and poplar (*Populus* species) at the margins. Building trackways on the wetlands below caused their own changes.

Oaks taken from woodland for the northern end of the trackways were aged about 400 years and those for the southern end were 100 years. Poplar grew only in the southern, holly only in the northern woods. Elm declined a little earlier near the southern end of the Sweet Track and did not regenerate. Analysis of hazel tree-rings showed they were being coppiced on a seven-year cycle. Lime trees failed to re-grow in places.

From such small beginnings, prehistoric farmers gradually changed woodland and reduced its proportion in the landscape.

In southern Britain woodlands were altered or eradicated extensively during the New Stone Age and following Bronze Age (4300–2700 BP). For instance, oak, lime and elm were removed from some Shropshire woodlands, leaving gaps where birches, ash and yew seeded in. Intriguingly, other sites in Shropshire have never supported woodland in the 3500 years since their trees were replaced by cereals.

Pastoral farming in the north and west reduced woodland unobtrusively. When sufficient density of domestic animals and wild herbivores constantly nibbled off tree seedlings (which grow from their tips), the next generation was eliminated and their parents were not replaced. Grasses still thrive when nibbled (as they grow from their base), so heavily-grazed woodland metamorphosed into floriferous grassland.

At Eskdalemuir, Dumfries-shire, the earliest effects of pastoral farming can be detected about 5100 BP. Pastoral farming continued intermittently for over three millennia, and after 2500 BP wooded landscapes rapidly became open country. In the Cheviot Hills on the Scottish–English border, large-scale woodland clearance was later, starting during the Iron Age (2700 BP–43 AD).

Tree-line

On mountains, trees and bushes grew to a natural upper altitudinal limit or 'tree-line', which varied with geology, aspect and slope. It hovered around 900 m

(3000 ft) in the Scottish mountains, but on Dartmoor it was about 610 m (2000 ft). Woody tree-line species like birches, Scots pine, juniper and hazel were susceptible to climate changes such as earlier spring frosts or natural fire, and to human interventions. Trees set less seed, produced few or no seedlings and were not replaced, so the tree-line 'moved' downhill. Dwarf shrubs such as cowberry (*Vaccinium vitis-idaea*) and cloudberry (*Rubus chamaemorus*) which grew above the tree-line could then 'move' downhill in their place. In the Welsh hills this happened during the Iron Age and coincided with abandonment of upland settlements.

Ecology

Woodlands were dynamic, with natural processes causing continuous change. Human activities altered the dynamics, species and structure of woodlands, as did climate or volcanic eruptions.

Human-induced changes included selection, which produced a greater proportion of younger, smaller individuals at the expense of older, bigger trees; it also altered the contribution of each species. Large woodlands were fragmented into discrete fragments so that 'woodland-edge' habitat lengthened substantially.

Bracken, heather and grasses increased at the expense of trees. Invertebrates of open ground increased at the expense of woodland species. Lighter canopies encouraged vernal herbs like bluebells (*Hyacinthus non-scriptus*) and primroses (*Primula vulgaris*). The genetic pool of species such as Scots pine and juniper contracted when woodlands became small and isolated.

Soil

When people opened up woodland canopies or put stock into glades to graze, they reduced woodlands' natural filtration properties: less canopy and undergrowth intercepted less precipitation; water reached soil and rivers more quickly and with greater force; rivers might have received heavier loads of sediment; flooding was probably more frequent. If all tree cover were removed, soil could disintegrate from the direct force of wind and rainwater, and be washed into valleys to collect as fans of colluvium. Nutrients in solution were washed out of treeless and harvested arable land, whilst nutrients from annual leaf fall would no longer be added to humus.

Climate changes could trigger soil degradation. In northern Scotland there was an enormous decline in Scots pine just before the Bronze Age, coincident with severe rain, storms and lower temperatures. Build-up of mineral soils was halted and blanket peat developed, exacerbated by people dismantling woodlands.

On hills of metamorphic rocks, brown earth soils were replaced by podsols and peaty podsols when tree cover was fragmented. Grassland, heathland and moorland ecosystems evolved on the denuded slopes. Early soils on the chalk hills of southern England varied from deep silty loams to shallower, chalky soils, but all derived from late-ice age, wind-blown deposits. Deciduous trees and Scots pine formed woodland mosaics.

Natural, small-scale creep took some soil downhill. But when prehistoric farmers cleared the woodland to cultivate the fertile soils, there was significant erosion which sent 'slurry' into the valleys, forming colluvial fans. Thin, calcareous, rendzina soils replaced the silty loams, still characteristic of the chalk hills. I have seen where ill-timed modern tillage produced deep 'slurry' from the same chalk hills.

Fire

Fire, both natural and deliberate, occurred frequently in British woodlands during prehistory, leaving layers of charcoal, a useful source of evidence of past tree species.

Conclusions

Seven decades of research have shown how woodland cover reached a maximum of 60–90% of the British and Irish landscapes about 7000 BP. Tree and shrub content varied with location, geology and soils. Landslips, windblow, flooding and wild herbivores produced mosaics of bare or grassy land where trees could re-colonise.

Woodland loss after human intervention started on a large scale in the Neolithic period and continued during prehistory. There are examples of local or regional regeneration, for instance a site in the Cheviot Hills, where farming declined after only a few decades, and young trees colonised the abandoned land.

Prehistoric woodland history in Britain and Ireland contrasts with prehistory in the native range of Sitka spruce: woodland and trees migrated and colonised but were not cleared on a large scale; soils were not changed so significantly by human activity; fewer anthropogenic ecosystems developed; farming of introduced crops and stock was minimal, though there was some gardening with introduced species; food webs remained complex, tree populations were not simplified by selecting large trees.

When Roman legions eventually appeared on the southern coast of Britain in 43 AD, they saw the results of three millennia of woodland decline and farming progress. They found frequent settlements in a mosaic of pasture, arable, waterway, marsh and fen, heathlands – and discrete woodlands. Few of them experienced pastoral Ireland: its life and landscape developed without military conquest for another six centuries.

The Romans needed good farmland for food and quality timber for armaments, vehicles, harbours, wells, boats and tools, for instance. So did their descendants and successors. The next chapter takes the story of woodlands in Ireland and Britain forward fifteen centuries to the time when the parlous state of woodlands finally dawned on those who needed them.

CHAPTER SEVEN

Woodland History and Britain's Need for Sitka Spruce

> At no time in the last two thousand years have there been woodlands in Ireland that matched in extent and density those of ancient times.
>
> Valerie Hall 2011

Romans needed trees

In 43 AD Roman legions landed on the southern shore of Britain in their first successful invasion. They found prosperous, trading Iron Age tribes-people who lived within a humanised mosaic of farmed arable and pasture, of marsh, acid bog and fen peat, heathland and woodlands.

Rome's military leaders viewed Britain as an attractive addition to their empire for its metal ores and grain. Ynys Môn (the Isle of Anglesey) was known for its wealth of metal ores including copper, lead and gold, Cornwall for its tin deposits, the South Downs for their grain surplus.

During their four-century occupation, Roman legions required timber and wood to build forts, camps and artillery, to smelt metal ores, to construct boats and landing craft to reach Ynys Môn. The circulating hot water under the floors of hundreds of civilian villas was heated by wood fuel.

They eradicated woodlands along their network of straight roads (such as Watling Street), clearing wide corridors free of new growth for military safety. Similarly, tree cover was removed either side of Hadrian's Wall, the high stone rampart which crossed northern England at the Empire's fringe.

After successfully defeating the native Ordovices and other tribes in Wales, the amount and structure of woodland there changed too. The ancient rocks of Wales held desirable metallic ores, but charcoal was needed to extract and smelt them. The Roman timber-using industry was well-organised, carpenters were skilful and their saws were better than Welsh saws. From Welsh oak trees they could produce posts one foot square and planks a foot wide.

The Romans *en masse* reached no further north than the Scottish Lowlands, where woodland cover had been reduced to an open landscape by Iron Age farming. They remained in Scotland for only a short time; in the Highlands, they and their industries had only localised effects. However, Highland woodlands continued to provide indigenous people with wood and timber of birch, hazel and oak particularly, as well as pasture for their domestic stock.

Ireland was barely colonised by Roman military and artisans. Trees spread over recently-abandoned, cultivated land in Ireland and some of the settlements attached to cattle pasture were abandoned.

Several food trees were imported to Britain during the Roman occupation: medlar (*Mespilus germanica*), mulberry (*Morus nigra*), Sweet chestnut (*Castanea sativa*) and walnut (*Juglans regia*) for instance. Species of edible animal, including the Fallow deer (*Dama dama*) and brown hares were also introduced and became feral. In four centuries Romans also added herbs and vegetables to the landscape and their own style of infrastructure: roads, harbours, towns, hamlets, and opulent villas built of clay tiles and bricks.

Later woodlands

Roman armies departed from Britain between 383 and 410 AD, leaving less woodland in England and Wales than when they arrived; in the Scottish Highlands woodland covered a larger proportion of the landscape than elsewhere.

When the empire declined, Britain was populated by Iron Age descendents such as the Picts in north-east Scotland and a mix of new 'Romano-British' people from four centuries of intermarriage. Germanic people (Angles, Jutes and Saxons) migrated into southern and eastern England and south-east Scotland. Norse people migrated into Ireland, northern and western Scotland. Small kingdoms formed throughout Britain.

Inter-tribal warfare in Wales contributed to neglect of hill-pastures, so that trees regenerated and woodlands developed. In subsequent good times, farmers re-stocked the steep hillsides with domestic animals and cleared woodland for crops. The pattern of increase and/or decrease in farm animals in the uplands was repeated over centuries. The time came, however, that trees could barely regenerate on what were now impoverished soils; much Welsh hill country became treeless. However, the lower slopes and valleys of Wales were still full of woodland at the start of the eleventh century and caused problems to the invading Normans for two centuries and to later kings of England who also wished to subdue Wales.

Scottish woodlands declined significantly in post-Roman centuries. In the uplands particularly, woodlands were important not only as a source of wood, but as pasture and shelter for domestic stock. They were deliberately burned; peasants' small sheep, cattle, goats and horses, as well as wild herbivores, grazed and browsed them. Tree regeneration waned in Scotland's short growing season, on its acid, peaty soils and in weather which brought frequent severe rains but the occasional drought.

In Ireland, nomadic pastoral life continued throughout the Iron Age and 'non-Roman' period. But, by 500 AD, human activity was suddenly on the increase even at higher altitudes. During the next three centuries, 30,000 to 50,000 cattle-rearing and dairying centres or 'raths' were constructed and the stock fed on extensive and expanding peat vegetation and lush grasslands.

Documentary sources

From Roman times onwards, documents and oral traditions, as well as archaeology, can give us information about trees and woodlands.

Land charters, poems and treatises, legal proceedings, probate inventories, statutes, tree sales, land leases, rent rolls, factors' reports, market prices, and work contracts are some of the useful types of document. In recent centuries especially, letters, maps, journals, photographs and film have left information for us. Artists such as John 'Warwick' Smith and John Constable painted many scenes of trees and woodlands.

Archives can be found in national and local records offices, castles, stately homes and museums, solicitors' offices and attics.

Map 8 gives the older names of counties in Britain and Ireland which are used throughout this book because they are the names used in pre-1970s documents.

Woodland resources 'in common'

Woodland and tree management had been codified by the post-Roman centuries. Strata of society had their own rights and responsibilities, with sharp differentiation between peasants, landowners and crown to woodland resources.

Rough, uncultivable land or 'waste' on poor soil was often held jointly or 'in common' by the tenants of the landowner. Tenants had rights, for example, to gather herbage, berries or reeds, to dig peat, turf or stone, to graze sheep and cattle. Typical rights on commons with trees ('woodland commons') were: to gather twigs, branches and coppice shoots, to bring pigs to gorge on oak acorns and beech mast in autumn ('pannage') and to graze stock for prescribed periods.

What they could gather or graze, *when* and *by whom*, were closely prescribed by custom and local courts.

Hunting parks and forests

Formal hunting was an important pastime, a demonstration of high status and a method of food collection for medieval nobility. 'Parks' and 'Chases' held 'beasts of the chase' like red, roe and fallow deer; 'game' birds like pheasants (*Phasianus colchicus*); 'wildfowl' such as bittern (*Botaurus stellaris*); and 'beasts of warren' like brown hares. They were normally enclosed by a deep ditch and a turf or stone bank topped with wooden palings.

Parks and chases also contained trees and bushes for shelter, timber trees to be taken for construction, and foliage as fodder for their inmates. Good visibility was necessary for hunting, so trees could be sparse. In winter, inmates provided desirable meat for their noble owners but were out of bounds to hungrier, lower-status people, whose rights were limited at the expense of the beasts of chase or warren.

The eleventh-century Godwinson family frequented hunting parks on their vast Sussex estates. A panel of the famous eleventh-century Bayeux Tapestry

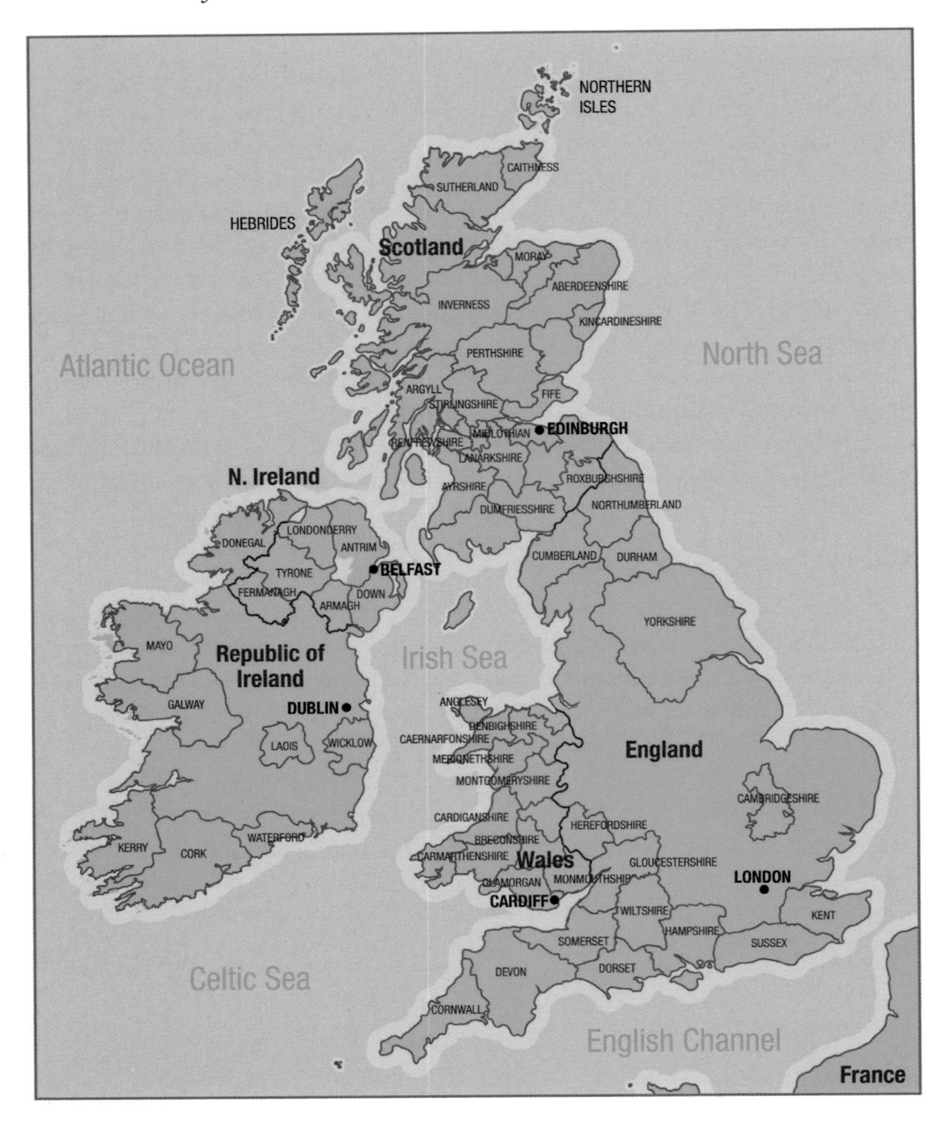

MAP 8. Pre-1974 counties mentioned, prepared by Susan Anderson (information from OS Map, *United Kingdom: Administrative Boundaries, Ancient or geographical counties (2001)*).

shows Earl Harold Godwinson (later the unfortunate King Harold II) at his Bosham estate in 1064 with his hunting falcon on his wrist.

'Royal Forests' were special hunting parks set aside – 'afforested' – by and for the Crown. Afforestation of Crown land often took in tenants' holdings, yet the aim was still to encourage deer, their primary resource and focus. Those tenants unfortunate enough to be *in situ* could continue pastoral farming under strict rules but could not enclose and cultivate, as it would restrict the movement of deer.

Harsh 'forest law' took precedence over 'common law' in forests, parks and chases. For unlawfully killing a wild beast you'd be hung, for killing a beast outwith forest boundaries you'd lose only your feet and eyes...

FIGURE 7.1. Rosemary is touching the Chirk park wall and standing in its ditch.

PHOTO: RUTH TITTENSOR, 2013.

The New Forest, founded by William the Conqueror by the time of Domesday Book about 1086, is a famous example. It took in 27,100 ha (67,000 acres) of existing Crown land plus nearby parishes, on infertile sand and gravel soils in Hampshire.

A cadre of professional 'Verderers' looked after all aspects of the Forest's resources and enforced forest laws. 'Forestry' was a paid game-keeping job carried out by a minor official. Foresters ensured there were enough trees, enough winter foliage for deer and consequently, enough deer to satisfy their paymasters.

The conflicting needs of tenants versus landowners for woodland products and pasturing – and the resulting disputes – have been part and parcel of woodland history through to the present day. Commoners needed, in particular, to graze their domestic animals on 'herbage' (grass and herbs); the Crown and landowners wanted deer, which, as well as herbage, needed 'vert' (green growth of trees and shrubs). Thus domestic stock and deer competed for vert, herbage and woodland habitat.

As centuries went by, the New Forest's harsh forest laws gradually eased. The roles of verderers and foresters gradually embraced the needs of commoners as well as deer protection; it became feasible for peasants to enclose small cultivation pockets called 'assarts' at the forest edge.

During the eleventh to fourteenth centuries, Anglo-Norman aristocracy were gifted Lordships in southern Scotland, where people of Brythonic and Anglo-Saxon origin lived. In 1124, King David I of Scotland granted the Lordship of Annandale to the Northumbrian-Norman, Robert de Brus. This lordship gave permission to enclose land for hunting parks and forests. Although the Lords de Brus granted land within the parks and forests to local peasants, it too was subject to forest law. However, forest law in Scotland was more flexible than in England.

Four hundred miles (*c.*640 km) to the north, the Royal Forest of Darnaway in Morayshire provided oak timbers for castle roofs until the end of the fourteenth century. There were many royal forests in Scotland, but they have not received the same detailed research as English royal forests. There were no royal forests in Ireland.

From the late eleventh century on, Anglo-Norman 'Marcher lords' also afforested on a large scale along the border of Wales with England. The incomers took over Welsh lands and imposed forest law, causing considerable resentment. Many woodland commons were taken over, further depleting tenants' resources. South Wales and the Marches hosted over one hundred forests and 21 deer parks during the fourteenth century.

Other crown demands eliminated Welsh woodlands. Hundreds of paid English woodsmen cleared them for military roads; to prevent ambush of incoming armies; to do away with robbers' hiding places; to keep communication routes and passes open; to build forbidding castles. Thousands of Welsh trees vanished into castles during five centuries.

Norman royals also granted well-wooded Welsh lands to French monastic orders to establish 47 monasteries. For four centuries after the 1100s, Cistercians rooted out woodlands for arable land, which became permanently devoid of trees. Their sheep ranching, often within woodlands, indirectly destroyed them.

However, monks managed chosen woodlands in an efficient and sustainable way for regular harvests. Timber, bundles of hedge rods and fuel wood, bark for tanning and wood for charcoal were produced – while still keeping 'capital reserves' of non-harvested woodland for future use. Trained woodsmen and foresters carried out the work. They managed woodlands by traditional medieval methods such as coppicing, pollarding, shredding and natural regeneration.

There are also extant written records of woodland management by Cistercian monasteries in the fifteenth and sixteenth centuries in Scotland. A local 'forester' was sometimes engaged to make sure that tenants kept to their agreements not to take or burn wood, not to graze their domestic stock, nor to cultivate in woodlands managed for sustainable harvesting. Agreements were, unsurprisingly, often infringed, but later on, flexible contracts gave the tenants more rights. For instance, *young* coppice re-growth might be protected against browsing by tenants' stock with enclosure fences; by and by, the fencing would be removed from around areas of *older, taller* coppice and opened-up for grazing by domestic stock.

Woodland management was a constant attempt at an equilibrium which satisfied harvests of big timber, smaller wood, commoners' rightful materials *and* woodland pasture for domestic animals – without destroying any one, or all, of the resources.

Transhumance

When farm families or herders walk with their grazing animals on a seasonal basis to another district, to feed, it is called transhumance. In Britain and Ireland transhumance originated in prehistory as a system for using resources throughout a region from sea-shore to mountains. Transhumance had profound effects on woodlands in some areas.

In Cornwall, archaeological and documentary evidence showed that from the Bronze Age until the sixteenth century AD, cattle and other stock were walked annually in spring from their lowland farms up to communal pastures on the acid granite of Bodmin Moor. They grazed on the high pastures up to 420 m (1300 ft) during the summer, and their herdsmen returned downhill with them to their permanent farms and homes in the autumn.

A different style of transhumance took place from the English Channel coast of southern England during medieval times. High sunshine, warmth, and excellent loam and silt soils on the coastal plain ensured that woodland had long been replaced by cultivation and pasture.

Every year in late summer, pigs – in particular – were driven northwards along 'green lanes' from the treeless coastal pastures to the well-wooded, clay region known as the Weald. Terrain with abundant beech and oak trees was chosen, where pigs could gorge on the mast and acorns. In early winter, drovers took the pigs southwards to coastal farms again. From this annual 'pannage' emerged seasonally-used, communal feeding glades or 'denns' which, after centuries, became permanent openings and settlements in the woodland or were formalised as 'woodland commons'.

An example can be found in a charter of 953 AD. It records a grant from King Eadred to his mother, of land at Felpham on the Sussex coast and its 'swine-pastures' 32 km (20 miles) to the north in the Manor of Bedham within the Weald:

> And here are the pastures of swine, four dwellings in the place which is called Boganora, at Hidhurst in the woodland and in the common-woodland-pasture which is called Palinga Schittas.
>
> Tittensor 1978

By 1300, these swine-pastures were a big woodland-common, named 'Mennesse', meaning 'common-land'. Even now there are 23 commoners who have inherited the 1000-year-old rights of pasture on 'The Mens' woodland-common.

'Pasture-woodland' is the umbrella name given to royal forests, parks, chases, denns, woodland commons and upland grazed woodlands – where domestic animals and woodland were expected to co-exist.

FIGURE 7.2. Chirk Castle and formal gardens, Wrecsam.

PHOTO: ANDY TITTENSOR, 2013.

Early woodland laws

The medieval 'Welsh Laws' written down in the time of Hywel Dda in the late twelfth century contain not only the earliest surviving picture of managed Welsh trees (coppiced and shredded) but details of how harvests of woodland products were regulated. Eleven tree species were listed and valued: alder, apple, ash, beech, Crab (apple), elm, hazel, oak, thorn, willow and yew. There was some 'secondary' woodland (naturally regenerated on cleared sites) especially in the Marches: raiding was common and agriculture was in decline. In South Wales, one stool of hazel coppice was worth more than 3.75 sheep, so rare was it as a commodity. Shelter, an important role of woodland in windy Wales, was also legally valuable. The Laws also defined intricate rules for hunting and falconry in forests and parks.

Trees were extremely significant in Ireland for their spiritual, historical, landscape and utility associations. They were important markers of land boundaries, providing permanence and definition to a site or area. Ownership and protection for all these purposes gave them legal status.

Seventh-century Irish laws, the *Becbretha* show that honey and beeswax from the native, black, woodland honeybee (*Apis mellifera*) were crucial woodland resources. Honey was a rare sweetener and beeswax was used for quality candles. People kept or guarded honeybee colonies in holes within wild trees and, by medieval times, in specially-built beehives.

FIGURE 7.3. Autumn parkland at Penicuik House, Midlothian.

PHOTO: RUTH TITTENSOR, 2013.

The eighth-century Irish *Bretha Comaithchesa* listed 28 tree species according to the value of their products, often fruit. Individual trees were owned and protected and there were harsh penalties for damaging or felling them and for taking the fruit. Oak was vital for tan bark and acorns for pig feed, apples for their fruit, ash for furniture.

Industry and woodland

Transport and travel along waterways and coast, and to reach the country's 6000 islands, needed boat-building skills from prehistoric times. A boat-building site, probably Bronze Age, was discovered by a now-dry lake near Monmouth in 2013.

Boat construction grew into an important industry around the 17,820 km (11,073 miles) of British coastline (current measurement) and along rivers. Wooden cradles or dry-dock hollows were cut into waterside ground where fishing, cargo and travel boats were built. John Constable's famous 1815 painting 'Boat-building near Flatford Mill' Suffolk, shows his father's wooden barge under construction.

Centuries of war between English, French, Spanish, Dutch, Scots, Irish and Norse had required warships. Each ship required a large quantity of straight, curved and kneed timbers cut from – particularly – oak trees.

Henry VIII's English warship *Mary Rose* was built at Portsmouth, Hampshire, in 1510 and needed oaks from the equivalent of 16 ha (40 acres) of woodland. By that time, large oaks were scarce and they were brought from many parts of southern England. You can visit the famous *Mary Rose* and *Victory* (a famous eighteenth-century warship) conserved in Portsmouth Dockyard and see for yourself just how many peculiarly-shaped timbers were needed to construct them.

The Weald and New Forest were handy to Portsmouth for ships' timber. Early enclosures for growing trees in the New Forest had been small, had relied on natural regeneration rather than deliberate planting. However, in 1698, the crown enclosed 809 ha (2000 acres), the first of a yearly intake of Forest land for tree-planting, where beech mast and oak acorns were sown. To favour trees, the crown took away its commoners' rights to pasture in the enclosed land – harsh but necessary.

Six centuries after Norman afforestation, a shortage of accessible trees with suitable timbers for naval shipbuilding was responsible for the value of *timber* to exceed the value of *deer* from the New Forest – silviculture at last took over from hunting.

The meaning of 'forest' had changed from deer-growing to tree-growing; the meaning of forester had changed from deer-preserver to tree-manager. But naval surveyors had to look further afield than the New Forest and Weald for trees suited to building warships.

To Scotland, with its indented 9910 km (6160 miles) of mainland coastline, fishing fleets and cargo boats were the most important craft, centred on the big estuaries such as the Firth of Forth. The Scottish warship *Great Michael*, twice the size of *Mary Rose*, was constructed and launched from a natural harbour there in 1511. Said to be 73 m (240 ft) long and with wooden walls 3.05 m (10 ft) thick, there were insufficient large, home-grown oaks within reasonable reach, so further timber was imported from Norway!

Boat-building was a common trade in Welsh estuaries such as Conwy and Aberdovey, using native hardwoods from forests such as Gwydyr.

Consequence of conflicting rights

A consequence of centuries of conflicting rights of both commoners and landowners to woodland produce is that royal forests, parks, woodland commons and upland pasture-woodlands contain veteran trees up to seven centuries in age. These escaped felling in the past because commoners had the right to take their branches or cut any re-growth for their own use, so the bole was short and misshapen.

Such venerable old trees provide specialised habitats for scarce bryophytes, lichens and invertebrates whose livelihoods require the specialist conditions of non-stop shelter, constant temperature and humidity, ridged bark or old roots.

Construction timber

A flurry of building monasteries, churches, priories and castles took place in medieval Britain. Early phases used timber, but later buildings included stone from home and Normandy. Scaffolding for building 'sky-high' stone cathedrals and minsters was also made from timber.

The timber-framed Octagon of Ely Cathedral rests on sixteen timber struts which the designer intended to be 12 m (40 ft) long by 34 cm (13.5 in) square. But there were not enough trees of this size available locally between 1328 and 1342 AD, so shorter, tapering timbers were fitted. Many timber-framed structures were constructed of numerous young (25–75 year old) oak trees – not few and big timbers – suggesting that big oaks were scarce.

Landowners used timber and wood for constructing their houses and farm buildings, for horse-drawn vehicles and boats. Peasants used split wood for lath panels in the walls of their dwellings; they used a variety of woods to make, for instance, implements, tools, utensils such as dairy buckets, also barrel-hoops, cart-wheels, gates and hurdles to enclose sheep.

There have been many detailed studies of the species, amounts and sizes of timber and wood which went into both vernacular and aristocratic houses of the twelfth to nineteenth centuries. At St Fagan's National History Museum, Cardiff and the Weald and Downland Open-Air Museum, Sussex you can visit timber vernacular buildings saved from demolition.

Other woodland industries

Oak bark was important throughout Britain in the medieval and later industry of tanning hides. The Montrose Estate woodlands along Loch Lomond were managed on a 21- or 24-year rotation as oak-coppice-and-oak-standards from the late seventeenth to early twentieth century. When the sap was rising in spring, people peeled off bark from the oak trees, rolled it, and sent it and the felled trees by boat along Loch Lomond and down the River Leven to the smelly tanneries of Glasgow. Small trees and branches were used locally to make charcoal and, later on, acids for Glasgow printing works.

Charcoal production gained value when medieval iron- and glass-making industries expanded, as in the Weald. Charcoal was produced within woodlands using small woody material. But by the nineteenth century, charcoal had been replaced by coke for many industrial purposes. This, the replacement of bark by chemicals for tanning, and changes in sheep-farming heralded the decline of coppice-and-standards woodlands.

Woodland resources in Scotland and Ireland

Despite warm weather during the eleventh and twelfth centuries, the upland topography, igneous and metamorphic geology, short growing-season and

FIGURE 7.4. Fallow deer in Petworth Park, West Sussex.

PHOTO: ANDY TITTENSOR, 2015.

acid, often peat, soils, made growing and maintaining woodlands difficult in Scotland.

Climatic deterioration during the Little Ice Age in the fourteenth–eighteenth centuries reduced vegetation growth generally (as well as animal productivity) in the north and west. At the same time, the ratio of sheep to cattle increased in Scotland, which reduced tree regeneration: cattle's treading produces ground conducive to seed germination, but sheep graze so closely that even tiny tree seedlings get eaten. I once found a 21-year-old, 15 cm (6 in) high, ash seedling in a sheep-grazed Scottish woodland.

On windswept land east of the Pentland Hills in Midlothian, farming between about 1100 and 1300 AD was sufficient to provide food for the peasant population and peat for fuel. But from 1300 to 1700 AD, farming declined, with cycles of crop failure and famine, possibly due to the onset of the Little Ice Age. The only woodland shown on seventeenth-century maps was within ravines along the Rivers North Esk and Black Burn, places difficult for stock reach.

Despite areas of treeless hillsides, Scotland, up to the 1450s AD, could supply its ecclesiastical needs and main burghs with some large, native-grown oak for construction. The oaks were usually 200–350 years old, that is from trees which started life in the tenth and eleventh centuries. An early fourteenth-century well in Elgin was lined with split-oak planks from trees aged between 235 and 355 years; the earliest timber in the well had been living in 886 AD. Such accurate dating is obtained by dendrochronology (tree-ring dating).

But by the middle of the fifteenth century, Scottish building timbers were much younger when felled. For example, timbers from Threave Castle, felled in

1446–1447 AD were only 100 years old. The most important buildings constructed between 1450 and 1600 AD were not of native Scottish oak, but Norwegian and Swedish oak. During the sixteenth and seventeenth centuries, imported Scots pine from south-west Norway supplied the bulk of Scottish timber needs – until that part of Norway was itself deforested!

During the eighteenth century, native pinewoods on some western Highland estates were clear-felled to provide much-needed income for their owners. In the eastern Highlands, some owners neglected their pinewoods because Scandinavian pine timber was so easily imported there. Scots pine woodlands of ancient lineage were gradually reduced in northern Scotland too.

Unenclosed native woodlands on Scottish hillsides often declined from long-term grazing, which curtailed tree regeneration. Social pressures, rising rural populations and extreme rural poverty between the seventeenth and nineteenth centuries caused displacement of rural communities from their homes and lands to coasts, urban centres and abroad. Landowners tended to replace farms, people and cattle with sheep. The beloved 'wild', 'natural', 'open' Scottish upland landscape had arrived.

The area of woodland in Scotland had dropped to below 10%. Surviving woodlands were usually well-managed during the seventeenth and eighteenth centuries and produced some construction timber, charcoal and tan bark. But, as in the New Forest, some estate owners and agents enclosed them, excluded tenants and denied them their rights.

In Ireland, several centuries of increasing warmth towards the end of the first millennium AD lowered water tables and reduced peat growth. There were six years when the harvest of oak acorns was recorded as exceptional. Expansion of agriculture reduced the area of Irish woodland and re-structured the landscape. Things then changed as the Little Ice Age started.

Black rats (*Rattus rattus*) were introduced into Ireland and Britain by the Normans in the twelfth century. During the fourteenth century rats, or rather their fleas, brought the plague microbe 'Black Death', which killed about one-third of the human population initially. Recurrent plague, frequent cattle disease and poor harvests allowed resurgence of woodland.

Irish landscapes changed again during the next two centuries. Soils deteriorated from overstocking, and eroded when large-scale cultivation replaced grasslands. English invaders forced a change from Gaelic to English culture and in the early seventeenth century, Scottish, Welsh and English people were deliberately 'planted' in Ulster, enticed by promises of land with commercially-valuable woodland. There was unrestrained felling at this time. Native hazel woodlands were destroyed almost countrywide to make room for more cultivation. Yet more trees went as the 'planters' cleared woodland for farming; the human population increased again. When trees had gone, fuel was obtained by digging peat blocks by hand from bogs and drying them in the wind.

Tiny remnants of native woodlands survived the seventeenth century but palaeo-environmental evidence shows that many species of flora and fauna dependant on native woodlands became extinct in Ireland.

Non-native tree species were first introduced into Ireland at the end of that century. The importance of trees and woodlands in the almost treeless Irish landscape and economy was realised at last.

Attempts to grow trees were hindered by another of the Normans' favourite animals. Rabbits (*Oryctolagus cuniculus*), native to Iberia, were introduced as breeding stock to Britain and Ireland in the late twelfth century for food and fur. Although kept in enclosed 'warrens', they escaped, right from the start, and ate their way through crops, grasslands and young trees. Documents show that when rabbit numbers rocketed during the eighteenth and nineteenth centuries, they caused severe distress to farmers from crop losses. In the nineteenth and twentieth century they were also a major forestry pest, requiring dug-in netting to exclude them from plantations.

Enclosure and segregation

Cultivation of food crops in Britain and Ireland was often carried out communally within unenclosed or partly-enclosed landscapes, depending on the region. Enclosure of land into single-ownership or single-tenancy units was piecemeal until the big 'Enclosure' and 'Improvement' movements of the seventeenth to nineteenth centuries (England and Wales) and eighteenth to nineteenth centuries (Scotland). It coincided with tenants' rights weakening but landowner powers increasing.

Enclosure produced and reinforced *static* landscapes: pasture, arable and fallow – even woodlands and forests – were sealed within defined boundaries of fence, hedge, dyke, ditch or wall. Landowners asserted rights to woodland resources except on common-land. Tenants were sometimes bought out and they moved to urban centres or abroad. Those left had to obtain woody produce from hedgerow trees on their tenancies.

Woodland management and farm management had become segregated into two different systems.

The ancient value of woodlands to fulfil a variety of needs for many people had been replaced by a narrow definition where timber and wood were the major products and for one section of the population only. Only on extant woodland commons and a few royal forests have both landowners' and communities' legal rights to woodland resources continued until recent times.

The end for some

Not only were native trees and woodlands depleted and confined, but native fauna had been deliberately reduced or forced into extinction. Loss of woodland habitat and deliberate killing were both factors. For instance, the brown bear and lynx (*Felis lynx*) were extinct by 1066 AD. The European beaver (*Castor fiber*) lasted until about 1200 AD, the Wild boar (*Sus scrofa*) until about 1400 AD. The insatiable Norman desire for hunting deer caused European wolves (*Canis lupus*) to be exterminated; a few survived in Scotland and Ireland until the eighteenth century.

A new relationship with nature

The new relationship with nature brought by farming is one reason for Britain and Ireland's modern need for Sitka spruce.

Farming was more extractive and manipulative than hunting and gathering. Soil was regularly scraped or turned-over (ploughed), exposing it to oxidation and erosion; its character was changed by adding fertilisers. Domestic herbivores were, and are, stocked at a higher-than-natural carrying-capacity. Farmers intervened in the environment by removing species ('weeds' and 'pests') which competed with crops and livestock. Domesticated species came from quite different environments than the cool, maritime conditions to which they were brought, so selective breeding was required to produce better-adapted types. Hunters and foragers ate a huge variety of wild foods from complex ecosystems; farmers worked with fewer species and gradually simplified their artificial ecosystems.

Farming had extreme and extraordinary influences on flora and fauna compared with hunting and foraging.

Woodlands were no longer widespread and contiguous, but became discrete units, tightly-held within boundaries. Their composition and structure were determined by human choice not ecological factors. Scarcity increased their economic value and community uses were restricted. Eventually, a caste of professional 'foresters' took on woodland management for owners, and developed rational, scientific methods of growing and harvesting trees.

Why Sitka Spruce was – and is – needed in Britain and Ireland

- Britain has few native tree species and Ireland has even fewer.
- After six millennia of post-glacial hunting and foraging, people's food procurement methods and diets changed drastically. Farming was bequeathed to all areas within five centuries.
- Increasingly intensive and sophisticated farming influenced native plants, animals and ecosystems. Change was continuous. Few natural systems remain: even rocks were quarried and mined in prehistory.
- Ireland and north-west Britain have a temperate maritime climate, unique at their latitude – which is the same as Labrador and Moscow.
- Native woodlands grew to a tree-line of at least 700 m (2297 ft), but climate changes and land uses altered it periodically.
- Climate changes and human activities intensified the growth of blanket peat over large tracts of north and west upland Britain and in Ireland.
- Millennia of both farming and peat growth reduced woodland regeneration in the uplands. Trees which grew again after early prehistoric use did so less as time passed by.

- Forests have been utterly changed and reduced to a small fraction of their post-glacial, Mesolithic extent.
- Many soils were degraded or lost due to climate changes and human actions. There are few native tree species which can grow to maturity and old age in the altered soils and climate of Ireland and north-west Britain.

Chapter Eight will discuss how changes in society brought new attitudes to trees, woodland and landscapes.

CHAPTER EIGHT

Realisation: New Trees for New Woodlands

As you know, the Grants have been at the forefront of innovative forestry theories and practices over the past 400 years ... we will be able to discuss some of my ancestor's achievements including planting 50 million trees and his interest in photography in the eighteenth century...

Sir Archibald Grant's welcome to the Royal Scottish Forestry Society 2009

Nadir?

In the sixteenth century, the landscape of Britain and Ireland and was still a humanised mosaic from sea-shore to mountain slope. Woodlands, which contributed only a small proportion, were managed to provide local and regional needs; some big construction timber was imported. Table 4 shows the approximate percentage area of their woodland between 1500 and 1700.

Early tree planting

In places as dissimilar as royal forests and tenants' yards, seeds and saplings of native trees such as oak and rowan got planted.

In the New Forest, the royal obsession for hunting deer gradually gave way to raising timber trees within its bounds. Merely encouraging saplings *in situ* was superseded by planting acorns and beech mast as a more certain way of growing trees.

Archives of Bedford Estate show that, in Bedford Purlieus (part of the ancient Rockingham Forest, Northamptonshire), people planted local species, for instance oak in 1457, 'quickthorn' (*Crataegus monogyna*) in 1663 and ash setts in 1701. After 1752, more species were planted: two 'pecks' (*c.*17.6 litres) of Crab apple (*Malus sylvestris*) kernels; oak, sloes (*Prunus spinosa*) and sallows as seeds or setts. Oak and Ash were deliberately sustained as the main species for four centuries; management was by coppice-and-standards between the sixteenth and nineteenth centuries. Non-native species like European larch and Sweet

England	4%	Walker and Kirby (1987)
Ireland	1%	DellaSala (2011)
Scottish highlands	8%	Smout *et al.* (2005/7)
Scottish lowlands	3%	Smout *et al.* (2005/7)
Wales	<10%	Linnard (2000)

TABLE 4. Late Medieval woodland: percentage area in Britain and Ireland.

chestnut came on the scene in 1801. Oak, ash and larch became the main timber crop after part of the Purlieus was grubbed out in the mid-nineteenth century.

The impetus for planting trees on *non-wooded* land was greater contact with Europe. Young aristocrats on their travels were impressed with the landscaped woodlands surrounding European mansions and took the ideas back to Britain and Ireland. Country houses of aristocracy were planted around with trees to form parklands and policies. They were for ornament, as a timber crop and – in exposed landscapes – to shelter farm crops too.

Crown and other landowners planted native broadleaves such as oak, elm and ash to start with. Scots tended to buy the big-acorn oak (*Quercus robur*) from England; the smaller-acorned *Q. petraea* was the commoner species in Scotland but its acorns deteriorate more quickly. In due course, landowners bought and tried out European broadleaves such as sycamore, Sweet chestnut and Horse chestnut.

In 1722, Cawdor Estate, Nairnshire took delivery of 100 Horse chestnut, Maritime pine (*Pinus pinaster*) and oak seeds from an Edinburgh firm; in 1725, a writer described the young plantation as 'handsome'. When an estate nursery was set-up soon after, tree-planting with a variety of native broadleaves, pine and European larch was possible. In 1698, judicious marriage had allowed Cawdor to acquire Stackpole Estate, Pembrokeshire, where eighteenth-century landscaping and tree-planting produced one of the most famous designed landscapes in Wales.

Welsh landowners planted significant numbers of trees in the early seventeenth century and by the middle of that century exotic conifers from Europe were grown for ornament. They included Norway spruce and European larch, 'cypress' and 'cedar' (probably *Cupressus sempervirens* and *Cedrus libani*). Scots pine, oak, elm and walnut were planted commercially at (for instance) Chirk in the Marches, Gwydyr Estate near Conwy, Margam and Trefloyne estates in south Wales:

> Master Thomas Bowen of Trefloine in the County of Pembroke ... had manie young and small plants of this kind [the Firre tree] brought him home by saylers from the Newfoundland
>
> R[ooke] C[hurche] 1612, 8, quoted in Linnard 2000

The smaller estates on Ynys Môn used European broadleaved trees, and Scots pine from Scotland for their commercial woodlands. By including it in their leases, Welsh landowners of the sixteenth and seventeenth centuries often forced their tenants to plant trees. On the Denbigh Estate in the 1560s, tenants were required to plant hundreds of trees each, a heavy burden.

In Scotland, from the 1730s, tenants in low-lying lands around the Moray Firth were given written contracts which obliged them to plant trees on the dyke (wall of turf or stone) around their yards. Saplings were supplied by estate gardeners from the proprietor's nurseries or grown from seeds supplied by commercial growers. Species included: ash, birches, gean (*Prunus avium*), holly, European larch, rowan, Scots pine and sycamore (*Acer pseudoplatanus*): native and European species. They were required by the landowner, whose trees they were, not only to plant them but to look after them too. Perhaps that is why some are still growing.

There were still sufficient existing small woodlands in mid-eighteenth-century Moray to supply local tanneries, make furniture, wheels and farm implements. Birch, hazel, willows and alder were coppiced regularly. Timber from small plantations or from 'standard' oaks in native woodlands supplied crucks to construct farm buildings. Imported European conifers such as larch, European silver fir (*Abies alba*) and Norway spruce were planted to supplement local resources.

Despite tree planting during the seventeenth and early eighteenth centuries, it contributed little to the overall wooded-ness of the UK landscape. Tree planting on new sites takes a lot of preparation: enclosure of the land to exclude stock; buying small trees or seed; awaiting their arrival from long journeys (by sea or on rough roads); training the labour force; awaiting the most suitable time of year (winter and early spring) for planting – and suitable weather!

Existing native woodlands were *still* on the decrease.

Age of Enlightenment and Improvement

But attitudes towards the rural estate and in society generally were changing.

During the seventeenth and eighteenth centuries, the 'Enlightenment' developed as a period of immense intellectual curiosity and creativity throughout Europe. Philosophy, science, medicine, engineering, political economy and agriculture advanced rapidly. Education expanded. Reading and discussion between interested people, led to the idea that 'evidence' and 'reason' should guide human thought and development for the benefit of both society and its individuals. It was an era of striving for knowledge and achievement.

The desire for personal improvement triggered travel as an accepted feature of British aristocrats' education. With Europe as a template, splendid new houses with designed gardens and woodlands were laid out on many British and Irish estates. The beautiful Stourhead in Wiltshire is a famous example. Landowners wishing to improve their estates, themselves and their standing in society transformed their landscapes. They sought new plants from far-away places which would grow satisfactorily, beautify the surroundings – and be the subjects of neighbours' admiration!

New methods in architecture, agriculture, arboriculture, silviculture and landscaping developed during the seventeenth to nineteenth centuries as part of this social change. The obsessive rural 'improvers' reckoned that any land which was not productive of food, trees, economy or obvious beauty was wasted. For them, tree planting and tree husbandry were as important as farm husbandry.

Thinkers and writers such as Jonathan Swift contributed their creativity and wit to Irish society. In England, Isaac Newton led the way in science, Francis Bacon and John Locke in philosophy. Aristocratic women, such as Elizabeth Montague, Elizabeth Vezey and their friends, known as 'Blue Stockings' from their informal attire, met regularly in their London homes for reading, discussion and intellectual improvement.

Scotland was also a hub of activity: think of Adam Smith, David Hume, James Hutton and Walter Scott. This may have been partly due to the population's high literacy level in the eighteenth century, estimated at about 75%, the result of a public education system which educated boys to the age of twelve. No other European country formally educated its population at this time.

Men met within the universities of Aberdeen, Edinburgh and Glasgow, in clubs and on the street, for discussions, learning and reading. After the union of Scottish and English parliaments in 1707, educated Scots travelled more frequently and further. In Europe, Scottish ideas and knowledge were held in high esteem. Scottish landowners in particular did the ground-work for deploying plant explorers to seek new species from afar and the Scottish diaspora took their ideas and achievements to other continents.

In Wales the improving estates were owned by descendants of Welsh-Norman and Anglo-Welsh families. In Ireland they were sometimes British who were given lands during the Plantations of Ulster or Irish and Norman-Irish families. In England, many big estates were still in the hands of descendants of Anglo-Norman or Anglo-Norse families.

The developments of the Enlightenment led, eventually, to the discovery, import and planting of Sitka spruce in Britain and Ireland. Why and how?

Need for imported timber

Timber was needed in Britain for ships, housing, bridges, castles, churches and harbours. Coppice woodlands could still provide wood for tools, utensils, vehicles and fuel and they were regulated to ensure optimum productivity.

Tree-ring-dating of timbers from Glasgow Cathedral and Caerlaverock Castle in Scotland, showed that their home-grown oak timbers felled and used in construction during the thirteenth and fourteenth centuries were between 200 and 300 years old, having grown up during the Medieval Warm Period. All pre-1450 timbers from other sites were the same. But timber felled during and after the later fifteenth century, was only 100 years old, for instance in Threave Castle near Dumfries. Smaller, younger home-grown timbers than previously were used when large timbers were not available. Sometimes they were cobbled together for the job in hand, as in the bell-tower of Ely Cathedral.

Timber had, of necessity, been imported for several centuries, as very few big trees were left in British and Irish woodlands by the fifteenth century. Home-grown timber – amazingly – was more expensive to cut and extract than buying in from abroad (it sounds familiar). Large-scale import of timbers had started by the fourteenth century in Scotland, coming from Germanic states, the Baltic and Norway. But after 1602, oak export from Norway was prohibited – it was needed for building the Danish-Norwegian navy!

During the Enlightenment, timber was imported to growing urban centres, for new country houses and the expanding merchant marine. Trees planted at that time would not provide timber for at least a century.

New tree species from Europe

Roman invaders and Cistercian monks had introduced a few foreign trees, mainly to start fruit orchards, which needed both good soil and frequent care.

Norway spruce was first recorded in Britain in 1548; it suited places where cool, moist conditions prevailed and late spring frosts were unlikely.

The beautiful *European silver fir* was introduced about 1603 and became popular as an ornamental right away, and still is. By the mid-nineteenth century it was planted as a forest tree. *Cedar of Lebanon* (*Cedrus libani*), native to mountains in the Levant, was introduced to Britain in 1683 for ornament.

European larch, native of the Alps and mountains of northern Europe, was introduced by 1629. In 1664 a tree of 'good height' was growing in south-east England. In Scotland, European larch was a favoured ornamental and as a commercial forest tree on estates such as Atholl.

Sycamore, a broadleaved tree of southern European mountains, was moved northwards very early because it germinates easily and grows well in windy places. Sycamore possibly reached Wales with prehistoric people sailing north along the west coast of Europe (see More and White 2003), though the earliest record is 1280; it was not common until the seventeenth century.

The earliest record of a non-native tree species in Ireland was in the late 1600s, when *Scots pine* was introduced four millennia after it had become extinct naturally. It was planted for ornament and in commercial plantations. Its wood was in great demand for commercial, ornamental and domestic use. *European larch* and *Norway spruce* were subsequently taken to Ireland. In 1682, a variety of introduced trees, including pine, cedar, lime, sycamore and beech were planted by Lord Granard on an estate in County Longford. Lime and beech reached Ireland by human agency.

Realisation dawns...

Landowners interested in tree-planting realised that, although native and European trees were fine around their mansions and on the fertile soils of lowland Britain, something else was needed if they happened to own land in north and west Britain or Ireland. European larch and Norway spruce would grow in some of their landscapes but these were continental species native to very cold winters which change quickly to very hot summers (no long-drawn-out spring and autumn). The maritime climate of upland Britain did not suit them well.

Landowners in the uplands concluded that Britain and Ireland had few or no native tree species suited to the poor moorland soils, soggy peat bogs, rocky outcrops, high rainfall and constant winds where so many upland estates were located. There was a decided dearth of useable native or European species for such regions.

Tree species as yet unknown, must be sought elsewhere and tested for their suitability in the difficult conditions bequeathed to them by history.

Help with the new forestry

Until the 1600s woodland management had operated for millennia with native species and in similar ways: by coppicing, pollarding, shredding, felling, layering, selection of seeds, saplings and timber trees.

But after that, unknown tree species were imported: how did you obtain seeds and saplings, how did you germinate them? How to plan, plant and grow the new trees? How to manage resulting woodlands? And how to find potentially suitable species from uncharted places?

There was no helpful government forestry agency, no university forestry departments nor forestry societies. It was a matter of self-help and the experience of friends.

Help came in several ways: practical *treatises* written by those who discovered by experience what to do; commercial *nurseries* which sold seeds and seedlings; existing estate *foresters* with basic but perhaps out-of-date expertise; *statutes* encouraging or forcing tree-planting; new *physic* and *botanic gardens* where imported seeds and plants were tested; *learned organisations* with enlightened landowners and intellectuals; finally, *explorers* who went on dangerous plant-hunting expeditions to bring back potential species for arboretum and plantation.

Treatises

The role of forester had changed from Norman deer-keeper to tree and woodland officer. There had been millennia of traditional woodland management using native tree species, so the principles were understood. Foresters learned their trade by starting with boys' work and undertaking all tasks, one by one, over the years.

A Forester's manual and accounts for 1269–1270 survives in the archives of Beaulieu Abbey, Hampshire. This *Tabula Forestarii*, described as the earliest forester's *vademecum* (Linnard 1979) show that his main forest products were 20-year-old coppice stems cut for firewood, hedging rods and charcoal. He occasionally sold small acreages of oak high forest. Monks, as well as their forestry staff, cut timber for the workshops, where carts and watergates were built. The Forester contributed to the Abbey its second-largest annual income of eleven departments.

When times changed and landowners bought 'new-fangled' tree seeds and seedlings, perhaps their gardeners and woodsmen were aghast at the prospect of trying unfamiliar species and learning new techniques?

To help them, men with or without experience of gardening and trees put pen to paper. In 1612 Rooke Church wrote and published:

> **An Olde Thrift newly revived,**
> Wherein is declared The Manner of Planting, Preserving, and Husbanding yong Trees of divers kindes for Timber and Fuell. And of sowing Acornes, Chesnuts, Beech-mast, The Seedes of Elmes, Ashen-keyes, &c. With the Commodities and Discommodities of Inclosing decayed Forrests, Commons, and waste Grounds. And also the use of a small portable Instrument for measuring of Board, and the solid content and height of any Tree standing. Discoursed in Dialogue betweene a Surveyour, Woodward, Gentleman, and a Farmer. Divided into foure parts, by R. C.
> R[ooke] C[hurche] 1612 quoted in Linnard 2000

Yorkshireman and explorer Christopher Levett was King James I's Woodward of Somersetshire in the early 1600s. In that capacity, he wrote, in 1618, a treatise on measuring and selecting trees for harvesting, used by the Royal Navy as its standard timber-measuring tool:

> **An abstract of timber-measures**
> Wherein is contained the true content of the most timber trees within the realme of England, which vsually are to be bought and sold. Drawne into a briefe method by way of arithmeticke, and contriued into such a forme, that the most simple man in the world, if he doe but know figures in their places, may vnderstand it, and by the due obseruing of it, shall be made able to buy or sell with any man, be he neuer so skilfull, without danger of beeing deceiued.
> By C. L. of Sherborne in the countie of Dorset, Gent…
> http://www.worldcat.org/identities/lccn-n88-32230

Englishman John Evelyn published his influential book *Sylva; or, A Discourse of Forest Trees and the Propagation of Timber* in 1664. It encouraged and exhorted people to plant trees and gave practical information which was used for the next two centuries.

In 1683, John Reid, an experienced worker on several estates in Scotland, published *The Scots Gardener*, which gave advice to Scottish landowners on growing trees to form woodlands and orchards: how to look after them in the nursery, when their seeds are ripe, when to sow, in what soils and how to transplant them. After all, people needed to know these basic items of information.

During the later seventeenth century, many more treatises about forest trees were available to landowners and foresters. Wales produced some after the mid-eighteenth century. For instance, William Watkins of Hay-on-Wye published a comprehensive *Treatise on Forest Trees* in 1753, which exhorted landowners to plant trees for posterity and gave much practical advice on individual tree species, on plantation and nursery management.

Tree nurseries

Plant nurseries already grew orchard trees and early forest nurseries were set up in the early eighteenth century throughout Britain. They collected or bought seed or plants of native trees and of species newly-arrived from Europe and North America. Nurseries germinated the seeds, grew seedlings to a suitable stage for transplanting and then sold them. In 1729, Archibald Dickson and his family founded a tree nursery near Hawick in Roxburghshire, and later one in Perth, Edinburgh and Belfast. The fifth generation of the Dickson family was still in business in 1835.

Until the end of the eighteenth century, there were no Welsh nurseries, so landowners frequently bought tree seed and plants from Dickson's, from English and Irish firms. Estates on Ynys Môn could use the short sea route from Ireland to obtain seed and plants. Some Welsh landowners found that long journeys over difficult terrain or round the coast had deleterious effects on their planting stock. This forced them to set up their own nurseries. The most successful of Welsh tree planters, Thomas Johnes of Hafod Uchtryd near Aberystwyth, started a nursery which produced nearly 1 million oak seedlings in the four years after 1798.

Some Scottish estates also set up their own nurseries, especially if they already owned trees from which they could collect their own seeds. The Clerk family, which owned the Penicuik Estate near Edinburgh were ahead, as stated by John Wilson in 1891:

> From a desire to improve the amenities of his estate Sir John began about the year 1703 to make nurseries for the propagation of young trees, and thereafter he started a regular system of planting. One of the very first strips laid down was on the south side of the mansion-house, near to the [River] Esk, covering the ground where there still exists an old coal-hole...
>
> Wilson 1891

The big Scone and Atholl estates in Perthshire also started their own tree nurseries – they needed huge numbers of little trees to fulfil their plans.

Commercial tree nurseries were later set up in Wales: for instance, between 1810 and 1815, Hindes of Newcastle Emlyn sold over half a million trees annually to Welsh landowners.

Statutes

Statutes on royal forests were in force from the eleventh century when they dealt with the promotion of deer and penalties for infringement. The first statute to encourage *trees* in the New Forest was in 1698, when an, 'Act for the Increase and Preservation of Timber in the New Forest in the County of Southampton' authorised immediate enclosure of 809 ha (2000 acres) for planting trees. Verderers were given powers to penalise people who broke down enclosure fences, stole wood or set fire to vegetation. Two further acts in the nineteenth century provided for yet more enclosure, which further constrained commoners' rights. Dis-afforestation of a royal forest needed a statute.

During 1883, Thomas Hunter published his interesting account of Perthshire in which he pointed out some Scottish statutes of past centuries which promoted tree planting where there had been no woodlands:

> Although the whole of the low-lying land is now well drained, the Carse of Gowrie, up till 1760, was, like many other tracts along the banks of the Tay, disfigured with large pools of water.
>
> So naked and open was the country at this time that its condition engaged the serious attention of Parliament, and various Acts were passed with the view of covering the nakedness of the land, – a state of matters which may have led the 'conjectural historian' into the error of supposing that there had at no time been extensive woods in the country. In the 'Acts of the Parliament of Scotland' (14th Parliament, James II, 6th March, 1457), we find it enacted 'Item, anent plantations of woods and hedges, and sawing of broom, the Lords thinks speedful that the King charge all his freeholds, baith spiritual and temporal, that in the making of this Whitsunday set, they statute and ordain that all their tenants plant woods and trees, and make hedges, and saw broom, after the faculties of their mailings in place condiment therefor, under sic pain as law and inlaw of the Baron or Lord shall modify'. In the 6th Parliament of James IV, 11th March 1503, it is also enacted, 'Item, it is statute and ordained anent policy to hehalden in the country, that every Lord and Laird make them to have parks with deer, stanks, cunningares, dowcatts, orchards, hedges, and plant at the least ane acre of wood, where there is na great woods nor forests'. Again, we find in the fourth Parliament of James V, 7th June, 1535, 'The above ratified; and that every man having an hundred pounds land of new extent, where there is no wood, plant, and make hedges and haining, extending to three acres, and that the tenants of every merk land plant a tree'.
>
> Hunter 1883

Three eighteenth-century acts for England and Wales listed severe penalties for damage, felling and burning timber trees. But five acts encouraged enclosure, planting and preserving orchard and timber trees. Fourteen broadleaf species and three conifers ('fir', 'cedar' and 'larch') were on the planting lists for timber trees.

Statutes were not particularly helpful to those wishing to plant trees on a large scale as there was neither advice nor financial incentives to accompany them. Trees got planted because some people loved trees, decided it was the right thing to do, was fashionable and acceptable and because their neighbours did it.

Physic and botanic gardens

Formal 'physic' gardens with plants grown for medicinal and educational purposes emerged during the seventeenth century. They broadened their scope and became 'botanic gardens' when more foreign plants were sent to them during the next century.

Two local physicians founded the Edinburgh Physic Garden in 1670 to cultivate medicinal and other plants to stimulate improvement of medical practice in the city. The Chelsea Physic Garden was founded in 1673 as the Apothecaries' Garden to grow medicinal plants and to train apothecaries in plant identification. In Ireland, physic gardens were most often within monasteries.

In Europe, the botanic gardens of Uppsala and Amsterdam were founded during the seventeenth century. French explorer and mapmaker Samuel

Champlain joined expeditions to eastern North America and laid out a garden near the new town of Quebec in about 1609. But the first real botanic garden in North America was developed after 1694 by Johannes Kelpius on the shore of the Wissahickon Creek, Philadelphia to grow medicinal plants, study and use them.

Botanic Gardens became important centres for receiving plants and seeds from abroad. They grew and tested plants and taught generations of gardeners, tree experts and explorers. In the modern world they are still important educational institutions and places of beauty.

The role of learned organisations

Intellectual societies were started by people who wanted to learn the ideas, discoveries and inventions in science and arts. The Royal Society of London for Improving Natural Knowledge became the Royal Society of London in 1663, its remit restricted to science. The Linnaean Society of London was founded in 1788 to further the study of natural history, following the establishment of an international system of naming plants and animals.

In the buzz of enlightenment Edinburgh, the Society for the Improvement of Medical Knowledge, started in 1731 and became the Royal Society of Edinburgh in 1783; it promotes science, arts and medicine.

The Highland Society of Edinburgh was formed in 1784 as a reaction to agricultural crisis and famine on some upland estates. In 1787 it became the Royal Highland and Agricultural Society, which encouraged knowledge of trees and tree-planting as well as farming.

The Welsh Enlightenment progressed differently, through the printed word, art, song and oral tradition. Some of its thirteen county agricultural societies, set up in the mid-eighteenth century, performed a practical function in supporting and enhancing rural life. Several included forestry in their activities and gave sought-after prizes for tree plantations and nurseries.

Landowners – if they wanted commercial plantations – needed *new trees* or their seeds, and *lots of seeds*. Learned and horticultural societies, botanic gardens, wealthy landowners, nurseries and royalty paid for their chosen plant hunters and explorers to travel afar in a quest for trees; for specimens and knowledge of plants of far-away places; to fulfil the growing enthusiasm for new plants, beautiful gardens, new woodlands and designed landscapes. The Horticultural Society, (Royal) Edinburgh Botanic Garden and Chelsea Physic Garden all sent explorers to far-off places. The British Royal Navy played its part too.

Tree-planting mania

Although landowners throughout Britain built fashionable mansions and beautified their estates with trees, it was in upland Britain and in Ireland that *afforestation* escalated during the eighteenth century. Landowners vied to plant the most trees. The competitive planting of huge commercial forests by landowners was described as 'mania'.

After a life in politics until 1732, Sir Archibald Grant decided to supplement the improvements started the previous century on his estate at Monymusk, Aberdeenshire. By 1754 he had planted 2 million trees, eventually topping 14 million trees!

The Earls of Moray planted 12 million trees on their Darnaway Estate in north-east Scotland during the eighteenth century. Most were Scots pine: seeds were collected from trees making up the estate's native pinewoods, then germinated and grown in the tree nursery and transplanted into plantations. Scots pine in mixtures was favoured for Scottish heathlands with the intention that at least one of the mixture would survive.

During the eighteenth century, the Atholl estates covered 141,000 ha (350,000 acres) of Perthshire. The hill land had been used as 'deer forest' since the sixteenth century and was an almost treeless domain where deer stalking took precedence, although a few tenants' sheep were allowed on the poor pasture.

The surviving Atholl native woodlands grew in the river valleys. They were managed traditionally as oak coppice-and-standards. The first European larch trees were planted on Atholl land in 1737 or 1738 (you can still see some of them). The second Duke of Atholl put more Larch trees in his policies in the middle of the century; the third Duke decided to plant conifers on the mountain sides in 1767 – a brave move. However, the fourth Duke really got the bit between his teeth, and prepared the land for millions of trees with miles of fencing and roads with bridges. During the later eighteenth century, he organised huge plantings of European larch. Between them, the Dukes of Atholl out-planted other landowners, with their 21 million trees on 6070 ha (15,000 acres) between 1730 and 1830. These were intended to become large commercial plantations, but beautiful landscapes and shelter for farming were as important as profit.

European larch, Scots pine, European silver fir and Norway spruce were the main conifers planted in the British uplands during the eighteenth century.

There was less tree-planting in the Scottish Lowlands. In the early 1700s the lands of Penicuik Estate in Midlothian were still described as barren bog cut through by treeless ravines. The Clerk family had bought the Estate in 1653 or 1654, intending to improve farms and plant woodlands. However, their plans did not reach fulfilment until after 1722 when Baron Sir John Clerk inherited it. John Wilson wrote in 1891:

> After succeeding to the Penicuik estates the Baron added largely to their amenities. He planted both sides of the [River] Esk from a point above the mansion-house down to the present serpentine walk. He also began the system of enclosing parks by double dikes of turf planted with thorns and hardwood trees... Not on his own property only were the Baron's enlightened views on arboriculture given effect to, but on that of the neighbouring estate of Newhall also.
>
> Wilson 1891

The word 'parks' here means 'fields', not deer parks. Up to 1782, trees were also planted as commercial plantations – even in those ravines! New plantings in the Scottish Lowlands often used oak coppice-and-standards, beech, willow, elm and ash as well as introduced conifers.

The Dukes of Argyll pioneered tree planting on the Scottish west coast in the 1730s; they used native and European broadleaved trees (particularly birch) and conifers such as Scots pine and larch.

In 1783, Colonel Thomas Johnes inherited the desolate, run-down estate of Hafod Uchtryd in the Ystwyth valley, Ceredigion. When he realised how much the woodlands had deteriorated from being used in lead mining and smelting, he determined to improve the farms and woodlands. His intention was to create an up-to-date example of good husbandry, woodland planting and an estate of great beauty.

He was and is regarded as the greatest Welsh planter of that period. Some of his planting records were burnt in an 1807 fire, so that estimates vary of the number of trees he planted. But at least one-and-a-half million trees were planted between 1785 and 1800, mostly European larch; the rest were alder, ash, beech, birch, Wych elm and oak. Ceredigion nurseries could not supply all his needs. So he collected and planted acorns direct; he dug up wildings and bought small plants from Liverpool and Scottish nurseries, but they were sometimes injured during the long journeys. He also started his own nursery at Hafod, and by 1802 had raised nearly one million oaks for planting out. James Todd, his Scots gardener, was in charge of the nursery.

The tree flora of Wales expanded considerably between 1750 and 1825: of eighteen conifer species planted, eight were from North America and eight from Europe. Twenty-five species of broadleaved trees were planted, of which twelve had been introduced from Europe and one from North America.

Something still missing

Growing and testing many new tree species was exciting, challenging and long-term. A new scenario was emerging in the landscape.

People were getting accustomed to European trees. Norway spruce, European larch and Corsican pine were suitable for lowland Britain's fertile soils and more-continental climate, and grew well in places like Bedford Purlieus and Goodwood Estate, Sussex. It took decades to discover that Norway spruce, European larch and Scots pine, planted in millions on some big upland estates had not always been planted on suitable land.

People were also getting used to species from eastern North America such as Balsam fir (*Abies balsamea*), Eastern hemlock (*Tsuga canadensis*) and Weymouth pine. These were suitable for parkland and gardens but not for forestry plantations, and certainly not in Britain and Ireland's oceanic uplands.

During recent millenia, both European and eastern North American trees grew in 'continental' climates, quite different from the maritime conditions of the coast and islands of north-west Europe, which receive the full force of regular low pressure systems from the Atlantic Ocean ameliorated only by the meandering Gulf Stream. The uplands of north and west Britain enjoy (!) regular, high rainfall, gales, sea mists, mild summers and cool winters.

It took time, of course, to discover whether a particular tree suited any part of Britain or Ireland. Seeds of introduced species were collected from their home territories and became increasingly available in Britain during the later eighteenth century. In 1765, the Society for the Importation of Forest Seeds organised three collections of tree seeds from eastern North America into Scotland. Nurseries could then grow seedlings in hundreds of thousands. It was not until individuals of a new species matured and produced cones (or not) that home-produced seed became available.

Western North America (it was so much further from Europe, with Cape Horn or trans-America to negotiate) gradually became better known by European explorers, traders and governments. Ships were sent from the UK, Spain and Russia to survey coastlines and report back. Pacific North America was where, during the late eighteenth and nineteenth centuries, the tree species really needed, were eventually discovered by plant hunters – Douglas fir, Lodgepole pine, Western hemlock, Western red cedar – and Sitka spruce!

These were exciting times for 'enlightened improvers' and 'tree-manic' landowners. Their progress was not replicated for people whose personal tenancies, as well as land held in severalty, were lost to enclosure and specialist use as arboreta, tree nurseries and plantations. Tenants and their domestic stock were usually excluded from the tree-planted lands by turf and stone walls or fences. Yet these might be the very places for which they had historic, communal rights for grazing or for collecting fuel. Estate owners planted trees on their tenants' peat-cuttings, on open moor, on anywhere that could not be improved for arable crops. The tenants of Archibald Grant of Monymusk were so outraged with such treatment that they cut down the young trees in the new plantations and put their farm stock back again.

In England and Wales enclosure and tree-clearance on many commons by Acts of Parliament also caused loss of tenants' historic right. In 1649, the 'True Levellers' or 'Diggers' cultivated some commons to grow food for hungry peasants whose land rights had been abolished. They published a pamphlet 'from the poor oppressed people of England' to the manorial lords who intended to fell trees on commons and exclude the commoners. Their protest was quickly squashed; loss of commons, their rights and their trees, continued. In the unenclosed uplands, native woodlands were lost insidiously when grazing regimes were changed from mainly cattle to sheep and deer which are good at eating naturally-regenerated tree seedlings.

How did the explosion of tree-planting with alien species along with enclosure of commons and loss of common rights affect the amount of woodland in Britain and Ireland, the remaining native woodlands, their structure and ecology?

And where does Sitka spruce fit in?

CHAPTER NINE

Ships, Surveyors, Scurvy and Spruces

> The observatories were taken ashore, and placed upon a rock, on one side of the cove, not far from the Resolution. A party of men was ordered to cut wood, and clear a place for watering. Having plenty of pine trees here, others were employed in brewing spruce beer. The forge was also erected…
>
> The trees of which the woods are chiefly composed are the Canadian pine, white cypress, and two or three other sorts of pine. The two first are in the greatest abundance, and at a distance resemble each other…
>
> Captain James Cook, Nootka Sound at 49° 33' N 233° 12' E, 8–26 April 1778

> We observed few vegetables of any kind; and the trees that chiefly grew about this Sound were the Canadian and spruce pine, some of which were of a considerable size.
>
> Captain Cook, Prince William Sound, 19 May 1778

> Before night, on the 13th, we had amply supplied the ships with wood, and had conveyed on board about a dozen tons of water to each. On the 14th a party was detached on shore to cut brooms, and likewise the branches of spruce trees for brewing beer.
>
> Captain Cook, Norton Sound, 14 September 1778
>
> (three quotes above from James Cook 1795)

Trees, ships and explorers

In 1998, three of us Women's Institute members were researching and preparing an exhibit, 'Trees, Ships and Explorers' for local display. We lived near Portsmouth dockyard in Hampshire and the well-wooded English Weald which had supplied timber to Roman and Germanic industries and to the navy. *Trees*, *ships* and *explorers* were part of our heritage and everyday life, while woodlands, archive offices, museums and Nelson's timber ship *Victory* were our main habitats at that time.

Background to plant exploration

Britain has been an archipelago since Doggerland and the Channel River were submerged by the North Sea about 8000 years ago. Rivers, coast and seas have been our high roads: people, their animals, plants and freight travelled across the English Channel, North, Irish, Celtic, and Norwegian Seas as well as to hundreds of islands. The coast bordering the Atlantic Ocean has been a route between Iberia and France, Cornwall, Ireland, Wales, Scotland, the Northern Isles and Faroe Islands.

Rivers, seas and coasts were also food sources, while tides provided a clock and calendar. To go exploring meant travel over water. *Everybody* lived less than 97 km (60 miles) from the sea, most much closer. *Everybody* lived on a small or very small island.

Christianity was brought to Ireland and Britain by missionaries in tiny boats; political wrangling required missives or royal brides to be sent over water. Enemies appeared on the horizon in boats, while navies left for foreign missions by ship. Battles took place on the sea or on the shores where enemies disembarked. Every coastal town had a port or harbour, fishing boats, fishermen and sailors. Many rivers were tidal far inland. The sea, rivers and estuaries affected everyone.

So when new ideas, new places, new discoveries – and new trees – were sought during the Enlightenment, people's quests started by boat. And Europeans came across the seas to participate.

Overseas travel fulfilled the quest for scientific knowledge. It also brought potentially-useful plants, animals and foods, other goods and – wealth. Trade was set up and trade routes across oceans were devised. Furs, bones, oil, ivory, spices, fruits, seeds, plants, animals and even people were imported.

Exploring, having adventure and seeking new lands across the sea was (and still is) a basic part of the British and Irish psyche. It brought rewards (or shipwreck more often) from political and financial masters. New Found Lands were desirable as allies or as territories to maintain supplies of new-found-goods for home consumption.

But the ethics of our ancestors' activities are nowadays subjected to debate.

Individuals, commerce and governments sent ships away with instructions to explore and survey coastlines and natural harbours of New Found Lands. Ships were fitted out for long voyages because they could be away for two to five years. Naturalists hoping to explore the flora and fauna of other continents found ways of getting passage on vessels. Exploring instincts of the insular British took ordinary individuals to new continents.

Early plant explorers

Jacques Cartier, an avid French explorer, sailed three times to (what are now) the Canadian Maritimes in the 1530s and 1540s. The conditions were severe and as he brought home plants not treasure, the French were deterred from exploring there again for some decades.

Map 9 shows the east coast of North America and sites relevant to the discovery of Sitka spruce by Europeans.

Englishman John Tradescant, born in the 1570s, was keeper of the royal gardens. He went adventuring in Europe, the Mediterranean and the Arctic, returning with seeds and bulbs, with which he and his son started a botanic garden and nursery. John Tradescant the Younger was sent to North America, and from his 1654 trip he introduced the Tulip tree (*Liriodendron tulipifera*) and Deciduous cypress (*Taxodium distichum*).

While at medical school, British surgeons received botanical training. On board ship they would need to deal with major wounds and amputations, which required medicines, salves and vitamins made from plants. They were often keen naturalists, collected specimens and wrote in their journals of the natural history they observed.

Scandinavian naturalist Carl von Linné (Linnaeus) was encouraged to explore little-known Lapland in 1732. Celebrated for his system of naming plants and animals, he spent the next 50 years teaching students field biology. He persuaded twenty or so of his young apostles onto dangerous natural history expeditions to five continents: eight never returned.

One apostle was academic Pehr Kalm, sent by the Royal Swedish Academy of Sciences to North America from 1748 to 1751. By sending him to a northern locality, they hoped he would return with plants suited to the climate of Scandinavia. Pehr Kalm was intrigued to discover that First Nations of eastern North America made a drink he called 'spruce beer' from the shoots of a conifer tree.

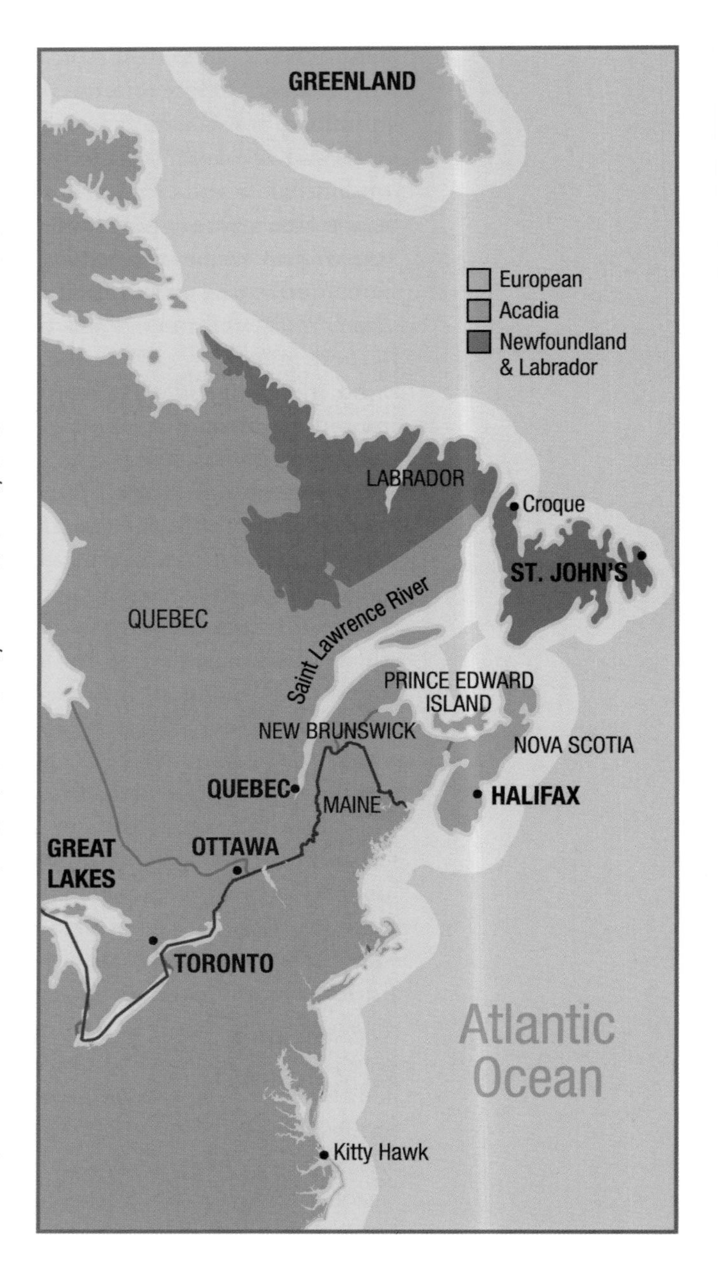

MAP 9. East coast of North America, prepared by Susan Anderson.

SOURCE: ATLASES.

Georg Steller and the Great Northern Expedition

Following their predecessors, British, French and Scandinavian explorers reached eastern North America from across the Atlantic Ocean. They sailed up the St Lawrence River beyond Quebec, to Nova Scotia, Newfoundland, Labrador and into Hudson Bay. To reach west coast North America they also sailed from the east, across the southern Atlantic Ocean, round Cape Horne and northwards.

Spurred on by their Tsars, the Russians reached the Pacific coast of North America first, but from the west. Before setting sail, they had to travel across Siberia – which took at least a year – and then to build their ships on the shores of Kamchatka Peninsula!

Russian naval ships often took foreign officers, scientists and surgeon-naturalists. Thus the two-ship Great Northern Expedition led by Dane, Vitus

FIGURE 9.1. A large Culturally Modified Sitka spruce showing scars from using an axe to get pitch (resin) from beneath the bark; Prince William Sound, Alaska.

PHOTO: JOHN F. C. JOHNSON, 2010.

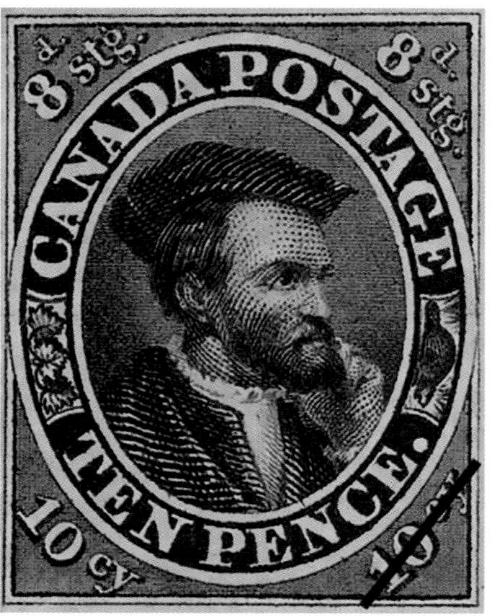

FIGURE 9.2 *(left)*. Jacques Cartier, whose sailors were saved from dying of scurvy in the Canadian winter of 1535.

PHOTO RETRIEVED FROM: HTTP://COMMONS.WIKIMEDIA.ORG/WIKI/FILE:10D_JACQUES-CARTIER_RISS_1855.JPG.

FIGURE 9.3 *(right)*. John F. C. Johnson, a Chugach Alutiiq ethnohistorian, who followed in the footsteps of Georg Steller when he visited Kayak Island to see modified Sitka spruce trees.

PHOTO: COURTESY JOHN F. C. JOHNSON, 2015.

Bering, took Georg Steller, a German physician and all-round naturalist, on its quest to find America from north-east Russia.

The crew of Bering's ship first set their pioneering feet on American soil in July 1741 on the small, steep Kayak Island on the Gulf of Alaska. Georg Steller wrote at length in his journal about this island's forests and signs of people. For instance:

FIGURE 9.4. James Cook, whose sailors were kept healthy with green plants and spruce beer.

PHOTO: LIBRARY AND ARCHIVES CANADA, ARCH REF. NO. R13263-94; PRINT OF ENGRAVING, 1750–1779.

> On Saturday 18th, we were so close to it towards evening, that we were enabled to view with the greatest pleasure the beautiful forests close down to the sea... We kept the mainland to the right and sailed northwesterly in order to get behind a high island which consisted of a single mountain covered with spruce trees only... I pushed on farther for about three versts, where I found a path leading into the very thick and dark forest which skirted the shore closely ... some tree bark, which was laid on poles in an oblong rectangle three fathoms in length and two in width. All this covered a cellar two fathoms deep in which were the following objects (1) Lukoshki, or receptacles made of bark... (3) different kinds of plants, whose outer skin had been removed like hemp, which I took for nettles which grow here in profusion and perhaps are used, as in Kamchatka, for making fish nets ... (4) the dried inner bark from the larch or spruce tree done up in rolls and dried; the same is used as food in time of famine not only in Kamchatka but all through Siberia...
>
> Golder 1925

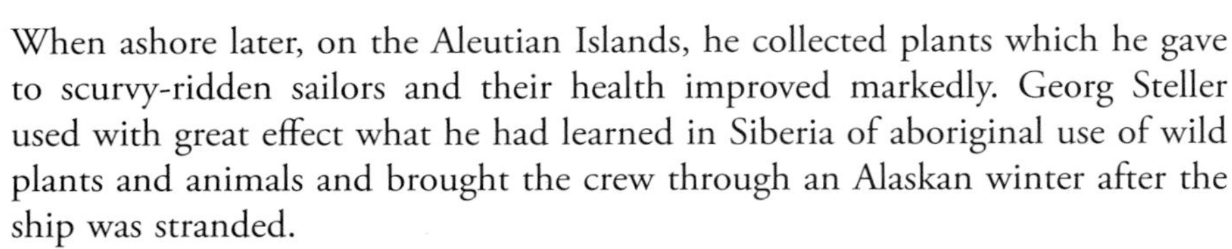

When ashore later, on the Aleutian Islands, he collected plants which he gave to scurvy-ridden sailors and their health improved markedly. Georg Steller used with great effect what he had learned in Siberia of aboriginal use of wild plants and animals and brought the crew through an Alaskan winter after the ship was stranded.

Chugach Alutiiq ethnohistorian John F. C. Johnson has confirmed that the Sitka spruce forests on Kayak Island still contain bark-stripped individuals. Georg Steller listed the hundreds of plants he observed throughout his travels, was the first European to study the natural history of Alaska and its oceans and to observe Sitka spruce in its native range.

James Cook, Joseph Banks and the Sitka spruce story

James Cook was born in 1728 and joined the Royal Navy in 1755 where he was taught surveying and map-making. After several naval battles in North America, he was then appointed to survey and chart the 9656 km (6000 miles) coastline of Newfoundland, which took from 1763 to 1766.

FIGURE 9.5. Joseph Banks, botanist and polymath, who recognised the importance of spruce beer.

PHOTO RETRIEVED FROM: HTTP://WWW.BIOGRAPHI.CA/EN/BIO/BANKS_JOSEPH_5E.HTML .

A quite different, young English aristocrat, keen botanist, frequenter of Chelsea Physic Garden, got himself and a friend two berths on a fisheries patrol ship, HMS Niger, sailing to St John's, Newfoundland in 1766. He was Joseph Banks, later a famous naturalist, visionary estate owner and polymath; in 1766 he was a beginner. The day after putting into St John's, he started to collect rocks and plants. The patrol ship later sailed along the east coast of Newfoundland to Croque and in due course to Labrador. Joseph Banks took the opportunity to go ashore, observe and collect at each stopping place. He wrote many observations on fish, fishing methods, wildlife and humans into his journal.

These two very different men, Banks and Cook, both learned something relevant to Sitka spruce during their voyages in eastern North America.

James Cook was fanatical about sailors' health: he had long been distressed at one dreadful disease suffered by sailors on deep-sea voyages; scurvy, a debilitating, fatal illness. It was taken for granted at that time that half a ship's complement would die from scurvy and dysentery on any long voyage, although it had been known for 150 years that lemons (*Citrus limonum*) could prevent and cure it.

Joseph Banks was fanatical about botany: he discovered while in Newfoundland that a drink he called 'Spruce beer' was made from the shoot tips of the Black Spruce which grows profusely in Newfoundland and northern Canada. He did not record whether it was made by First Nations Beothuk people, Acadians or other Europeans. On 12 May 1766, at St John's he wrote:

> ...the Country is Covered with wood fir is the only tree which can yet be distinguished of which I observed 3 sorts (1) Black Spruce of which they make a liquor Calld Spruce Beer (2) white Spruce & (3) weymouth Pine...
>
> O'Brian 1987

He put the recipe in his journal – it came in useful by and by:

> Take a copper that contains 12 Gallons fill it as full of the boughs of Black Spruce as will hold them down pretty tight Fill it up with water Boil it till the rind will strip off the spruce boughs which will waste it about one third take them out & add to the water 1 gallon of Mellases Let the whole boil till the Melasses are dissolved Take a half hogshead and out in 19 Gallons of water & fill it up with the Essence, work it with Barm or Beer Grounds & in less than a week it is fit to Drink...
>
> Justice 2000

It was coincidence that both James Cook and Joseph Banks were in St John's and met briefly in October 1766, but it contributed to the discovery of Sitka spruce.

The famous *Endeavour* voyage

James Cook's exceptional skills in mapping the Newfoundland coast so accurately and beautifully were recognised by the Admiralty. A couple of years later, he was given a *much* larger task. Scientists knew that the planet Venus would cross in front of the sun (a 'transit') in June 1769 but would be visible only from the Southern Hemisphere.

Lieutenant Cook received instructions from the Admiralty in 1768 to sail to the southern Pacific with a team of specialists to observe Venus; then to seek and, if found, survey the continent of 'Terra Australis Incognita'. He was then to proceed to New Zealand and survey its coasts before returning back to Britain round Cape Horn or the Cape of Good Hope. He was instructed also to organise the collection of specimens of rocks, flora, fauna, as well as 'seeds of trees, fruits and grains'. Ninety or so men and various pet and food animals prepared to set off in a small, ungainly ship HM Bark *Endeavour*.

Meanwhile, Joseph Banks heard about this special voyage and used his influence to get a position on board the *Endeavour*. He took Daniel Solander, a friend from the British Museum (another apostle of Linnaeus) as professional botanist. Sydney Parkinson, a talented natural history painter was engaged to illustrate flora and fauna. Joseph Banks took several other professionals, four servants and two dogs! In fact, the officers and professionals on board got on well and Joseph Banks' bubbly personality helped morale for 3 years on a tiny, crowded ship.

Lieutenant Cook's concern about sailors' health, and scurvy in particular, ensured that, during this global voyage, his uncooperative sailors ate sauerkraut, malt and citrus syrup with their normal diet. The sauerkraut had been prepared beforehand and stored in barrels. Cook and Banks also knew that by adding green plants and tropical fruits to the sailors' diets, scurvy could be reduced or prevented. However, wild plants and lemons could only be collected or bought at coastal stopping places, so it was back to sauerkraut on the high seas!

Sailors ate what they were told to eat, or were flogged. There were *no* deaths from scurvy during that three-year voyage, which was amazing! James Cook became well-known for the clever crew diets and drinks which he designed and for the unheard-of on-board cleanliness he put into practice.

On returning from this voyage, James Cook, Joseph Banks and their helpers were fêted for the discoveries and surveys of new lands, the huge collection of geological and natural history specimens, Sydney Parkinson's paintings and their detailed journals.

Joseph Banks then settled into London and his Lincolnshire estate. He had plenty to occupy him cataloguing his collections, preparing a book and promoting the development of science.

Using spruce beer

French explorer Jacques Cartier learned the value of 'spruce beer' the hard way. While overwintering in the St Lawrence River 1535–1536, his ship was stuck in the ice at Stadaconé (now Quebec city). After two months, most of his crew were ill or dying of a disease we now know was scurvy. Iroquoian First Nations people made them a 'spruce' drink from the leaves and bark of an evergreen conifer they called 'annedda'. Most of the sailors were miraculously cured, so naturally they called it 'Arbor-vitae'; we know it as White cedar (*Thuja occidentalis*). It is not surprising that he took arbor-vitae back to France in 1536 along with Sugar maple (*Acer saccharum*).

'Spruce beer' made by extracting and fermenting liquid steeped in leaves and young shoots of Norway spruce, was probably a drink in Baltic and eastern Europe. It could be made alcoholic or not depending on the amount of sweetener added. Recipes were published in Spons' Encyclopaedia of the Industrial Arts of 1879. In Russia, it was made from the same species and the closely-related Siberian spruce (*Picea obovata*).

Spruce beer was not produced in Britain and Ireland because there were no native spruces or suitable conifer foliage. A 'Dantzig' beer from Prussia was imported to Scotland from the fourteenth century: it was a thick, brown beer probably made from wort containing treacle, with essence from spruce foliage and bark added.

As Norway spruce is native to a small area of France, spruce beer was unlikely to have been a common drink there: it was obviously unknown to Jacques Cartier. Surprisingly, Pehr Kalm the Swedish plant hunter, was not familiar with spruce beer either and wrote about it with great interest to the Swedish Academy of Sciences in 1752:

> Among other liquors commonly drank in the European plantations in the North of America, there is a beer which deserves particular notice; it is brewed from a kind of pine that grows in those parts, and is by botanists called Abies Piccea foliis brevitus conis minimis. Rand. Mill. Gard. Distion. Spec. 5, – The French in Canada call it Epinette, and Epinette Blanche; the English and Dutch call it Spruce. This sort of pine is pretty common in Canada, and differs so little from that of Sweden, that at first sight one would think it was the same; but the cones of the American are by several degrees smaller. In the English plantations in North America, this tree is pretty scarce; for it requires a colder climate, and in those near the South it is hardly ever seen, except on the top of the highest mountains, or on that side of them which faces the North, where the snow remains, in the spring, a great while longer than in the adjacent country, and returns earlier in the autumn; but in Canada this pine grows in the same soil, and in the same manner, as in Sweden.
>
> This liquor is chiefly used by the French in Canada … As there is a great resemblance between the pine and that which is common in Sweden, it would be worth while to try out whether ours could be made use of in the same manner.
>
> Kalm 1752

He was describing White spruce to his mentors; he provided a recipe, not given here.

Acadians, settlers of French descent, who, since 1604, had lived in the Maritime region of eastern North America, made spruce beer during the mid-eighteenth century. It is likely that they obtained a recipe from Mi'kmaq First

Nations people who drank an infusion of spruce leaves during the cold, snow-covered winters when herbaceous green foods were covered with snow. Acadians drank spruce beer while under siege in Fort Louisbourg by the British in 1758 – perhaps James Cook learned of spruce beer as a member of the siege fleet.

The British army in North America probably learned about spruce beer at the same time. Major-general Jeffrey Amherst, commander of British forces in North America from 1758 to 1763 had the task of capturing Fort Louisbourg with his 140,000 siege troops. In 1760, he wrote down in his journal (yes, he had a journal too) a recipe for spruce beer. He was insistent that it was made in large quantities for his troops:

> **1760 General Amherst's Spruce Beer**
> Take 7 Pounds of good spruce & boil it well till the bark peels off, then take the spruce out & put three Gallons of Molasses to the Liquor & and boil it again, scum it well as it boils, then take it out the kettle & put it into a cooler, boil the remained of the water sufficient for a Barrel of thirty Gallons, if the kettle is not large enough to boil it together, when milkwarm in the Cooler put a pint of Yest into it and mix well. Then put it into a Barrel and let it work for two or three days, keep filling it up as it works out. When done working, bung it up with a Tent Peg in the Barrel to give it vent every now and then. It may be used in up to two or three days after. If wanted to be bottled it should stand a fortnight in the Cask. It will keep a great while.
>
> Anon 2013

During his second global voyage in 1773, Captain Cook, ever watchful of his sailors' health, would go ashore and collect, or order to be collected, edible wild plants. In April 1773, from Queen Charlotte Sound, New Zealand, he wrote: 'Also began to Brew Beer with the leaves and branches of a tree which resembles the Americo black Spruce Inspissated Juce of Wort and Melasses…' (Victoria University of Wellington Library 2014).

On another occasion, the sailors on the sister ship *Adventure* refused to eat their greens. With scurvy disabling and killing the crew, he ordered them to: 'brew beer of the inspissated juice of wort, essence of spruce and tea plants…' (Edwards 2003).

But during that voyage, the two ships had not sailed through any region where even one spruce species grew. So where did they get their spruce essence? They probably got it from three conifers whose juvenile forms have needles very similar to spruces, Kahikatea (*Dacrycarpus dacrydioides*), Matai (*Prumnopitys taxifolia*) and Rimu (*Dacrydium cupressinum*).

Captain Cook's *Method of Making Spruce Beer*:

> We at first made our beer of a decoction of the spruce leaves; but, finding that this alone made it too astringent, we afterwards mixed with it an equal quantity of the tea plant (a name it obtained in my former voyage, from our using it as tea then, as we also did now), which partly destroyed the astringency of the other, and made the beer exceedingly palatable, and esteemed by every one on board. We brewed in the same manner as spruce beer, and the process is as follows. First make a strong decoction of the small branches of the spruce and tea-plants, by boiling them three

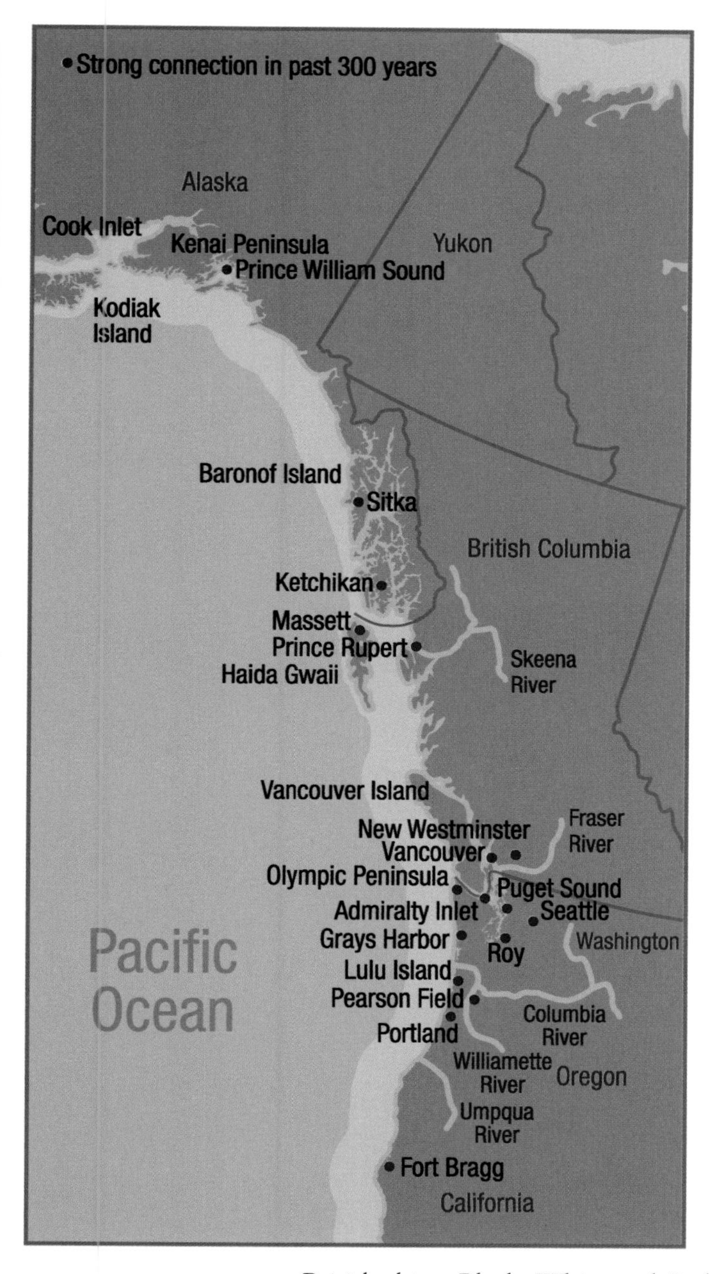

MAP 10. Sites associated with Sitka spruce in North America, prepared by Susan Anderson.

DATA COURTESY OF ALAN FLETCHER, SYD HOUSE, ELIZABETH ROBERTS.

or four hours, or until the bark will strip with ease from the branches; then take them out of the copper, and put in the proper quanitity of molasses, ten gallons of which is sufficient to make a ton, or two hundred and forty gallons of beer. Let this mixture just boil; then put it into casks, and to it add an equal quanitity of cold water, more or less according to the strength of the decoction, or your taste. When the whole is milk-warm, put in a little grounds of beer, or yeast if you have it, or anything else that will cause fermentation, and in a few days the beer will be fit to drink.

Anyone who is in the least acquainted with spruce pines will find the tree which I have distinguished by that name. There are three sorts of it: that which has the smallest leaves and deepest colour is the sort we brewed with, but doubtless all three might safely serve that purpose.

Victoria University of Wellington Library 2014

Mention of Black spruce in his 1773 recipe suggests that the spruce beer of Captain Cook's later voyages was concocted as a result of his time in Quebec and Newfoundland and his friendship with Joseph Banks. And his journals demonstrate that he was always willing to learn from First Nations. Joseph Banks had written down the recipe for spruce beer using Black spruce during his botanical trip to Newfoundland in 1766 and even attempted to concoct it.

Spruce beer using fresh foliage could only be produced when the voyages reached coasts with spruce growing nearby, when ships reached their voyage-end at spruce-covered coasts or harbours. Newfoundland and Nova Scotia were frequent termini for British ships; Black, White and Red spruce which could all be used for spruce beer grow there. Of course it was also necessary for someone who recognised a spruce tree to be on board.

When spruce-tree coasts were not on the itinerary, shoots from trees which looked like spruce were used. A drink flavoured with conifer shoots was probably far more acceptable to British sailors than sauerkraut!

A manuscript of Joseph Banks' Newfoundland journal with lists and paintings of plants and birds can be consulted at the library of McGill University, Montreal. Lysaght (1971) has an account of his 1766 Newfoundland visit based on his manuscript diary and herbarium specimens (held in the British Museum and National Museum of Natural Sciences, Ottawa).

Captain James Cook organised spruce beer while surveying the Pacific and Bering Sea coasts of North America during his final voyage in 1778. As usual, he was on the look-out for 'spruce' for his sailors to collect and make beer. He mentioned 'spruce' six times in his journal for autumn 1778, on the journey from Vancouver Island to the Aleutian Islands, Norton Sound and back.

Sitka and White spruce occur north of Vancouver Island, but Sitka spruce is the predominant *coastal* species. With Western hemlock it forms the coastal forests north from Vancouver Island to Prince William Sound, but hemlock goes no further west than the Kenai Peninsula. From there, Sitka spruce reaches further west to Kodiak Island and Cape Kubugakli (longitude 155° 6' W). White spruce overlaps with Sitka spruce in a few places on the very northern British Columbia coast and in the Kenai Peninsula but is the only spruce to grow naturally north and west of Sitka spruce in Alaska.

Captain Cook ordered spruce beer to be made at six places between Nootka Sound on Vancouver Island and Norton Sound in Alaska. The 'spruce' could only be Sitka in Nootka Sound (unless he mistook hemlock for spruce, which is unlikely). It could only be White spruce in the Norton Sound area. He likely used Sitka spruce at the other four sites because it is the only littoral spruce of British Columbia and Southeast Alaska. He may, of course, have given the name 'spruce' or 'spruce pine' to any needle-leaved conifer, but given the time he spent with Joseph Banks on the three-year *Endeavour* voyage and his interest in plants and scurvy – his botanical skills were probably sufficient to recognise the needle-leaved trees he really wanted. And in his journal he specifically differentiates several conifers by names we recognise.

Captain Cook and his crew were the first Britons to observe Sitka spruce in its native range on the North American Pacific coast and the first Europeans to collect its foliage for spruce beer to prevent scurvy.

Sending the Spruce beer recipe on naval voyages

The botanical interests of Joseph Banks and the dietary interests of James Cook fortuitously combined and gelled during the famous global voyage of 1768–1771. This ensured that green plants and fruits – and spruce beer or spruce essence – would be important ingredients of at least some future naval voyages. But it was not until 1795 that the British Admiralty *officially* recognised the importance of green plants for its sailors.

Meantime, Joseph Banks used his influence to persuade the Admiralty that *his* chosen botanists should be taken on future naval voyages of discovery and survey. He commissioned trained surgeon-botanists and naturalists as professionals on board government survey ships.

One of these was Archibald Menzies, a Scottish gardener who had the luck to be taken on by the new Royal Botanic Garden, Edinburgh. Following classes in science, he studied medicine and became a surgeon, typical of contemporary surgeons in his botanical knowledge. He joined the Royal Navy in 1782 and from 1784 to 1786 he was posted to Halifax, Nova Scotia. He was given plenty of time for study and to collect plants avidly, sending seeds to Joseph Banks for Kew Gardens and to John Hope, one-time boss at the botanic garden.

From 1788 to 1789, Archibald Menzies took a civilian job as surgeon on the commercial ship Prince of Wales which sailed to the north-west coast of America, where the crew trapped sea otters for their pelts. He collected plants, for instance, during a month moored at Nootka Sound. The *Prince of Wales* was not a naval ship so the spruce beer recipe was not on board: the crew suffered scurvy by the time the ship reached there.

Archibald Menzies was later contracted as naturalist (not surgeon) on HMS Discovery commanded by Captain George Vancouver (veteran of James Cook's final voyage). This was a four-year, two-ship, round-the-world expedition, from 1791 to 1795, taking in the Pacific coast of America. The instructions given Captain Vancouver by the Admiralty included: to survey the American west coast from 30° N (Guadeloupe, Mexico) to 60° N (Cook Inlet, Alaska) to note navigable rivers flowing from the interior, to observe and note native peoples, their customs and landscapes.

Under sufferance, Captain Vancouver took along the well-connected Archibald Menzies who also had his orders: to seek and enumerate new plant species (including mosses and ferns), to find their native names, to bring home seeds and living plants and to write a diary. When the ship's surgeon was sick, he had to attend to men's health too!

Joseph Banks made sure that his spruce beer recipe was taken onboard *Discovery* with both Archibald Menzies and George Vancouver – just to be certain! Archibald Menzies instructed the sailors who did the brewing, but first they had to find spruce trees of some sort.

George Vancouver and his surveyors managed to survey the coastline from San Francisco to Cook Inlet near Anchorage, Alaska. This was an enormous task, which we can barely imagine, and done mainly from very small, open boats called yawls and pinnaces.

Sitka spruce and spruce beer

Not surprisingly, by the time they reached the Pacific coast of North America, the sailors of *Discovery* showed signs of scurvy: they were far from the tropics, from fresh fruits and greens. So the hunt was on for spruce trees along the North American coast.

While the surveyors were working from their open boats, Archibald Menzies and his helpers went ashore, integrating the search for spruce trees with more general plant-hunting. In summer 1792, they came across large trees of Western hemlock. Menzies mistook it for Eastern hemlock and called it *Pinus canadensis*. British Columbia archives department has transcribed and made available his manuscripts covering North America.

On 6 June 1792, while in Admiralty Inlet (latitude 48° N, near modern Seattle, see Map 10) he wrote about the trees he saw:

> The Woods here were chiefly composed of the Silver Fir – White Spruce – Norway Spruce and Hemlock Spruce together with the American Arbor Vitae & Common Yew; & besides these we saw a variety of hard wood scattered along the Banks of the Arms, such as Oak – the sycamore or great Maple – Sugar Maple – Mountain Maple & Pensylvanian Maple – the Tacamahac & Canadian Poplars – the American Ash – common Hazel – American Alder – Common Willow & Oriental Arbute, but none of their hard wood Trees were in great abundance…
>
> Newcombe 1923

What he called 'Norway spruce' was almost certainly Sitka spruce. On 31 July 1792, further north in the Broughton Archipelago, he wrote again about their search in coastal forests for spruce beer ingredients:

> Early in the morning of the 31st two Boats were equipped and sent off under the direction of Mr. Puget & Mr. Whidbey up the Western branch on a surveying expedition, while Cap* Vancouver & Mr. Broughton went off at the same time in the Pinnace with intention to accompany them further up with the Vessels, & if so, to settle on a place of rendezvous.
>
> A party began to Water & another to brew Spruce beer, but after erecting the Brewing Utensils on shore, they brought me word that there was none of that particular Spruce from which they used to Brew to be Found near the landing place, on which I recommended another species (*Pinus Canadensis*) which answered equally well & made very salubrious & palatable Beer.
>
> Newcombe 1923

The sailors made spruce beer from the (Western) hemlock on several occasions along the north Pacific coast. Archibald Menzies had noted that its needles were a similar length to those of Black spruce and thought it would suit the purpose.

He collected many other tree species new to European science, for instance Coast redwood, Douglas fir, Chilean pine (Monkey-puzzle tree) and Western red cedar. He wrote the first description of [Sitka spruce] in his journal, recognising it as a distinct type of conifer, but not giving it a name. Herbarium specimens of it and many other plants were packed and sent back home, to be deposited in the

FIGURE 9.6. Friendly Cove, Nootka Sound, 1792. Volume I, plate VII from *A Voyage of Discovery to the North Pacific Ocean and Round the World* by Captain George Vancoouver. copied from NOAA, USA.

RETRIEVED FROM HTTP://COMMONS.WIKIMEDIA.ORG/WIKI/FILE.VANCOUVER.FRIENDLY-COVE.JPG.

FIGURE 9.7. Puget Sound, Washington, famous from early European explorers, surveyors and plant collectors; in modern times a source of lumber and outdoor recreation.

PHOTO: SYD HOUSE, 2004.

British Museum. Unfortunately many of his living plants on the ship's deck were ruined were ruined in the cold.

At the end of the voyage Archibald Menzies wrote up his journal and catalogued his specimens. His life's work drew attention to the very rich botany of Pacific North America. He saved hundreds of sailors' lives by following Cook's and Banks' methods of making spruce beer from conifer needles. During the four-year Vancouver voyage, spruce beer ensured that *no sailors died from scurvy* – quite an achievement! He also collected at least 400 plant species new to science during his lifetime's voyages.

For enjoyable and fascinating stories of the life and explorations of a surgeon-naturalist on a British naval ship at the end of the eighteenth century, try reading some of Patrick O'Brian's novels about Captain Jack Aubrey and his Surgeon-naturalist Stephen Maturin.

It was not until 1831 that some *seeds* of Sitka spruce first reached Britain and Ireland from North America. Imagine the excitement and tension as landowners and tree-planters awaited conifer seeds collected by expeditions to the Pacific coast of North America.

Chapter Ten will describe how Sitka spruce participated in both tree-growing trials and fast-changing landscapes during the nineteenth century.

CHAPTER TEN

Journeys and Experiments for Seeds and People

> A handful of Scots ... have hunted out and introduced into the West more plants from around the world than probably all the other European nations combined. Every garden in Britain, and those of most of our European neighbours as well, contains plants originally brought back to Europe by a Scot. From the early eighteenth century onwards, Scots' expertise, cunning, curiosity, intelligence, adventurousness of spirit, scientific knowledge and business acumen dominated the horticultural scene not only in Great Britain, but around the world.
>
> Ann Lindsay 2008

Early travellers to North America

Humans lived on the land mass of Beringia about 15,000 BP, fishing and hunting small mammals for food. The urge to move on and explore, for whatever reason, took some people along an ice-free coastal zone in easterly and south-easterly directions from Asia into North America. Some of their descendents reached Southeast Alaska by the time Sitka spruce was infiltrating Lodgepole pine woodlands about 10,000 BP.

The earliest known written records show non-indigenous, European, people arrived in *eastern* North America about 1000 AD. It took another several centuries before Europeans explored *western* North America.

The Norse 'Vínland' sagas tell us that Eirik the Red was outlawed from Iceland to Greenland in the late tenth century AD with his family. His son, Leif Eiriksson, heard tales from a Norwegian explorer of a new country sighted to the west. So, in 1000 AD, he and a ship's crew sailed away westwards. After 1328 km (825 miles) of sea, they discovered (what are now) Baffin Island and Labrador, which they named 'Helluland' and 'Markland'. Sailing on southwards, they found pleasant, fruitful lands between the Gulf of St Lawrence and New Jersey, where they spent a season. German-born Tyrkir happily recognised plants and fruits resembling grape-vines and grapes from home, so Leif Eiriksson called that country 'Vínland'.

They took home to Greenland a boat-load – about 22 tonnes (24 tons) – of timber and a trailer-boat of berries! Subsequent Norse expeditions also took away boat-loads of timber and berries after short visits, until the last-recorded Norse voyage to North America in 1347. Timber and berries, scarce in Greenland and Iceland, were valuable resources.

FIGURE 10.1. Botanist Karl Mertens, who sent many plant specimens from Sitka Island to a colleague in Russia.

PHOTO RETRIEVED FROM: HTTP://FLORAWONDER.BLOGSPOT.CO.UK/2015/05/OLD-PLANT-BOOK.HTML .

During the 1480s and 1490s, merchants from my home port of Bristol, in the west of England, often sailed across the Atlantic Ocean to Newfoundland, but left no journey logs. The famous merchants John and Sebastian Cabot also sailed from Bristol in their little ship *Mathew* and reached Newfoundland in 1497.

Their stories of 'more fish than water' (Roberts 2007) caused French, Portugese and Basque fishing fleets to chase across the North Atlantic a few years later – and Europeans continued to scoop huge cod out of the Grand Banks until a moratorium on fishing was imposed by Canada in 1992 AD. By then, there was definitely more water than fish.

In 1534, Frenchman Jacques Cartier sailed up the St Lawrence River to the Great Lakes and, after some weeks stuck in winter ice, took several local tree species home to France. Early the next century, Samuel de Champlain followed his route, reaching Quebec where he established a botanic garden. The era of Europeans exploring North America for new plant species was under way.

A few Russians, after long land journeys across Siberia, sailed to Alaska in the seventeenth century as surveyors and traders in pelts. An official exploration led by Vitus Bering and Aleksei Chirikov, landed in southern Alaska in 1741. Subsequently a strong Russian presence along coastal Alaska developed trading and warring relationships with indigenous people. Naval expeditions carried out scientific surveys from Alaska to California during the next 100 years; for instance the 1826 to 1829 voyage of *Senyavin* had an ornithologist and mineralogist as well as botanist Karl Mertens. They discovered thousands of unknown (to Europeans) species of plants and animals.

The North American continent, which had been colonised from Asia 9000 years before, was squabbled over for exploration, trade and removal of natural resources. After the Wars of Independence, the 1783 Treaty of Versailles established the new multi-state nation of America. A unified Canada was formed in 1840. People of European descent put themselves in political control of the continent and its First Nations.

Hence the numerous eighteenth-century explorers who sought new plants and coastal spruce-beer trees were individuals in an eight-century time-line of elsewhere-people seeking new resources in North America. But the continent's natural resources were already integral to the lives and ethos of its indigenous inhabitants.

The Scottish context

Scotland was the country from which the most – and the most prolific – plant hunters set out during the eighteenth and nineteenth centuries. It had a unique role in finding, collecting and growing new species.

Scottish Presbyterian society promoted a strong work ethic. Young gardeners, educated to primary level, could read and write. They were trained in rural crafts as well as gardening; life in the inclement Scottish weather ensured toughness; work was usually hard to come by, so travelling to find work and adventure was common.

From the late seventeenth century onwards Scottish gardeners ran numerous horticultural enterprises. For instance in 1722, Philip Miller took charge of the new Chelsea Physic Garden, central to British plant collecting and growing. During the nineteenth century, the famous Veitch family ran successful nurseries in Devon and London and sent out 23 plant explorers.

Scottish arboriculturalists and foresters were the only members of their profession in Britain with long experience of growing *conifers*: the native Scots pine. This and their reputation as good plantsmen ensured them positions on many Welsh estates, where Scots pine was grown from Scottish seed during and after the seventeenth century. Examples of Welsh estates which grew Scots pine are Faenol (by Peter Balden) and Penrhyn (by Angus Webster), both in Caernarvonshire and St Fagan's (by Robert Forrest) in Glamorgan. Local Welshmen were the woodsmen doing the hard work though.

Walking across America: David Douglas

After the Vancouver expedition arrived back in Britain in 1795, Archibald Menzies caused great excitement among botanists and landowners with his herbarium specimens and information on rich forests with new trees, shrubs and herbs. Yet it was not until 35 years after Archibald Menzies' visit – 1825 – that another British plant explorer arrived on the Pacific coast. Time was needed to raise funds for a long, complex expedition: nine months were needed for just a one-way journey there from Britain.

The Horticultural Society of London, formed in 1804, sent several men plant-hunting to other places during the 1820s. David Douglas, a humble gardener from Scone, Perthshire worked in Glasgow Botanic Garden by age 21. His life as an explorer started in 1823 when his boss recommended him to the Horticultural Society. He was sent to collect plants and seeds from Upper Canada: the Niagara Falls, Lake Erie and south to Philadelphia. This trip was successful for both plants and for his impeccable behaviour. So the Society chose him to follow Archibald Menzies' work with an expedition to western North America. He went first to meet Menzies and to study his herbarium specimens in London. In July 1824 he set off from Gravesend, Kent, on the Hudson Bay Company's ship *William and Ann*. The ship reached the mouth of the Columbia River in

FIGURE 10.2 David Douglas, Scottish plant explorer, who sent the first seeds of Sitka spruce and many other trees to the UK from North America.

PHOTO: COURTESY SYD HOUSE.

April 1825. One of the first trees he observed on the shores was named *Pinus taxifolia* at that time – known to us as Douglas fir.

David Douglas first explored onshore from the ship then changed his base to the new Fort Vancouver, upriver. His explorations really started when he travelled on foot, horseback and by canoe with employees of the Hudson Bay Company in the catchment of the Columbia River. They slept rough and caught much of their food. He collected specimens and wrote his voluminous observations when the weather allowed. Fruits and seeds were first dried over an open fire at supper-time and then kept dry by regularly changing the blotting paper in which he pressed them. They were then labelled (absolutely vital) and put within an oilskin to keep out rain and river water (also vital). He needed to dry, press and pack each specimen sufficiently for it to survive the rough horseback journey back to Fort Vancouver and the long sea passage to Europe.

I know how difficult it is, after a day outdoors, to tiredly identify specimens and write up one's notes even in a warm field station; it needed motivation and devotion to do it outdoors in a damp climate.

Despite his care, he sometimes lost his specimens in lakes, rivers or from mould. He was fortunate to have assistance from First Nations and Canadian helpers in every aspect of his work, which meant he could collect, pack and write more fully – and their knowledge of the country helped enormously. David Douglas recognised First Nations' huge knowledge of their native plants and animals, and was appreciative of their help.

While in London preparing for his trip, he had seen a herbarium specimen of an un-named conifer collected by Archibald Menzies from Puget Sound. One day he recognised it *growing* on the north shore of the mouth of the Columbia River. He collected his first specimens from there in autumn 1825. In his detailed written description of the tree and its habitat, he gave it its first Latin name *Pinus menziesii* to honour his predecessor, and wrote effusively about it:

> It may nevertheless become of equal if not greater importance [than Douglas fir]. It possesses one great advantage by growing to a very large size ... in apparently poor, thin, damp soils... This unquestionably has great claims on our consideration as it would thrive in such places in Britain where P. sylvestris finds no shelter. It would become a useful and large tree... This if introduced would profitably clothe the bleak barren hilly parts of Scotland ... besides improving the beauty of the country.
>
> Douglas 1914, quoted in House and Dingwall 2003

David Douglas perceived immediately that this new conifer might suit Scottish conditions. His knowledge of weather, soils and plants back home in Scotland and his acute observations of the tree's distribution on another continent coalesced into an understanding of the potential of [Sitka spruce] for his home country. An explorer from lowland UK brought up in its more amenable climate and good soils might not have made this connection.

To actually get seeds of [Sitka spruce], he had to laboriously pick up scattered cones from the forest floor, shoot cones down, or, more often, climb the tall trees and pluck them. The cones were also dried over an open fire so that the seeds could be shaken out. For a good herbarium specimen, he collected foliage, bark, cones and seeds, which he placed in a plant press.

Botanising for 1825 stopped when winter came: time for rest, writing and planning the next year's work. Contemplate the difficulties facing an explorer. There were few roads and travel was by foot, boat or on horseback. While away from a Fort, they slept on conifer foliage or the ground. Maps were basic; distances were enormous; native peoples might welcome strangers or not. When David Douglas hurt his knee, there was no penicillin to prevent infection, when he had malarial fevers he took a palliative and rested. Protein food had to be shot, prepared and cooked. A gun was also necessary to defend them from some mammals, to dislodge cones from trees and to loosen plants from cliffs. Seeds could be collected only from ripe cones, so sometimes he had to return later, remembering the locality in his mind (no maps or GPS!).

In autumn 1826, David Douglas trekked to the catchment of the Umpqua River south of Fort Vancouver, where he hoped to find Sugar pine (*Pinus lambertiana*) a tree with giant cones. While swimming in the Umpqua he happened to observe [Sitka spruce] again, growing along the banks: what a lovely way to botanise! Actually, the expedition members were extremely tired and had stopped for a break.

At the end of his expedition, he travelled with the annual Hudson Bay Company overland expedition from Fort Vancouver to York Factory before taking a Company ship home. When he set sail from Hudson Bay in September 1827, David Douglas had walked, canoed or ridden on horseback over 11,265 km (7,000 miles) in North America.

In 1830 David Douglas was in North America for the third time. That year he sent back more [Sitka spruce] seeds from the Puget Sound, the inlet which had been mapped meticulously in 1792 by Peter Puget and George Vancouver. He boated and walked through its temperate rainforests and keenly observed the growth and distribution of Sitka spruce and other flora.

In 1834, he was killed accidentally in Hawaii. He introduced over 200 species of plant to Britain from North America, his journal was written painstakingly, a bonus to his other accomplishments. He travelled thousands of miles, worked well with diverse people, kept going despite bouts of poor health and added immensely to scientific knowledge.

Sending conifer seeds across the ocean

To ensure that at least some of his specimens reached London, David Douglas sent several parcels via the different overland and shipping routes available to him. Even so, the first Sitka spruce seeds, despatched to Britain in 1825 from Fort George, never reached London.

His second, 1826, collection of Sitka spruce seed arrived in Britain later that year or in 1827. His final, 1830, parcel of Sitka spruce seeds, collected from Puget Sound, arrived in London during 1831 and the contents were distributed amongst Horticultural Society members and Scottish estate owners. Some germinated into the first Sitka spruce to grow in the United Kingdom.

Collecting conifer seeds in bulk, maintaining their condition and getting them back home were all difficult to accomplish. Small quantities arrived in Britain to start with so early handouts, were of few seeds of each species; packets were sent to the societies, estates, arboreta and gardens which had contributed expedition funds.

FIGURE 10.3 *(below left).* Young Sitka spruce planted in 1834 at Keillour, the first Scottish estate to plant Sitka spruce seeds.

PHOTO: COURTESY ROYAL SCOTTISH FORESTRY SOCIETY.

FIGURE 10.4 *(below right).* A Sitka spruce planted in 1897 in the policies of Murthly Castle, Perthshire by Colonel Walter Steuart Fothringham; in 2007 it was 49 m (161 ft) tall and had a girth of 6.20 m (20 ft).

PHOTO: ANDY TITTENSOR, 2013.

First attempts at growing Sitka spruce

Landowners hoped that imported evergreen trees would enhance gardens and arboreta during the UK's dreary, grey, damp winter. They also hoped for species to utilise the degraded, exposed uplands so the land would be more productive.

We can guess that there was great excitement with the news that packets of seeds had arrived from North America. Perhaps there was even more excitement when smaller packets were received and opened inside mansions, nurseries and societies and a few unfamiliar seeds were found inside.

In 1834, Keillour Estate in Perthshire, Scotland, was first to plant Sitka spruce seeds. Murthly Estate, Perthshire planted its first Sitka spruce seeds in 1846; Drummuir on Speyside in 1850; Drumlanrig Arboretum, Dumfries-shire in 1850 and Culzean Estate on the Ayrshire coast in 1851 (Map 11).

In Ireland, the Curraghmore Estate, Co. Waterford, ancestral home of the Marquesses of Waterford since the twelfth century, planted some of David Douglas's seeds in 1834 or 1835. Tender loving care and the sheltered planting position in a hollow near the River Clodiagh, ensured that one is still alive today. At 55 m (179 ft) tall, this beautiful specimen is the second tallest tree in Ireland.

The beautiful Abbeyleix Estate, Co. Laois received some of David Douglas's original Sitka spruce seeds between 1829 and 1836. A specimen tree from those original seeds still survives in the arboretum, having reached 38.5 m (126 ft) tall.

In England, there are fewer records of early Sitka spruce planting. However, at Pencarrow in Cornwall, a seed planted in 1842 grew into a tree which was 31 m (102 ft) tall in 1970, but has since died. In 1860, Sitka spruce was planted in the Bolderwood enclosure in the New Forest, and in 1861 some were planted in the beautiful grounds of Stourhead Estate in Wiltshire.

The Penrhyn Estate, Caernarvonshire, planted Sitka spruce seeds soon after seed became available, probably when Penrhyn Castle was redesigned and expanded in the early 1830s. An 1880 list of specimen trees included a Sitka spruce 16 m (53 ft) tall, with its original name *Abies menziesii*. It had grown over one foot a year. In 1959, Penrhyn boasted a Sitka spruce from original seed at 35 m (116 ft) tall, possibly the same specimen.

The big Welsh estates acquired seeds of all the available nineteenth-century Pacific coast trees. After growing European and eastern North American conifers the previous century, they were well equipped to experiment with newly arriving species. Some estates recorded their trees' height and girth as they grew, as well as their reaction to different soils and climatic conditions. Early ornamental plantings were useful trials for later plantation forestry: foresters discovered that Douglas fir, Noble fir, Japanese larch, Lawson's cypress and Sitka spruce were suitable for Welsh conditions.

Many Sitka spruce trees grown in Britain and Ireland from early seed were alive and of exceptional height and girth in 1970 when they were measured by Alan Mitchell (see Mitchell 1972).

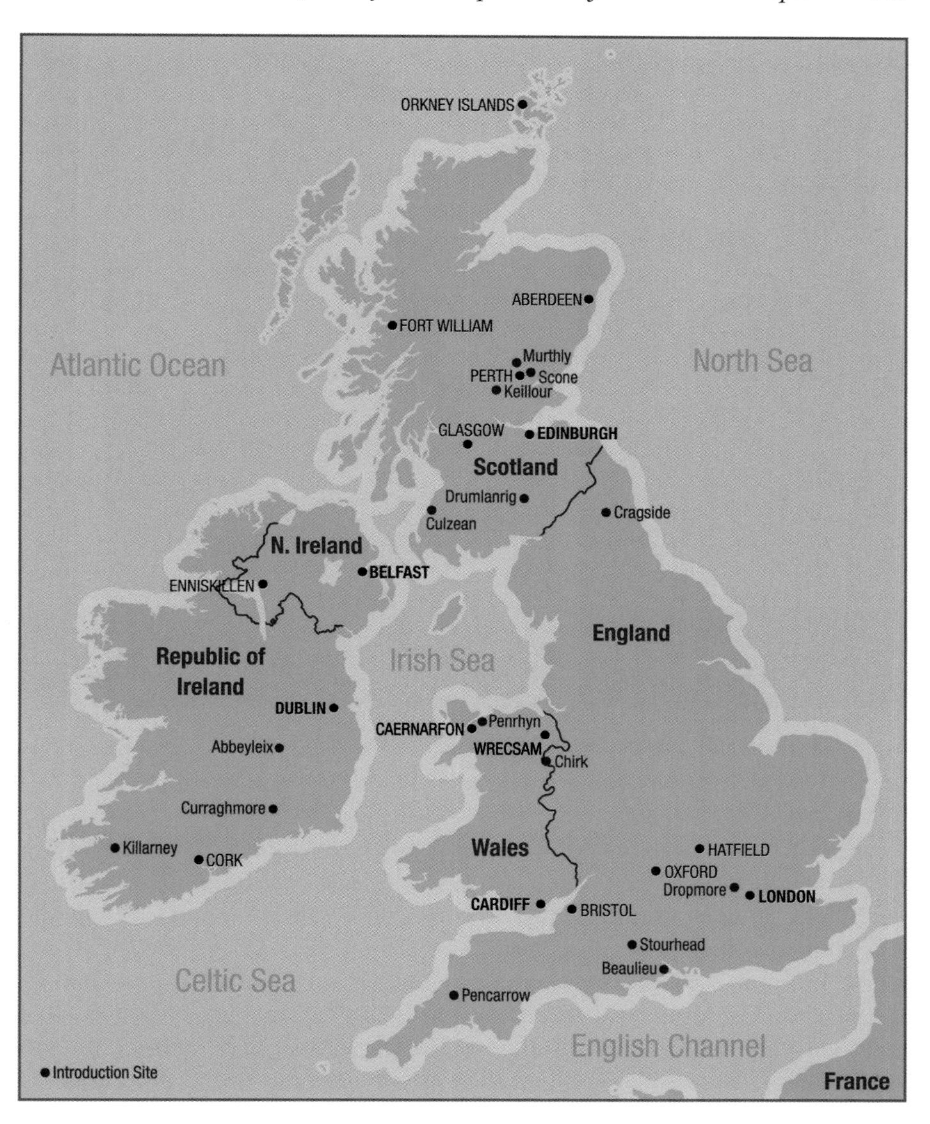

MAP 11. Some introduction sites, prepared by Susan Anderson.

DATA: SEE BIBLIOGRAPHY FOR CHAPTER TEN.

The estates I have mentioned have historic castles, houses, parkland, arboreta and gardens open to the public and I recommend a visit to them for their architecture, historical interest, horticulture, internal ornament and Sitka spruce.

The *very* first

It comes as a surprise to discover that it was Russians who first planted Sitka spruce outside its native range. Several Russian expeditions noted the treeless-ness of the Aleutian Islands at the turn of the eighteenth to nineteenth centuries. The Aleutian Islands support herbaceous tundra vegetation apart

FIGURE 10.5. *(left)* This beautiful Sitka spruce in the arboretum of Curraghmore Estate, Co. Waterford, Republic of Ireland, started as a seed planted about 1835.

PHOTO: ANDY TITTENSOR, 2013.

FIGURE 10.6 *(centre)*. One of the earliest Sitka spruce in Ireland, grown from seed collected by David Douglas, in the gardens of Abbeyleix House, Co. Laois.

PHOTO: SYD HOUSE.

FIGURE 10.7 *(right)*. An early Sitka spruce in the UK, planted in the arboretum of Drumlanrig Castle, Dumfries-shire in 1850.

PHOTO: RUTH TITTENSOR, 2013.

from low-growing alder, dwarf birch, willow and heathers. Colonisation by non-indigenous people would be difficult without accessible timber and fuel wood despite driftwood washing ashore.

About 1805 the Tsar's representative in north-west America, Nicolaĭ Petróvich Rezánov, ordered that small trees two to three years old (called *Abies* but they were Sitka spruce) should be taken from Sitka (now Baranof) Island and transplanted onto the small, volcanic Amaknak Island in a bay of Unalaska Island; more were planted later on nearby Expedition Island. These were small, sheltered islands, subject to fog and were about 837 km (520 miles) south-west of the nearest native Sitka spruce (or any forest) on Kodiak Island. Unalaska's first priest, Father Veniaminov reported that 24 were still alive in 1834 but growing out not up: the tallest were 2 m (7 ft) with stem diameter of 46 cm (18 in). He transplanted more Sitka spruce from Baranof Island onto Unalaska Island in the 1830s and 1840s and the trend continued on several Aleutian Islands until the Second World War.

Visitors described the slow-growing Sitka spruce: in 1903, when the tallest was 7.62 m (25 ft) high, they were compact, mat-shaped, with lower branches along the ground. Nevertheless, some survived and the tallest were 8.5 m (28 ft) high in 1958; they had produced cones and there were seedlings on nearby disturbed ground. When surveyed by American foresters in 1987 some of the 1805 Sitka spruce were still alive and are there today in Sitka Spruce Park.

Many tree species were planted in the Aleutians during the twentieth century but Sitka spruce survived the best.

FIGURE 10.8. 'The Forest' in 1899 looking east: Sitka spruce planted in 1805 on Amaknak Island, Unalaska, Aleutian Islands.

PHOTO: E. S. CURTIS (THROUGH THE OFFICE OF HISTORY AND ARCHAEOLOGY, ALASKA DEPARTMENT OF NATURAL RESOURCES).

FIGURE 10.9. Sitka spruce plantation alias 'The Forest' on Amaknak Island, Unalaska, 1975, planted in 1805; designated as a USA National Historic Landmark.

PHOTO: RICHARD W. TINDALL, 1975 (THROUGH THE OFFICE OF HISTORY AND ARCHAEOLOGY, ALASKA DEPARTMENT OF NATURAL RESOURCES).

FIGURE 10.10. Sitka Spruce Park, Amaknak Island, Unalaska, 2014, showing the remaining 1805 trees and their vigorous offspring.

PHOTO: ALBERT H. BURNHAM, RECREATION MANAGER, CITY OF UNALASKA.

Oregon Botanical Association: conifer seeds in bulk

Botanists, horticulturalists and foresters in the UK were anxious to discover more plants. They hoped that larger quantities of conifer seeds could be collected and brought back to undertake trials, but it took twenty years to raise enough finance for more botanical exploration.

In November 1849, the 'Oregon Botanical Association' was formed in Edinburgh by people keen to promote continued exploration in the area called 'Oregon Territory': the Pacific coast and inland to the Rocky Mountains of today's Oregon, Washington and southern British Columbia. Oregon Territory was claimed variously by Britain, Spain, Russia and America during the earlier nineteenth century. However, with the Oregon Treaty of 1846 its land south of 49° N would be American, north of it would be British, and Vancouver Island, Canadian. £5 shares in the Association were bought by estates, nurseries, horticultural societies and the Royal Botanic Garden, Edinburgh.

In May 1850, with sufficient money raised, John Jeffrey, a young worker at the Royal Botanic Garden there, was contracted to take on the task of continuing the work of David Douglas. He started out around Hudson Bay in August 1850, traversing a changed landscape from David Douglas two decades earlier. On his first day he collected 54 cones of *Picea rubens* and 22 cones of *P. glauca*.

His first letter and box of seeds did not reach Britain until late autumn 1851, and the Oregon Association was disappointed with their paucity. Perhaps members misunderstood how difficult it was to collect fertile cones in bulk, extract and dry the seeds thoroughly in difficult conditions. But a few seeds were allocated to some subscribers in spring 1852.

By January 1854 John Jeffrey was in San Francisco and had sent back to Britain ten consignments of specimens and seeds – but he disappeared mysteriously soon afterwards. Yet in three years he had sent back material from *35 species* of North American conifers, including Sitka spruce. He discovered new six conifer species, including Western hemlock and Foxtail pine (*Pinus balfouriana*).

Murthly and Scone estates in Perthshire, Scotland, were members of the Oregon Association; they planted their second batch of Sitka spruce seeds in 1852. Glenalmond Estate, Perthshire still grows several spectacular Sitka spruce planted from Oregon Association seeds at this time.

With so many conifer species imported during the nineteenth century, the UK landscape began to change dramatically. Competition between growers caused a 'conifer rage'!

Contemporary botanical exploration

John Jeffrey was not the only plant collector in North America: nurseries, horticultural societies, botanic gardens and estates in Britain and Europe contributed to the costs of plant and seed collectors who travelled there and to other continents. Nurseryman Thomas Drummond and his brother William went to the Arctic and Texas in 1827 and 1831 respectively; William Murray was sponsored by nurserymen Peter Lawson & Son who had seen the future commercial value of North American species. John Fraser is famed for his flowering shrubs from the eastern USA, Newfoundland and Russia. Irishman doctor-botanist Thomas Coulter collected plants while attending to people's health in Mexico, then visited Monterey and finally the Arizona desert. The Welsh were not avid plant hunters.

Nurseries were flourishing: 26 sponsored the Oregon Botanical Association. The Veitch family not only ran exceptional nurseries but went plant hunting and sent out explorers such as the Lobb brothers to South America and Ernest Wilson to China.

Writers prepared treatises on forestry and textbooks on plants. After their eighteenth century botanical expeditions, French botanists André Michaux and son Francois produced an inventory of trees in eastern North America, *Histoire des arbres forestiers de l'Amerique septentrionale* in 1812. And Briton, Edward Ravenscroft is famous for his 1884 tome *Pinetum Britannicum*, which described, with lithograph illustrations, all the exotic conifers growing in the UK at the time.

First Sitka spruce plantations and experiments

Conifer seeds arriving in Britain from western North America were germinated and grown in nursery beds or gardens. Each one which grew into a big-ish tree by about ten years old would be moved to its final position in an arboretum or park. But it was not known, yet, how long each species would take to mature and produce cones.

Conifers' attractive shapes and foliage colours were appreciated. Many grew into beautiful specimens – but which would produce good timber in the United Kingdom's varied geological, soil and weather conditions?

Douglas fir, Western red cedar, Western hemlock, Grand fir, Lodgepole pine and Sitka spruce all grew well. Sitka spruce did not attract much attention but found its way to some welcoming landowners.

Inevitably there was something of a time lag before landowners had confidence to start plantation experiments. Eager owners waited to see how each tree grew. After a wait of 20–25 years, their Sitka spruce trees probably produced fertile cones and they could then collect and germinate the seeds to produce offspring in bulk.

It was necessary for each owner to assess a tree's timber qualities in his or her particular growing conditions before collecting or buying quantities of seeds or transplants. Enclosing land and planting thousands of trees was expensive, and even more so if the trees failed to reach timber size and quality:

> The ostensible object in planting pinetums, apart from ornamental purposes, is to test the qualities and hardiness of the different kinds for ultimate timber trees in Britain, or, in other words, to find out if any of them, by their quality of timber and rapidity of growth, will replace some of our present timber trees, and return a greater amount of profit to the planter.
>
> France 1869

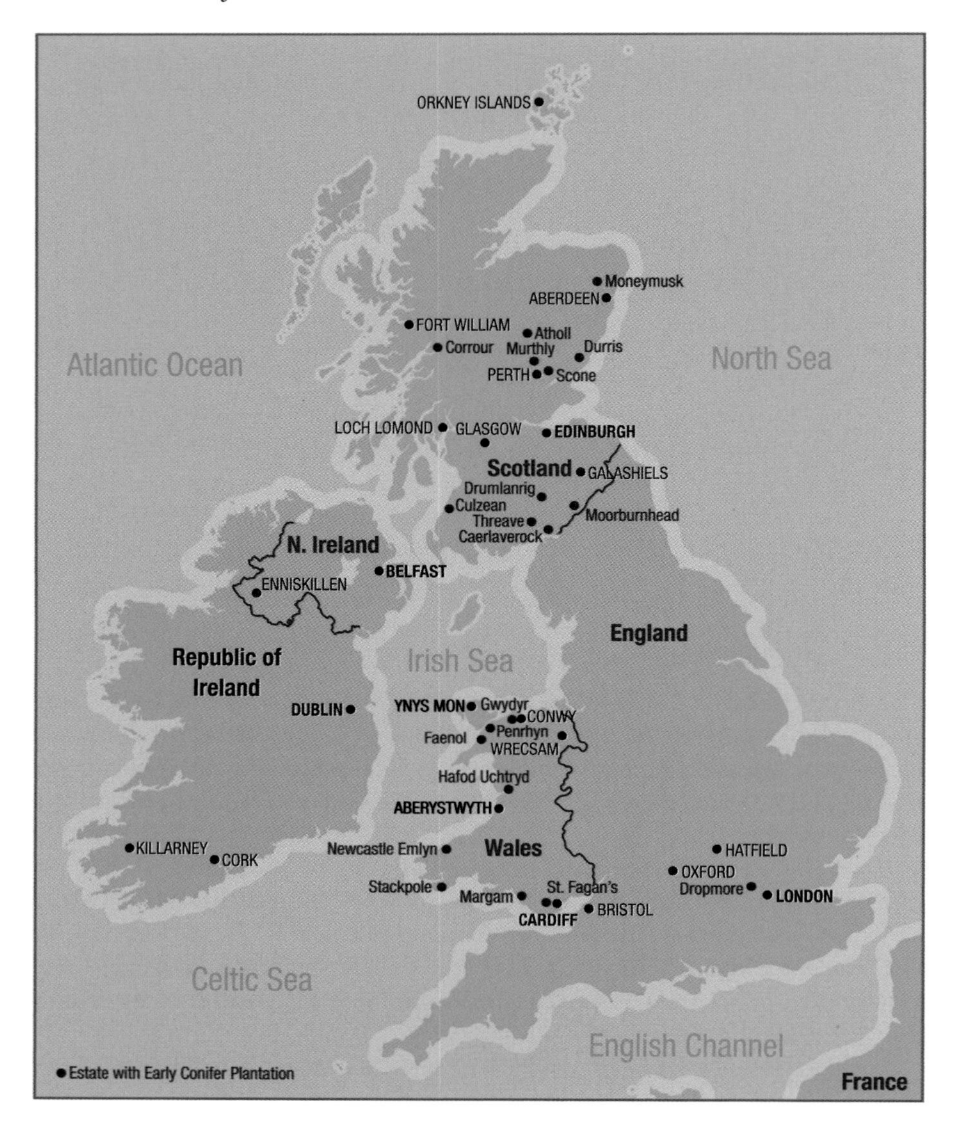

MAP 12. Private estates with early conifer plantations, prepared by Susan Anderson.

DATA: SEE BIBLIOGRAPHY FOR CHAPTER ELEVEN.

A few estates experimented with plantations during the mid-nineteenth century. At Wooplaw, Roxburghshire, Sitka spruce was planted at 274 m (900 ft) altitude in 6–9 m (20–30 ft) spacing – and they grew from 1866 until 1956.

Records show that Henry Robert Baird established plantations of Sitka spruce and Scots pine on his difficult land at Durris, Kincardineshire from 1873 to 1882, planting them 1.8–2.7 m (6–9 ft) apart. He also experimented with European larch, Norway spruce and Douglas fir. In 1931, Sitka spruce was the only species still growing. The Duke of Buccleuch started an experimental plantation on his Drumlanrig Estate, Dumfries-shire in 1900 (Map 12).

Following techniques used in Germany, some estates transplanted Sitka spruce from nursery to plantation at one to three years old, and close together

LOCALITY AND TREATMENT DATA. PLOT 13

rdnance Survey Sheet No. Kincardineshire 10 S.W.
" " " " Scotland 66.
er H.R. Baird of Durris.
at Charles Stewart, Forester, Nursery Cottage Durris, Drumoak.
ations 800' Aspect N.
sure Mod. Exposed. Slope
gy and Soil 1"–3" mixed humus on a layer of fine peat varying in depths from 3" to 6"–9" in the southern half of the t where it has been cut. Under the peat is a layer of e grey clay with stones and rotten rock. Boulder clay over Felspathic Gneiss.
tation Nil.
ment Established to augment the data on the yield of Sitka ruce in Scotland. Thinned on establishment and ain in 1925 to rather more than a B. Grade.

LOCALITY AND TREATMENT DATA. PLOT 14

dnance Survey Sheet No. Kincardineshire 10 S.W.
" " " " Scotland 66.
r H. R. Baird of Durris.
Charles Stewart, Forester, Nursery Cottage, Durris, Drumoak.
tions 800' Aspect S.E.
sure Mod. Exposed. Slope Gentle.
gy and Soil 1"–2" raw humus on 1–3' fine grained peat owing presence of some mineral soil, eg. mica. ck occurs at 1'–1½' in the N. position and at over 3' the S. position. Peat and boulder clay over Felspathic Gneiss.
ation Nil.
nent Established to add to the data on the yield Sitka Spruce in Scotland. Thinned to a B. grade. The plot was wrecked by the gale of Jan. 1927.

Plot No. 13. Species Sitka spruce. COPY
Site Strathgyle Wood. Durris Quality Class III (80')
Area .425 acre. Thinning Grade B.

RECORD OF PERIODICAL MEASUREMENTS PER ACRE.

Year of measurement	Age of crop (yrs.)	Main Crop: Number of trees per ac. after thinning	Height: Average of largest trees (ft.)	Height: Average of crop (ft.)	Form Factor	Girth at 4' 3" (ins.)	Basal area per ac. after thinning (sq. ft.)	Vol. per ac. (under bark) (cu. ft.)	Bark (%)	Intermediate Yield: Number of trees	Average Height (ft)	Girth at 4' 3" (ins.)	Basal area per acre (sq. ft.)	Vol. per ac. (under bark) (cu. ft.)	Total Crop: Basal area (sq. ft.)	Vol. (under bark) (cu. ft.)	Increment Periodic: Basal area (sq. ft.)	Vol. (cu. ft.)	Periodic Mean Annual: Basal area (sq. ft.)
9/1920	38	430	–	62	·394	32½	249	6085	8	55	46	21	14	270	263	6353	-	-	-
9/1923	–	–	–	–	–	–	–	–	–	5	–	–	3.5	95	–	–	-	-	-
9/1925	43	372	74½	69½	·420	35½	256	7468	8	49	58½	26½	19.0	470	275	7943	29.5	1955	5.9
5/1920	47	372	82½	78	·400	37	280·1	8867	8	-	-	-	-	-	280·1	8867	24.9	1399	6.2

Plot No. 14. Species Sitka Spruce. This plot was blown in 1927. COPY
Site Strathgyle Wood. Durris Quality Class III (80')
Area .336 acre. Thinning Grade B. Grade.

RECORD OF PERIODICAL MEASUREMENTS PER ACRE.

Year of measurement	Age of crop (yrs.)	Main Crop: Number of trees per ac. after thinning	Height: Average of largest trees (ft.)	Height: Average of crop (ft.)	Form Factor	Girth at 4' 3" (ins.)	Basal area per ac. after thinning (sq. ft.)	Vol. per ac. (under bark) (cu. ft.)	Bark (%)	Intermediate Yield: Number of trees	Average Height (ft)	Girth at 4' 3" (ins.)	Basal area per acre (sq. ft.)	Vol. per ac. (under bark) (cu. ft.)	Total Crop: Basal area (sq. ft.)	Vol. (under bark) (cu. ft.)	Increment Periodic: Basal area (sq. ft.)	Vol. (cu. ft.)	Periodic Mean Annual: Basal area (sq. ft.)	V (cu)
8/1920	38	570	–	64½	·419	28	249	6420	8	90	50	17	14	365	263	7025	-	-	-	-
9/1923	41	510	70	68	·425	30	253	7311	10	65	63½	21½	16	415	269	7726	20	1006	6.7	3
9/1926	44	503	73	71	·414	30½	261·9	7696	10	6	65	25	2·1	54	264	7750	11	439	3.7	1

FIGURE 10.11 *(left)*. Detailed 1920s plantation notes for Plots 13 and 14 on the former Durris Estate, Kincardineshire.

REPRODUCED COURTESY OF MORAY AND ABERDEENSHIRE FOREST DISTRICT, FOREST ENTERPRISE SCOTLAND.

FIGURE 10.12 *(right)*. 1920s data on the growth of Sitka spruce in Plots 13 and 14 planted between 1873 and 1882 in Strathgyle Wood, former Durris Estate, Kincardineshire.

REPRODUCED COURTESY OF MORAY AND ABERDEENSHIRE FOREST DISTRICT, FOREST ENTERPRISE SCOTLAND.

at 10,000 per ha (4000 pre acre). With experiments on different sites, climate, altitude and types of ground in Scotland, a picture of the growth of conifers there began to form in landowners' minds. They discovered that, without the tender loving care given in an arboretum, Sitka spruce in young plantations would 'check' or grow very slowly to start with. However, after a few years in check, the little trees grew rapidly even on peaty land in high rainfall areas. When they were planted in thick heather, however, growth remained poor, a problem not resolved until the later twentieth century.

Drumlanrig Estate had produced Sitka spruce 27 m (88.5 ft) tall by 1946, and at Wooplaw they were 33–37 m (110–120 ft) by 1956. Of course the girth and trunk shape are also important when assessing trees' usefulness: Sitka spruce turned up trumps in these features too. Not all landowners could wait half a century to assess their plantations, but they soon realised that Sitka spruce surpassed their other trial species in plantations on a variety of sites.

Irish doctor Augustine Henry led two expeditions to China in the late nineteenth century, concentrating on plant collecting, not people's health. Emotionally affected by severe deforestation in China, he decided to study for a French forestry degree and then took up academic life at Cambridge. By the early twentieth century, he was back in Ireland, as professor of forestry in Dublin, where he worked to evaluate the growth of several North American conifers. His research indicated Sitka spruce's potentially important role in Irish afforestation.

Welsh estates grew many conifers and broadleaved trees during the nineteenth century. To Welsh landowners, Sitka spruce was not initially the most obvious candidate for afforestation, because they were interested in many species. From earlier planting as ornamentals, Douglas fir, Lawson Cypress (*Chamaecyparis lawsoniana*), Noble fir (*Abies procera*), Sitka spruce and Japanese larch (*Larix kaempferi*) were found to be satisfactory.

Penrhyn Estate had maintained traditional deciduous and Scots pine woodlands since at least the sixteenth century: the large variety of trees there were tall and good quality. During the nineteenth century, these species were needed for farm tools, boat-building, turnery, wood-turning, charcoal, tanning leather and fish weirs. In 1880, the estate hired a new forester, Angus Duncan Webster, a Scotsman. He looked after the deciduous woodlands and new conifer plantations of Scots pine and larch, bequeathed from earlier in the century. Fortunately for us, he also wrote numerous forestry treatises, as well as publications about his estate work at Penrhyn. Other big Welsh estates, such as Stackpole Court, Gwydyr and Margam also combined traditional woodsmanship and experimental plantation forestry.

FIGURE 10.13. Augustine Henry, Irish forester and polymath who evaluated tree species in Ireland and recommended Sitka spruce for its superior growth.

PHOTO: COURTESY OF THE NATIONAL BOTANIC GARDENS, GLASNEVIN, DUBLIN.

Long-term plantation experiments at Corrour

In 1891, aristocrat and politician Sir John Stirling-Maxwell bought Corrour Estate, Inverness-shire, which boasted the highest altitude loch and inhabited house in Britain at 396 m (1300 ft) above sea level. You might not think this would be the best place to experiment on tree species' responses to plantation life on poor soils in the Scottish Highlands, nor to assess their silvicultural properties and potential commercial usefulness.

That is just what he did for the next 40 years.

On inhospitable, infertile, peaty land between 381 m (1250 ft) and 477 m (1565 ft) and in rainfall of 1778 mm (70 in) annually around the north-east of Loch Ossian, he gradually planted 243 ha (600 acres), starting in 1892. He tried over 50 native and imported trees, of which 24 were conifers. His Sitka spruce seed came from Alaska and Haida Gwaii.

FIGURE 10.14. John Stirling-Maxwell in 1895; he experimented with trees on his estate at Corrour, Inverness-shire.

PHOTO RETRIEVED FROM: HTTP://COMMONS.WIKIMEDIA.ORG/WIKI/FILE-JOHN_STIRLING_MAXWELL2.JPG.

Sir John tried new methods: the land was enclosed and drained; the soil and turf cut from drains was thrown down to make lines of – drier – mounds on which to plant the trees; two handfuls of basic slag and sand per tree were added as fertiliser. These methods meant he could afforest what was very soggy land. He compared the growth of *seedlings* with *transplants*, and varied planting densities from 3750–5000 trees per ha (1500–2000 per acre).

He found that Sitka spruce was wind-fast and snow-tolerant, grew quicker than Norway spruce and most other trial species and produced good timber. He published his detailed experimental results for all the tree species so a wide audience could benefit.

Timber imports to the United Kingdom

It was not only seeds and plant specimens that were exported from North America to Europe.

Scandinavia and the Baltic region regularly exported timber to the UK, but by the later nineteenth century it could come from Canada too. Governments welcomed low-priced timber from vast natural forests in the 'colonies' for its Industrial Revolution: for coal mines and smelting iron, for bridges, for house and factory construction, paper and fuel.

But warships were now constructed from iron and steel not big, home-grown oak trees.

Some broadleaved woodlands were grubbed-up, left derelict or, like Bedford Purlieus, inter-planted with larch. Some estates such as Tintern and Gwydyr still found markets for smaller oak trees, to build merchant ships, for charcoal and tan bark.

Welsh and Scottish foresters (in particular) were becoming experts in growing a variety of exotic trees, in afforestation of bare uplands and in plantation forestry. But there was no large-scale, home timber-processing industry to utilise and profit from their new skills. Governments' forestry policies pursued cheap imports not a thriving home-grown timber industry.

Timber import duties, which had been raised during the eighteenth century to increase government revenues, were increased again during the early nineteenth century, except for imports from colonies which were protected to encourage timber trade between North America and the UK. After protests by Baltic countries, the differential was reduced until, in 1866, timber tariffs were removed altogether. This meant that imports of conifer timber – called 'softwood' – from Canada increased throughout the century. Sitka spruce was one of them.

There was negligible government interest in a home-grown timber industry or in financial support for forestry and its new plantations, during the nineteenth century. So by 1900, the United Kingdom imported 90% of its timber and forest products.

The free market for timber is still with us today and timber is the UK's second largest import after oil.

This has been a very big mistake in the long run…

However, passionate tree planters throughout the UK carried on with their long-term experiments with many tree new species, and with moral support from like-minded people. Naturalists, intellectual societies, plant nurseries and explorers continued with their expeditions, education and encouragement of others. Experimenters continued to plant Sitka spruce, knowing their findings would one day be needed…

Sitka spruce in nineteenth-century North America

Many residents of North America expanded the horizons of natural sciences there. For example, in the mid-eighteenth century, John Bartram visited other states to collect plants, grew them on his Pennsylvanian farm, sent specimens to Linnaeus, kept journals and was instrumental in new species reaching the UK as seeds. He is known as the 'Father of American Botany'. Doctor and botanist John Torrey studied the plants of north-eastern USA. He published his *Flora of the Northern and Middle States*, a *Flora of New York State* in 1836, and he worked on a *Flora of North America* published 1838–1843.

Hardly had wonderful new trees been discovered and named than they were regarded as available for the taking by Canada and the USA – and vicariously by the UK. Forest soils, trees' dependent organisms and connections between aquatic life and forests were little understood except by First Peoples who were not, at that time, consulted.

From the mid-nineteenth century, felling ('falling' or 'logging') whole trees and watersheds opened up the forests of western North America, starting in the accessible hills, estuaries and coasts of Washington and Oregon.

Clear-fell programmes were carried on with the expectation that they could continue in perpetuity. Behemoth Douglas fir, Sitka spruce, Hemlocks and Western red cedar disappeared.

Rainforest conifers were felled for housing, furniture, railways, bridges and paper pulp, although the less abundant Sitka spruce had few markets and was considered a nuisance. However, in 1903, the frame of the 'Wright Flyer', the first plane to fly successfully, was built of ash and several hundred pounds of Sitka spruce, now known for its strength per unit weight. By 1911, 'Curtiss Pusher' biplanes were built in quantity from Sitka spruce and flown from the polo field of Vancouver Barracks, USA.

Keeping forestry going in the UK

Passionate tree-planting landowners in the UK continued to enrich their estates despite lack of tariffs on imports and a small home timber-industry.

The Wood of Cawdor, Nairn had been planted with 285,656 trees by 1818, with 17 species including European silver fir, Norway spruce, Weymouth pine and seven broadleaved species; Scots pine made up half the woodland. Tree planting is recorded on the estate until 1854 using trees from the estate nursery. Records of fencing, beating up and early thinning show the woodlands were well looked after.

Extant archives record that about 5 million trees had been planted on 486 ha (1200 acres) of Thomas Johnes' Hafod Uchtryd estate by 1816. His lands were stony, steep and at high altitude, received high rainfall and had derelict mines. His tree-planting was a magnificent effort and the beautiful wooded landscape can still be enjoyed.

Famous author and polymath Sir Walter Scott loved trees and helped his workers plant them on his new Abbotsford Estate, Roxburghshire. In the 1820s 162 ha (400 acres) of trees were planted. He wrote down his methods, judged their effects on wider land use and bequeathed posterity a newly-wooded landscape along the River Tweed. And lots of famous novels.

The Atholl Estate, Perthshire, still employed 120 men in its forestry department at the century-end, despite a slow-down of tree-planting.

In lowland England, with better soils and climate, landowners preferred to plant beech, larches, Corsican and Scots pine, and Douglas fir. Small numbers of rarer species, such as Monterey pine (*Pinus radiata*) grew well on warm coasts.

Ireland experienced a big population increase between the 1680s and 1840s. This was followed by the infamous Potato Famine during the mid-1840s, when about two million people died. Farms were neglected or left derelict, so trees regenerated on abandoned land. Ash and sycamore trees sprung up in hedgerows; many oak trees in the modern landscape date from that period.

This was not a time when Sitka spruce and exotic conifers took precedence in Ireland.

During the nineteenth century life as a farm tenant in Ireland was difficult. Poverty and famine there forced the UK government to make changes to landlord–tenant legislation (Ireland was now part of the UK). Between 1870 and 1909, the Ireland *Land Acts* gave tenants better conditions and the possibility of buying their holdings with financial assistance: numerous small farms and crofts were formed from some big estates.

With the Ireland *Land Acts* in force, landlords sold farmland to their tenants, but to gain additional finance, they sold the farms' woodlands first. And the new owners (ex-tenants) felled what trees remained on their land because this type of 'improvement' no longer incurred increased rent – and the cash was handy. It was not until after a government forestry department was set up in 1903 that the tide of woodland loss in Ireland started to reverse.

Rural England experienced large-scale land enclosure accompanying the agricultural revolution. The New Forest was still crown land and in 1851, the *New Forest Deer Removal Act* gave greater powers for silvicultural enclosure but surrendered the crown's right to keep deer. Commoners' land was segregated from Crown land, with a statutory area of about 18,211 ha (45,000 acres) of unenclosed land where they could exercise their rights.

In the following 20 years, 4856 ha (12,000 acres) of Crown enclosures of the New Forest were planted with at six conifer species, which altered the contemporary oak-beech-holly landscape. Sitka spruce was one of the chosen species – but it was not well suited to the dry heathland soils. The new conifers diversified the overgrazed, deciduous woodlands, but deer broke the new law by resisting removal.

The learning curve

With meagre government interest in forestry, some landowners sought additional sources of income, which included rearing and shooting game birds, deer-stalking on treeless hillsides, and extensive sheep-farming.

Old ways continued too, especially in the deciduous woodlands of lowland Britain. Foresters and woodsmen on British and Irish estates were trained by their seniors for several years of apprenticeship. In 1907, fifteen-year-old Jim Shaw met the factor of Killearn Estate near Loch Lomond. The factor stated:

> I expect you know what it would mean James? You would work for half a crown a week for three years. You would be shown everything and taught everything that might fit you to become a woodman, a forester like your father, in due time. Whenever there might be courses that would benefit you you would be sent on them. You would be expected to read and educate yourself in forestry and sylviculture. At such times as the Home Farm needed help at hay or harvest you would be available to help, and you would, on those occasions, get another shilling and sixpence a week. Do you think this is what you want?
>
> Niall 1972

Jim learned weeding in the nursery, clearing between rows of trees, pruning, thinning, hedge-trimming, tree-planting, fencing, helping with woodland deer drives, how to recognise insect pests and of course much country lore. It stood him in good stead because, in 1924, he took charge of Gwydyr Forest.

Landowners decided to provide their own support: the Scottish Arboricultural Society was founded in 1854, the Irish Forestry Society in 1901, the English Arboricultural Society in 1881.

They participated in species trials, gave silvicultural advice to foresters, held meetings and published journals to promote knowledge of forestry. Journal articles were directed towards people working at the grass roots. Sitka spruce was a frequent subject of articles.

Meetings and published articles stimulated professionalism: good seed sources were sought; tree species were chosen according to site; land was prepared for tree-planting by draining, cultivating, controlling weeds or fertilising; tree growth was stimulated by thinning out young trees in plantations; more treatises on forestry were published.

Starting in 1899, governments financed forestry lecturers at Edinburgh, Aberdeen and Bangor, which was first to offer a degree in forestry and to produce a woman graduate. Silvicultural methods based on forestry in India and Germany were inculcated into students' heads.

Landowning aristocrats lobbied on behalf of forestry in the House of Lords, London. With cheap and plentiful timber imports from North America and the Baltic their pleas fell on stony ground:

> If adequate measures were taken to try and grow the eighteen million pounds' worth of pine and fir we now import, and which imports may become greatly increased in value within a comparatively short period, a vast economic field, now left neglected and uncultivated, could easily be made to yield a golden harvest … the apathy shown towards forestry in this country is one of the things it is impossible to understand…
>
> Heroic measures to replace the woodlands destroyed can only be undertaken on a sufficiently large scale by receiving considerable encouragement and assistance from the State, whose attitude has hitherto been extremely unsympathetic in this respect.
>
> Nisbet 1900

FIGURE 10.15. Hafod Fawr, Merionethshire, first state forest 1899–1900, showing its mountainous site and difficult ground.

PHOTO: © IAN MEDCALF, 2014, CREATIVE COMMONS LICENCE.

John Nisbet, ex-Indian Forest Service, favoured a national approach to afforestation; he also pointed out to an Irish forestry committee the potential of Sitka spruce over the better-known Norway spruce.

Changes

At the end of the nineteenth century the UK's three Commissioners of Woods, Forests and Land Revenues (who managed crown forests) charged their officials, the Office of Woods, with acquiring 554 ha (1369 acres) of hill land near Ffestiniog, Merionethshire. The nine hill-sheep farms, including Hafod-fawr isaf and Hafod-fawr uchaf, were situated on a steep slope above the Afon (River) Cynfal, a slate-mining area. The intention was to remove all sheep, enclose the land and plant conifers for timber.

A change of heart?

It was poor land: north-facing, exposed to strong sea winds, with an average rainfall of 2032 mm (80 in) a year, heavy winter snow, at altitude from 213 m to 556 m (700–1823 ft), with acid clay and peat soils. Almost the worst land one could find.

Small-scale afforestation started in the 1899–1900 planting season. Norway spruce, Sitka spruce, Scots pine, European and Japanese larch, Common alder and some Corsican pine, European silver and Douglas fir were chosen by Commissioner E. Stafford Howard. A manager, a woodsman and a boy carried out the hard graft, coping with awful weather, long hours, farmers who objected to eviction and the change from farming to forestry. Luckily there was also a wife who cooked them a hot meal every mid-day.

Hafod Fawr Forest turned out to be another useful experiment in the growth of tree species on upland sites and demonstrated to officials the daily problems of abysmal land and working conditions, the agility of Welsh mountain sheep in jumping over walls and the costs of labour and trees.

The difficult decades experienced by landowners with forests were summed up half a century later:

> It is understandable that during the nineteenth century those anxious to encourage forestry operations could find few sympathisers in high places. Imports were adequate to meet all demands. And there was no major war in which Britain was engaged to cut off her ships from the world routes or otherwise exercise an unexpected drain upon her timber resources. A few warning voices were heard; but the effect of these eloquent pleas was negligible. Some brave, unsung British silviculturalists experimented with the introduction of new species and the so-called 'exotic' conifers such as Douglas fir, Sitka spruce and Japanese larch were tried out and when acclimatised, were found to be eminently suitable to the soil and climatic conditions of this country. In later years the experience of these unknown pioneers was of immense value in establishing forests which not only suit these islands but are of rapid growth and produce timber of excellent quality and supplement the indigenous Scots pine and the better known foreign stocks such as Norway spruce and European larch. Against the work of those who advanced silvicultural knowledge in this bleak period must be set the almost complete failure of the State to grapple with a problem which became daily more pressing, but it was not until the 1914 War that the real seriousness was fully realised – and by then it was too late.
>
> Jefferies 1945

FIGURE 10.16 *(opposite)*. One of the largest-girthed Sitka spruce in the UK, at Drumtochty, Kincardineshire, planted in the 1850s.

PHOTO: RICHARD MARRIOTT, 2015.

The real nadir?

Afforestation was small-scale and left to some passionate landowners during the nineteenth century. Annual taxes on woodlands possibly reduced enthusiasm.

Nationally, woodlands and trees were in a parlous state but society did not realise it. The landscape was very far from the 60–90% native tree cover of Mesolithic times.

The total area of woodland continued to fall, up to and beyond the end of the nineteenth century, except in Wales where it increased slightly from 3.02% to 4.22%. Table 5 shows the treeless-ness of the United Kingdom at the start of the twentieth century. The total area of tree cover was about 1,618,743 ha (three million acres).

More would yet be lost.

The role of Sitka spruce in making good some of the loss will be described in Chapter Eleven.

England	5%	Linnard (2000); Smith (2001)
Ireland	1.2–1.3%	Boylan (2010); Joyce and OCarroll (2002)
Scotland	3%	Smout (2003)
Wales	4.2%	Linnard (2000)

TABLE 5. Percentage area of woodland in Britain and Ireland in 1900.

CHAPTER ELEVEN

From Rare Ornamental to Upland Carpet

> That legacy of introducing new trees and developing new silvicultural systems was a vital heritage for the nation's foresters when the challenge of establishing a national timber estate was thrown down in the aftermath of the First World War.
>
> Syd House and Christopher Dingwall 2003

> The Sitka spruce is planted mostly on cold wet grassland and peat moors; many of these sites have been impoverished by prolonged overgrazing or burning, and often altered by mining, erosion, peat cutting, or lack of drainage. Few other trees would tolerate such places... In western Britain it will survive anywhere, from the highest tree-line down to sea level, where its tolerance of high soil-water sodium levels is outstanding.
>
> John White 1995

Waiting

At the start of the twentieth century, conifers from western North America had contributed to the UK landscape for about 70 years. Many grew satisfactorily in arboreta and gardens, but were not chosen for commercial plantation forestry. By the end of the century, there were *hundreds of millions of Sitka spruce trees* growing in the north and west of Britain and in Ireland. These millions had been planted on land thought of as *marginal* for farming, as *unproductive* or as *wasteland*.

Why and how did this happen to Sitka spruce?

What about Lodgepole pine, Western red cedar, Western hemlock or Douglas fir? Or Norway spruce, Japanese larch or the beloved native Scots pine? These were planted in a variety of places but in fewer millions.

Political and economic considerations, the wishes of those landowners with forestry interests, land availability, tree species' ecology, the farming industry and history all affected the choice of tree species, while the First World War determined the speed of change.

Estates in difficulties

The repeal of tariffs on imported grain (the *Corn Laws*) in 1849 had a negative impact on domestic agriculture. Grain imports were now cheap and prices for home-grown cereals dropped alarmingly. Stock farming increased at the expense of arable, many farmers and their labourers moved to towns or emigrated. Farm tenants could manage only low rents so landowners' income reduced.

The repeal of tariffs on timber imports in 1866 had a similar impact on tree planting by landowners. There was no financial assistance from governments towards tree planting or management; landowners still had to pay an annual land tax; heavy death taxes meant that estate land might be sold.

Atholl Estate, Perthshire had planted significant areas with Larch to supply timber to the navy. Imports of cheap timber from Scandinavia, North America, Russia and the tropics produced a glut in good-quality timber. There was little incentive for Atholl and other estates to continue planting trees as a crop.

The rate of afforestation declined. Woodlands were not worth the cost of management or harvesting; the home-grown timber industry struggled against cheap imports. Native woodlands were often neglected because their traditional products were no longer needed. Some estates converted their coppice-and-standards to high forest (tall trees) by the end of the century.

Tintern Estate, Monmouthshire was an exception to the trend as some of its coppices survived until after the Second World War. Gwydyr Estate continued to manage native woodlands for local markets and started new plantations. It planted blocks of Norway and Sitka spruce, European larch and Scots pine, as well as mixed plantations of conifers with beech, oak, sycamore and ash: they were still standing in 1920.

A bad gale in October 1881 blew over many of the Penrhyn Estate's mature trees. In the following twenty years forester Angus Duncan Webster planted conifers along the Penrhyn Quarry Railway and at altitudes of 305–366 m (1000–1200 ft) despite buffeting by Irish Sea winds. He also had growing plantations to look after, which in 1890 were Scots pine and nearly-100-year-old larch.

Arboriculturalist Malcolm Dunn persuaded colleagues throughout the country to measure the height of their specimen conifers. Sitka spruces, very tall compared with other species, suggested its potential for forestry:

> *Abies Menziesii* (Menzies' Spruce) – or, as it is now called by botanists, *Picea sitchensis* – is still another of the giants of the forests of North-west America, which in suitable soils of a moist, cool nature has made remarkable progress in Britain, particularly in Scotland, and in some parts of Ireland, for which it seems specially well adapted...
>
> All these dimensions show a great production of timber in the period, and the excellent quality of the wood places this tree at the head of the Spruces, and the most valuable of that tribe as a forest tree in soils and situations where it thrives.
>
> Dunn 1891

This echoed David Douglas's field observations in North America in 1825.

A tree at Dropmore, planted in 1841, was 22.2 m (73 ft) tall in 1891; a Penrhyn Estate tree had grown to 3.15 m (10 ft 4 in) girth by 1904; the Curraghmore Sitka spruce had reached 32.3 m (106 ft) high by 1905, an average growth rate of 1.24 m (4.08 ft) per year; an Abbeyleix Sitka spruce planted in 1834 or 1835 was 27.4 m (90 ft) tall in 1910.

When Henry Baird's 1870s conifer plantations at Durris were measured in 1912 the Sitka spruces, had grown to 15 m (50 ft), taller than all other species. By 1922 they were 22 m (72 ft) tall, but unfortunately blew down in 1927.

Practical as well as financial problems beset landowners wanting to grow woodland. The decline of medieval rabbit warrens and a long-term depression in farming paved the way for colossal populations of feral rabbits which nibbled small trees – and almost all other vegetation. Sheep, cattle and hares did the same: walls, fencing or netting to exclude animals were costly to build and difficult to maintain.

In Scotland, some landowners had converted their high, treeless country into 'Deer Forest' for stalking red deer. Remnant native woodlands suffered browsing and in the long-term they were unintentionally eliminated.

The view from the factory floor

John McEwan and Jim Shaw had trained as woodsmen at the turn of the century and both became professional foresters in due course. Their descriptions of countryside in the early twentieth century illustrate both busy and declining estates.

John McEwan's work linked the two centuries in his roles as estate worker, contractor, wartime woodsman and senior forester. Jim Shaw, a bit younger, had a similarly varied career. Both men had hard young lives as woodsmen. They experienced long hours, tough work in all weathers, poor housing, accidents, illness and the harsh attitudes of estate factors. It was the hard graft of men, lads, horses and sometimes oxen which managed the plantations and carried out the landowners' experiments with new species.

John McEwan's first job in forestry started in January 1905 on the Seafield estate, at Cullen, Morayshire: he learned hedging, fencing, clearing, and 'snedding' (cutting off side branches) blown Beech trees for the estate sawmill. Tree-planting did not feature in his work there. But at the nearby Altyre Estate, he was delighted to plant anew: each worker put 100 plants a day into a slit in the turf, without any ground preparation. His all-round experience there included measuring and marking trees, felling trees, cross-cutting after felling and general estate work. On the Duke of Montrose's Buchanan estate at Drymen, by Loch Lomond, he carried out similar tasks but also had the unenviable task of being the *underneath* man in a typical, sunken, sawpit.

After a spell in the timber-processing business, he worked in Glasgow parks and then Edinburgh Botanic Garden. He attended evening courses and obtained qualifications the hard way, vital to men who wanted to work their way up the career ladder. Jim Shaw had been an apprentice woodsman on Killearn Estate, Stirlingshire in the early years of the twentieth century. He also worked in the Edinburgh Botanic Garden and attended evening classes to improve his prospects.

Despite their wide experience of estates, neither man saw much *afforestation* in the early twentieth century. Their work was mainly with existing woodlands, not initiating new ones. Their experience of working as woodsmen on feudal estates and of lack of progress in expanding tree cover fostered both men's interest in a potential state forestry agency.

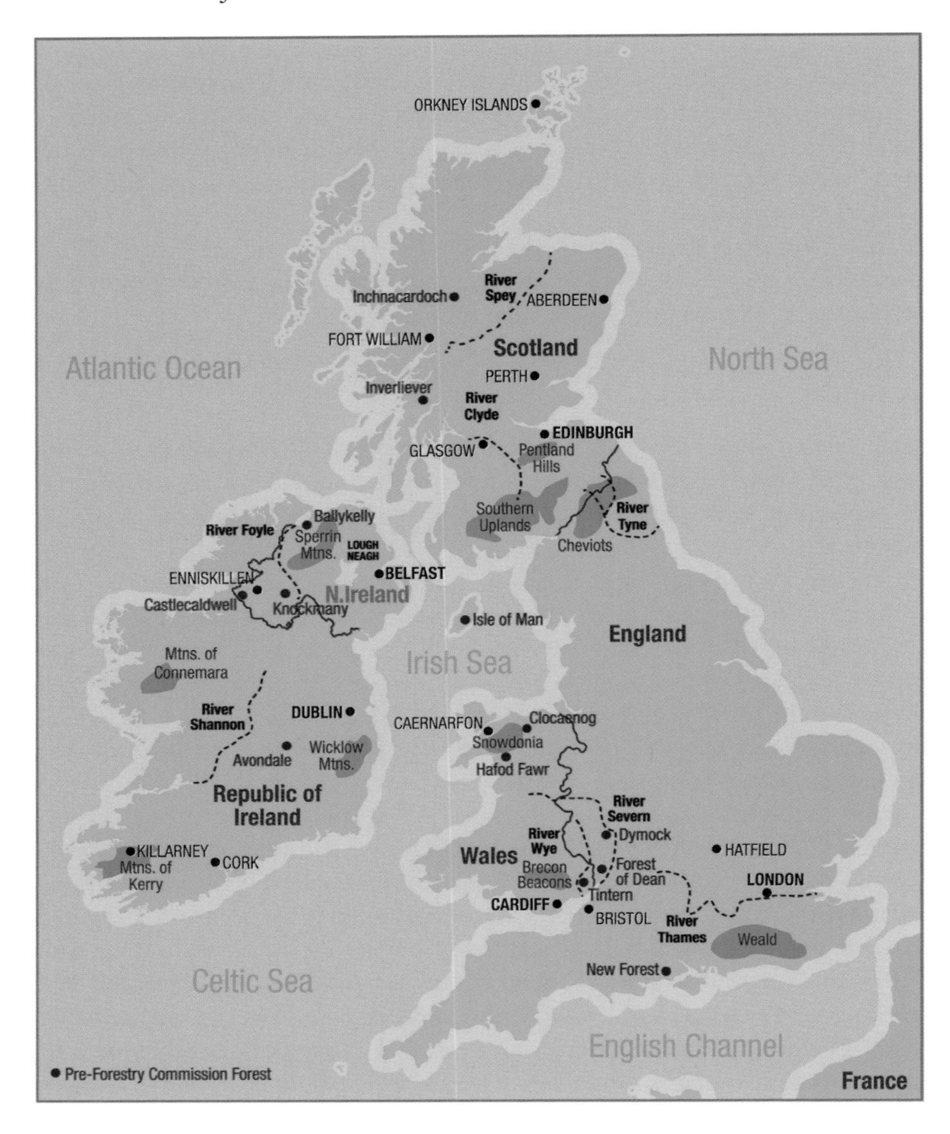

MAP 13. Pre-1919 state forests, prepared by Susan Anderson.

DATA: SEE BIBLIOGRAPHY FOR CHAPTER ELEVEN.

Education and experiment

In 1883 the first significant afforestation on crown land was carried out on the Isle of Man; then Hafod Fawr was bought by the Office of Woods in 1899 (see Chapter Ten). In 1901 it bought Tintern Estate, Monmouthshire, with its 1619 ha (4000 acres) of woodland, and the small oak forest of Dymock in Gloucestershire which did not participate in the Sitka spruce story. High moorland on Mynydd Hiraethog, Denbighshire was bought in 1905 for Clocaenog Forest (Map 13).

Landowners and foresters lobbied governments for a state forestry agency. Angus Duncan Webster wrote a strong case for *national* afforestation to the

FIGURE 11.1. Sitka spruce which were planted in 1916 in Inverliever Forest when it was a Crown Estate property.

PHOTO: RUTH TITTENSOR, 2013.

Board of Agriculture in 1905 and pointed out that Britain consumed the largest amount of timber in the world. He noted that some cities had already afforested their water-catchments and reiterated that landowners could no longer sink cash into a crop which would not be harvested for at least 25 years.

The 4856 ha (12,000 acre) sporting Inverliever Estate was bought by the Office of Woods for £25,000 (equivalent to 48,300 CAD, $38,000) from the Poltalloch family in 1907 which had experienced difficult times. Consisting of seven miles of steep slopes along the west side of Loch Awe, Argyll, its maritime climate receives 2032–2159 mm (80–85 in) of annual rainfall.

Inverliever's tenanted hill-sheep farms consisted of extensive grazing land and arable 'inbye' to grow hay and grains. Together, the farms supported only 6650 sheep and 30 cattle (less than two stock animals per acre) and were considered 'marginal'. But Red and Roe deer, Red and Black (*Tetrao tetrix*) grouse and a few Pheasants gave it sporting value. The poorest land was covered in shallow peat but some areas had better, mineral soil. When the estate was sold, the farm tenants presumably had their tenancies concluded.

Why would the Office of Woods buy poor land? Why did the government's first attempt at afforestation use marginal land? Why was fertile land in lowland Britain not bought and afforested? Why was this new industry initiated in the worst possible conditions, with little infrastructure, a mainly untrained workforce, housing in dormitory bothies and at long distances from potential markets?

One reason was large-scale *experiment*: Inverliever was considered typical of large areas of 'waste' land in western Scotland which awaited what was called 'rational use' – something more economic than hill-sheep. The priority was to investigate the growth of Norway and Sitka spruce and other species on the almost treeless hillsides up to 348 m (1142 ft).

Moorland

'Moorland' is used in Britain and Ireland to describe a particular upland landscape, such as the high land of Inverliever when it was surveyed in 1923.

In this book, I use the terms 'moor' and 'moorland' ('muir' in Scots, 'rhostir' in Welsh, 'mónteach' in Irish) to describe upland landscapes which are unenclosed, uncultivated but grazed by domestic stock and deer. Moorlands develop recognisable vegetation on hillsides or plateaux. They formed in the past from forest or forest edge when climate became wetter, humans cut trees and burned vegetation, and peat formed; they are maintained by human activities such as burning and stock grazing. Moorlands usually develop above 302 m (990 ft) altitude, are treeless, bear soils of acid peat, gley or peaty gley and support heathers, grasses, sedges, rushes and patches of bog moss (*Sphagnum* species).

Simmons (2000) gives an interesting historical account of English moorlands.

Experimenters carry on

At Corrour, Sir John Stirling-Maxwell carried on with his experimental plantations on the mountain sides of Inverness-shire. After two decades, he knew that Sitka spruce could grow at least one foot a year. Cold spring weather reduced its growth; trees might 'check' for four to fifteen years, but even so, top-dressing with phosphate fertiliser could make a difference of 8 m (25 ft) to their height.

Information in forestry journals encouraged other landowners to plant Sitka spruce, increasing its contribution to new forests until the First World War.

Afforestation slowed on the Atholl Estate but existing plantations contributed to the local economy. At the end of the century larches were still the most important species. On other Scottish estates, Norway spruce, Scots pine and larches continued as staples of conifer plantations.

The Earl of Plymouth was an outstanding experimenter at St Fagan's, Glamorganshire, at the turn of the centuries. But Sitka spruce was not one of the many tree species he tested in his plantations. Forests on the Penrhyn and Chirk estates were offered to Bangor college for research and to educate foresters.

The Manx Arboricultural Society planted trees on the Isle of Man in 1906 but it is uncertain whether Sitka spruce was included.

Sitka spruce in Hafod Fawr and Inverliever forests

A forward-looking Scottish forester, W. D. Crozier, who had worked in Ireland too, collated and published data on Sitka spruce in 1910, yes, *1910*! He noted its North American ecology and pointed out that introduced Sitka spruce was already larger and more vigorous than all the other Pacific Coast conifers, producing timber at high elevations and in wet soils. It exceeded Norway spruce, Scots pine and larches too. He had seen that Sitka spruce grew best in the Highlands of Scotland, in Wales and Ireland; he observed that it needed moisture-full soil and atmospheric moisture, preferably 2032–2540 mm (80–100 in) of rainfall annually, though 762–1347 mm (30–53 in) would do.

Crozier proposed planting densities, thinning practice, volume production, and gave information on coning and seed ripening, good nursery practice and suggested that the planting altitude could be raised to 305 m (1000 ft) in places. He noted that markets for Sitka spruce were opening up in Scotland, suggested a rotation of 100–120 years for structural timber, and a rotation of fourteen years for pulp, which would be in demand in due course. For *1910*, he was far ahead in his data and predictions for Sitka spruce. He deprecated the general lack of interest in the species:

> It is a matter of regret that, while Sitka spruce has, for a period of nearly eighty years, been known and appreciated in pineta and pleasure grounds, on account of its ornamental value, so little has been done in a practical way, to ascertain its commercial value; but probably, as in the case of the Douglas fir, a prejudice, based on insufficient knowledge of the real value of its timber, may have accounted for the lack of interest shown in the tree.
>
> Crozier 1910

At Hafod Fawr progress was slow and careful to make sure tree-planting was successful. Grown in its nursery, 8750 treelets per ha (3500 per acre) were squeezed into the Forest. Sitka and Norway spruce, and a few other conifers managed to grow in its difficult conditions. Despite new fencing and existing walls, feral rabbits and domestic sheep pushed through and nibbled even prickly Sitka spruce. So replacement or 'beating up' was constantly needed.

Each of the 735,000 little trees had the benefit of an individually-dug hole to give its roots a good start. This was *very* labour-intensive and time-consuming when a mere slit in the ground was usual. By 1915, several English and Welsh managers had worked at afforestation and community liaison at Hafod Fawr and they had overseen the planting of 85 ha (210 acres).

At Inverliever, a nursery was laid out at Ford, bothies were built for the workers and new planting areas were fenced to keep out sheep. In 1909–1910, the first trees were planted on the steep sides of Loch Awe: European larch, Douglas fir, Norway spruce, Western red cedar and a few Sitka spruce on the lower slopes.

William Mitchell, grandfather of current woodsman Charlie Mitchell, was a member of the workforce of over forty which had planted two-and-a-half million trees by 1914. Mackenzie's Grove, a short distance from the Cruachan car park contains Sitka spruces planted in 1915 to 1916 in a ravine; nowadays they impress people with their size and beauty. Mairi Stewart's booklet *Smell of the Rosin – Noise of the Saw* includes memories of people connected with Inverliever Forest.

Criticism

In 1912, Roy Robinson, a senior civil servant trained in forestry, visited both Hafod Fawr and Inverliever Forests. He was disappointed with the patchiness and growth of their trees. What did he expect? The Office of Woods had bought

FIGURE 11.2. Avondale House, Co. Wicklow, Irish forestry school and research site after 1906.

PHOTO: COURTESY COILLTE, CO. WICKLOW.

marginal land on which to grow trees. Many species had been planted to measure their success in the difficult conditions. It might take a decade or half a century to discover which survived, grew *and* produced marketable timber.

However, his disappointment pushed him, by and by, to set up research groups to study tree establishment.

Other forests

The woodlands of Tintern Forest were famed for their position along the gorge of the River Wye, Monmouthshire. Situated on a boundary of 'lowland' and 'upland' Britain, with rainfall of only 813 mm (32 in), Tintern had land suited to the growth of broadleaf trees such as oak, beech and Small-leaved lime. Records show there was regular coppicing from the thirteenth to the twentieth century. European larch, oak and beech planted in the mid-nineteenth century were well grown by 1901. European and Japanese larch, and Douglas fir were planted anew soon after. In 1913, a compartment of Sitka spruce and larch was planted at Bargain Wood on the slopes above the river, but the Wye valley is not really Sitka spruce land.

Ireland and the Avondale experiments

Despite the government's attempts to solve the 'Irish Land Question', assist big landowners with famine-caused debts and encourage more crofting tenants to

FIGURE 11.3. The Great Ride at Avondale, Co. Wicklow showing a towering Sitka spruce planted in 1920 at the far end, left; the other trees are the experimental plots.

PHOTO: MICHAEL CAREY, 2009.

acquire land, there was less change in Ireland than hoped. However, in 1899 a Department of Agriculture and Technical Instruction was set up to take responsibility for forestry. In 1904 it acquired Avondale House and its wooded estate, Co. Wicklow. By 1906 it had opened as the state forestry school and as a research site, with A. C. Forbes as head. Meanwhile the Irish Forestry Society had been formed in 1903.

Arthur Forbes laid out 49 test plots of tree species at Avondale. Sitka spruce was planted in two plots in a 50:50 mixture with Japanese larch. After a decade, the apparently dead Sitka spruces found new life when the larches were thinned. They put on 0.6–0.9 m (2–3 ft) of new growth annually after that!

Observations obviously took time, but by 1923, Arthur Forbes knew of Sitka spruce's power of recovery. In the late 1920s he concluded that Sitka was the fastest-growing spruce he knew, but needed nurse trees in frosty places. By then, it was grown widely on damp soils in Ireland.

A Forestry Agency for Ireland

A programme of state forestry started in Ireland in 1910 to provide a sustainable supply of home-grown timber by increasing its 1% of woodland area. Early acquisitions included Ballykelly (Co. Derry), Knockmany (Co. Tyrone) and Castlecaldwell (Co. Fermanagh) between 1910 and 1913.

To start with, afforestation was allowed on marginal and sub-marginal land only. In due course, conifers from north-west America gave high yields of

FIGURE 11.4. Some of the early Sitka spruce in Knockmany Forest, Co. Tyrone, one of the earliest Irish government acquisitions, about 1910.

PHOTO: ANDY TITTENSOR, 2013.

quality timber: Douglas fir, Lodgepole pine, Sitka spruce, Western hemlock and Western red cedar.

Sitka spruce was planted in mixtures with European larch and Norway spruce in parts of Ireland, but this confused foresters. It was not until they removed all the trees except Sitka spruce that it recovered from check. Giving rock phosphate fertiliser to the dead-looking specimens would usually start them growing.

Ireland recognised that it could have no forestry industry without introduced species, unless trees were grown on good farmland. But this was too valuable for food production to be used for trees. State forests had to be established on sites where conditions were such that its few native species just would not grow.

After 1925, Sitka spruce's potential for the country's forestry was acted on and it was given the chance as a single-species plantation crop. It was not used in places with lower rainfall, freely-draining soils, very acid peat, frost or heathers.

The First World War

During the First World War enemy submarines severely restricted timber imports to the UK. But coal-mines needed pit props, factories and farms needed wood for tools, machinery and fencing; homes, schools and work-places needed furniture; aeroplanes were built of wood. In 1916, the government realised, at last, that its apparently limitless supply of imported timber was coming to an end.

Any woodland was liable to be felled and the timber used in the war effort. There was no time to tidy up and replant. 'Devastation' was the contemporary description. The government sought help with its felling programme because young men were scarce on the home scene. Thousands of militarised lumberjacks were sent by the Canadian government to fell woodlands. They settled into rural camps, felled trees, extracted them to a roadside, cut them into logs, pit props and planks, then took the products in lorries to their destinations.

Royal Forests were not exempt. Heavy felling in the New Forest produced 226,750 tonnes (250,000 tons) of timber for the war effort. Hafod Fawr and Inverliever Forests had barely got going by the First World War. But the 1954 National Forest Park Guide for Snowdonia noted that in Gwydyr Forest:

> Many of the slopes had been wooded in the past, but very heavy fellings took place during the war of 1914 to 1918, and when the Forestry Commission took over the first sections of the area in 1920, these earlier woods had been devastated.
>
> Forestry Commission 1954

FIGURE 11.5. Soldier loggers of the Spruce Production Division at their logging camp, 1918.

PHOTO BY DARIUS KINSEY, COURTESY WHATCOM MUSEUM (1978.84.1751), BELLINGHAM, USA.

FIGURE 11.6. An important First World War training plane: 8340 of AVRO504 planes were built of Sitka spruce felled by American soldier-loggers.

PHOTO COURTESY TANGMERE MILITARY AVIATION MUSEUM, SUSSEX.

By the end of the war, 182,000 ha (450,000 acres) of UK woodlands had been clear-felled, about 14% of the total three million acres (1,214,100 ha) in the country.

North American spruce in the First World War

Sitka spruce played a vital role in North America's efforts during the War. It provided frames and sometimes bodies of military aircraft. The tree's high strength to weight ratio was the important factor. In 1903, the young William Boeing bought forests around Grays Harbor on the Olympic Peninsula and made good profits from trading timber. In 1910 he bought a boatyard on the Duwamish River near Seattle, where he built his first plane and furniture factory, close to the coastal rainforests.

Despite its relative rarity within the rainforests, Boeing used Sitka spruce to construct some of his early planes, such as the Boeing B&W in 1915. Sitka spruce was critical to pre-war and First World War planes:

> By 1911 the Curtiss Pusher biplane, made from spruce, had become the popular choice among pilots ... our Sitka spruce of the Pacific Northwest, which covered the hills around Olympia, Vancouver, and Pearson Air Field became a pivotal player in the Allied Command. The US Army Signal Corps was assigned to build and run what became the world's largest spruce-cutting mill at Pearson. The Army sent its servicemen ... to log the voluminous Sitka spruce from mountains and transport the logs to Pearson via truck and rail. During World War I, Pearson Air Field produced 71 million board feet of aircraft spruce, which was shipped down the Columbia River to airplane factories throughout the states... Few trees are left.
>
> Hansen 2004

Although training-planes such as the Curtiss Jenny were manufactured in North America, much of the lumber was exported to Europe to build thousands of fighters and bombers, including the AVRO 504, the de Havilland DH2 and DH4, the Bristol Fighter F.2B and the Sopwith Camel.

FIGURE 11.7. The Silver Dart biplane, designed by Alexander Graham Bell and colleagues, had a wooden framework of Sitka spruce. Its 1909 maiden flight was off the ice of Bras D'Or Lake, Nova Scotia and was the first powered flight in Canada.

PHOTO OF REPLICA 1959: RCAF PHOTOGRAPHER, COURTESY ATLANTIC CANADA AVIATION MUSEUM, NOVA SCOTIA.

It was also a busy time for logging huge Sitka spruce on Haida Gwaii. In a 1919 poetry book its importance for aeroplanes is stressed:

Sitka Spruce
Sitka Spruce is fine of grain,
And Sitka Spruce is tough,
To carry weight and stand the strain
There grows no better stuff;
It thrives upon Queen Charlotte Isles
And lifts its head on high,
When summer's sun upon it smiles
Or winter rages by.

When summer's sun upon it smiles
Or winter rages by.
Sitka Spruce is straight and clear,
And Sitka Spruce is light,
That aviator knows no fear
It girds into the fight;
For borne on wings that tire not,
He hurtles on the foe
Until he finds a vital spot
And sends him down below.
Until he finds a vital spot
And sends him down below

Sitka Spruce the allies need,
And Sitka Spruce must get;
The loggers answer: 'With all speed
This need shall now be met'.
And when the logger speaks his mind
It is not empty boast –
The Allied nations soon shall find
The thing they need the most.
The Allied nations soon shall find
The thing they need the most.

D. E. Hatt MA, Secretary of the YMCA, Moresby Island Administration Camp of the Imperial Munitions Board, Department of Aeronautical Supplies, Thuration Harbor, Queen Charlotte Islands, British Columbia.

Hatt 1919

A Forestry Agency for Britain

What would happen if there was another war?

The private woodlands of the country were now well and truly denuded while the few state forests were still young.

In 1910, the Board of Agriculture had requested the aforementioned civil servant, Roy Robinson, to investigate potential afforestation. He already networked with landowners and experimenters and was familiar with the interest in a potential forestry agency. In the war, he was secretary to a government sub-committee which considered potential future forestry. By 1918 it was obvious to members that the country needed a strategic reserve of home-grown timber.

In July 1918, the Duke of Buccleuch – owner of huge estates in Scotland – asked an official question about Inverliever Forest in the House of Lords:

> To ask His Majesty's Government what has been the annual and total expenditure by the Crown on the estate of Inverliever since its purchase in 1907; how much of that expenditure has been incurred in planting and replanting new ground; what is the total area of land so planted, and what number of plants have been used up to the present time; how has the total expenditure been apportioned under such heads as the following: (1) fencing against cattle, sheep and rabbits; (2) draining; (3) plants and planting; (4) beating up and establishing crop; (5) cleaning and bracken cutting; (6) nursery; (7) cost of clearing sheep; (8) general management and miscellaneous charges *etc.*;
>
> ...This scheme at Inverliever was started, I think, about ten Years ago-in 1907. It is a most important development, and to a certain extent, it is looked upon, I believe, as the start of a possible scheme of State afforestation...
>
> Hansard 10 July 1918, Lords Sitting

The forestry sub-committee advised setting up a forestry agency with a programme to afforest 60,703 ha (150,000 acres) within ten years and to assist private landowners to afforest an additional 20,234 ha (50,000 acres). But the long-term aim was to afforest 809,371 ha (nearly two million acres). This would increase the country's forest area to 1,618,743 ha (nearly four million acres) a massive task. Under the *Forestry Act* 1919, the Forestry Commission, an agency, not a ministerial department was set up.

This initiated the biggest change in UK land-use of the twentieth century.

It was managed on hierarchical lines by professionals trained in the style of the Indian and German forest services. Some experienced forest workers like Jim Shaw and John McEwan, familiar with the domestic industry, obtained posts as Forester for an individual forest. They found the new system was as hierarchical as the estates of their young days.

The agency was to provide for the nation a strategic reserve of two years' timber. But even if land was acquired and planted quickly, a war within 80 years would have insufficient timber supplies because the usual age at which broadleaved trees were harvested – the rotation length – was 80 years. The Forestry Commission bought war-felled woodlands and treeless property to increase the nation's trees. Private landowners, having sacrificed their woodlands to save the country from disaster, could obtain only minimal financial assistance to restock their own woodlands.

Land acquisition

What type of land was available?

The tree species planted depended on the land available to the Forestry Commission. Britain is a small, quite urban country with mountains mainly in the north and west, but not high by world standards: Ben Nevis is 1344 m (4409 ft), Yr Wyddfa (Snowdon) 1085 m (3560 ft) and Scafell Pike 978 m (3209 ft). However, they are rather bare of trees and exposed to maritime winds. They consist of mainly acid igneous and metamorphic rocks often with superficial ice age deposits. Many upland soils were degraded during and since prehistory. Above 305 m (1000 ft) there are nowadays few settlements and farming is difficult. There is a big climatic, cultural and ecological divide between the upland north and west of Britain and its lowland south and east.

The Irish Free State (formed in 1922) was an even smaller country with a maritime climate and considerable marginal land of low-lying peat and mountains exposed to the Atlantic Ocean. Many big estates had been broken up after the Irish *Land Acts*.

After a severe war, the importance of farming guaranteed that acquisitions for afforestation *had* to be poor, non-agricultural land. State afforestation was therefore restricted mainly to localities in the upland regions of north and west Britain. Thetford Forest is one example of afforestation in the dry-summer climate of southern England: East Anglia's 'Breckland' where soils had been impoverished by centuries of rabbit farming in large warrens.

Deciduous trees, such as oak, beech and lime, which the public are fond of, grow well on the fertile loams, clays and limestones of the lowlands. But their growth is sub-optimal in much of the north-west, so timber for modern needs cannot be guaranteed.

I emphasise, as foresters have often emphasised to me, that 'good', 'agricultural' land was forbidden to state afforestation.

Finding land for new forests

The new Forestry Commissioners and other significant landowners fell over themselves to get their own property sold and afforested. It was a way of dealing with their poorest lands, which also brought in the lowest rents.

By the end of 1919, lands at Borgie (Sutherland), Eggesford (Devon) and Rendlesham (Suffolk) had been acquired by the Forestry Commission. The Duke of Fife sold 1214 ha (3000 acres) for Monaughty Forest, Moray in 1920–1921 (Map 14).

The Duke of Buccleuch was quick off the mark when he sold 1416 ha (3500 acres) at Newcastleton, Roxburghshire in 1925. In 1925–1926, the Church Commissioners sold 1012 ha (2000 acres) just into England; the Duke of Northumberland sold 19,020 ha (47,000 acres) of his Kielder Estate in 1932. These blocks formed Kielder Forest – more of its story later.

2428 ha (6000 acres) in the eastern foothills of Yr Wyddfa (Snowdon) were acquired from Gwydyr Estate and the first trees were planted in 1921. Unemployed slate and lead miners provided the labour. Later, 2330 ha (5759 acres) were acquired from the Penrhyn Estate and added on. Over the years more blocks were added and by 1949 it was 7284 ha (18,000 acres).

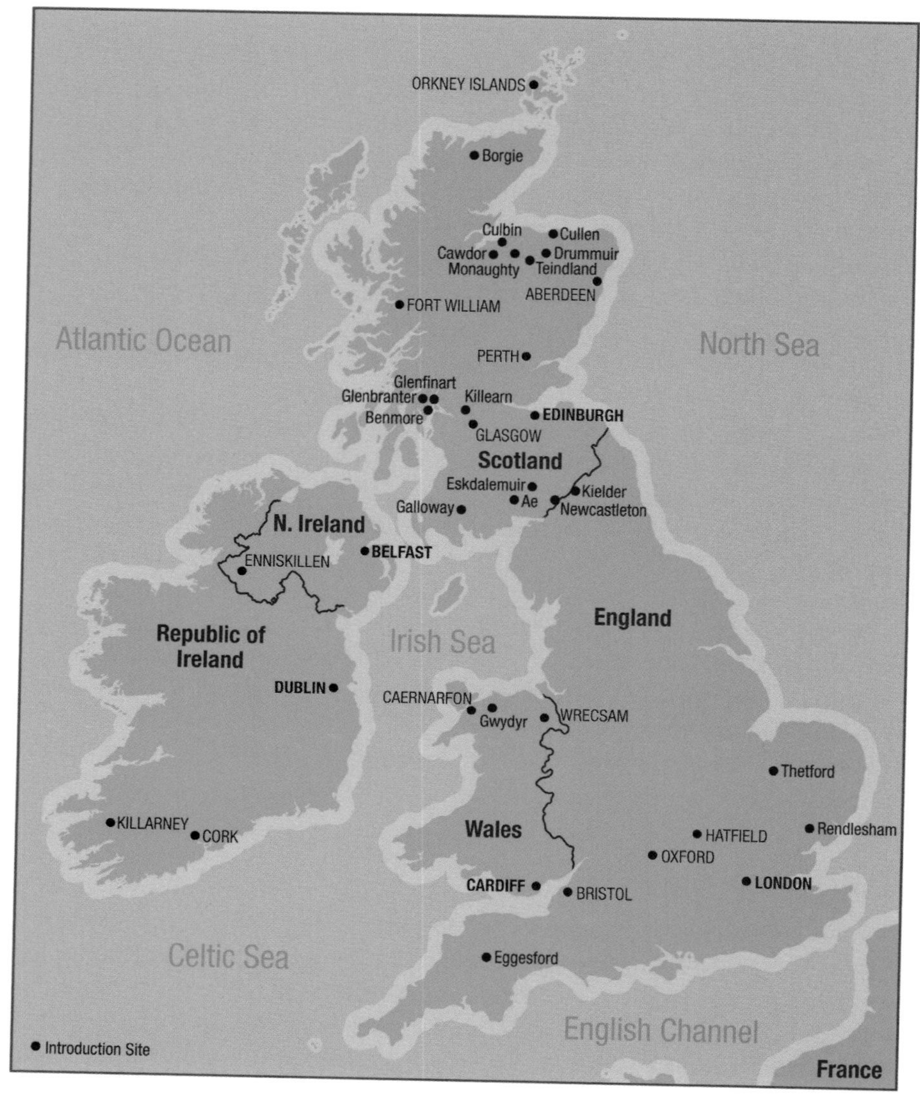

MAP 14. Some early Forestry Commission forests, prepared by Susan Anderson.

DATA: SEE BIBLIOGRAPHY FOR CHAPTER ELEVEN.

In 1923, the Forestry Commission took over the remaining royal forests (New Forest and Forest of Dean) and Hafod Fawr, Inverliever, Tintern and Dymock. When the easy supply of their lordships' land parcels declined, 'Acquisition Officers' were trained to locate and acquire suitable land. Oral history contributor Bill Sutherland explained:

> As Acquisitions Forester I became part of a very efficient team based at 25, Drumsheugh Gardens, Edinburgh, under the charge of the Chief Land Agent (Geoff Forrest, later to become Senior Officer for Scotland)... My job was to inspect properties, which they had found through adverts in papers and local knowledge. I would meet the owner (seller or tenant) and survey the property assessing its value for tree planting and prepare a report on my findings. My Acquisition Report would be examined by the Lands Staff of the Department of Agriculture and Fisheries for

FIGURE 11.7. Monaughty Forest, Moray was bought by the Forestry Commission in 1920 and John McEwan, a lover of Sitka spruce, became its first forester; it is still producing excellent Sitka spruce.

PHOTO: ANDY TITTENSOR, 2015.

> Scotland. These were qualified agriculture degree personnel… They decided if the land or how much of it could be transferred from Agricultural use to Forestry. And the Forestry Commission had to abide by their decision or we held mutual site meetings to negotiate the final outcome.
>
> Sutherland in Tittensor 2009

Goodwood Estate in Sussex and Gwydyr Forest were acquired differently, with a 999-year lease. Compulsory purchase was not used: 'Land is only acquired by negotiation with willing sellers, and so can only be bid for when it comes onto the market' (Forestry Commission 1974).

In 1922, the Northern Ireland Forest Service was formed within the provincial agriculture department. Large estates were not a major source of early forest land: it was a matter of buying farms and gradually adding more parcels. In later decades, larger areas of marginal land in the uplands were available, often cliffs, rocky areas, overgrazed hill slopes and peat (Noel Melanaphy pers. comm.).

In the Irish Free State, acquisition parcels were small because it too lacked big estates. John McEwan, who worked on acquisitions in Dublin, noted that there was never any timber to be valued on land changing hands, as it was sold by the vendor to raise cash.

Landscapes of acquisitions

The proposed Forest of Ae near Dumfries would be remote, with no roads or settlements. Rounded hills of Silurian rocks with mainly acid soils carried heather and grasslands, and many burns drained into the Water of Ae and thence the River Annan.

Valleys were considered too fertile to plant trees, while the highest hills (up

FIGURE 11.8. The Conwy Valley with forests planted on the slopes between valley-floor cultivation and hill-top moorland.

PHOTO: COURTESY VISITWALES. © CROWN COPYRIGHT (2014) VISITWALES.

to 696 m (2285 ft) on Queensberry Hill) were very exposed and carried poor soil – though sheep and cattle grazed up there. It was the lower hills and slopes up to 366 m (1200 ft) which were to be planted:

> Before the Forestry Commission started work in 1927 most of the land was used for sheep grazing. There was little woodland except a few patches of oak and alder in some valleys … the higher ground, most of which is covered with peat, is too poor and too exposed for the satisfactory growth of trees … but this land is not altogether waste because sheep and cattle can be run on it in summer. Planting has been confined to the lower slopes.
>
> Forestry Commission 1948

This illustrates both the poverty of the land and the rule that only marginal land could be afforested. The choice of tree species for planting was restricted to whatever would grow there.

Gwydyr Forest varied from 152 m to 305 m (500–1000 ft) altitude and average annual rainfall was 1016–2012 mm (40–80 in). Silurian igneous and metamorphic rocks formed the mountains which hold 11 lakes. Pre-afforestation, the high, exposed, peat moorlands supported one sheep to 1.2 ha (3 acres), a very low stocking density.

The rugged plateaux moorlands with vegetation of heathers, Western gorse (*Ulex gallii*) and grasses such as *Nardus stricta* were too difficult to afforest as there was not the necessary machinery or expertise at that time. The fertile lands of the Conwy Valley and its many tributaries were retained in farming when

Gwydyr Estate was acquired in 1920. It was the slopes, the 'ffridd' lands of acid grassland, gorse and bracken, in-between the valley fields and moorlands, which the Forestry Commission was allowed to afforest.

Nuts and bolts of early afforestation

John McEwan was appointed Forester for Monaughty Forest in 1920. At last he was able to establish plantations for the future. During his first year he got 202 ha (500 acres) planted and:

- Cleared branches left from wartime felling.
- Organised draining.
- Took on a trapper to control hares and rabbits.
- Obtained tools.
- Acquired labour (unemployed, untrained in woodsmanship).
- Cleared bracken from drier areas to plant larch.
- Organised two squads of 30–40 men and a draining squad.
- Got each man to plant 650 plants per day, carefully.
- Planted blocks of rectangles of up to 20 ha (50 acres), easy to measure.
- Built a wet-weather 'Rusky' for his men.

Choice of species

A variety of trees had been tested during the previous 90 years. But the new government foresters 'had their hands tied' because the land available to them was unsuitable for many species. They had been asked, told even, to grow trees as fast as possible, to ensure that Britain never again suffered such a timber shortage as during the recent war. 'Plant, plant, plant', they were told. Fast-growing conifers – which attained height and produced a harvestable crop quickly on the allowed land – were the pragmatic choice.

Foresters in the uplands tried Scots pine, Norway spruce, European larch and conifers from North America, particularly Western hemlock and Western red cedar. There was historic affection for native Scots pine and it was well-known how to grow Norway spruce and European larch. European silver firs were put in more amenable sites. Despite Crozier's informative 1910 paper most foresters took little interest in Sitka spruce.

In lowland Britain, Scots and Corsican pines and larches were commonly used, often in mixtures with Beech or oak. Douglas fir was planted on deeper, fertile soils and Monterey pine in warm coastal areas.

When, in 1922, Jim Shaw was appointed Forester for Tintern Forest he cleared-up after wartime felling and thinned the remaining woodlands, then collected acorns and restocked with oak and other broadleaved trees in the good soils. He planted larch, Douglas fir and other conifers in groups on varied sites to ascertain which grew best.

Obtaining plants

In 1920, John McEwan had to get transplants for Monaughty Forest from anywhere he could, many of poor quality. He started with Norway spruce, Scots pine, larch and probably Sitka spruce. Looking back in later years, he felt that he should have used only Sitka spruce:

> Norway Spruce was a European species in use in Britain for many years; the sitka spruce was not so well known at that time. If I were planting today, there would be a big difference in this selection. I would use practically no larch, which was one of the main species used at that time. It would have been much easier to lay out the ground, as the great bulk would have been in sitka spruce. Scots pine, too would not have been planted so generously as it was at that time.
>
> McEwan 1998, but recorded in 1984

Gradual increase in Sitka spruce

Through several decades of journal articles on the reaction of Sitka spruce to different ecological conditions, foresters learned of its ability to grow in conditions unsuitable for many other species.

When the Forestry Commission took over Hafod Fawr and Inverliever forests in 1923, experiments had been in progress for 24 and fifteen years respectively. Of the eleven species tested at Hafod Fawr, Scots pine and European larch failed to grow satisfactorily, but Japanese larch and Sitka spruce grew successfully in its difficult conditions. Hafod Fawr thus became the template for afforestation in Wales and Sitka spruce became the commonest afforestation species in its uplands.

When the first Sitka spruce of Hafod Fawr were felled, they had grown to more than 30 m (100 ft) tall and over 1.8 m (6 ft) girth – enormous for such environmental conditions.

So much for Roy Robinson's worries in 1912.

At Inverliever Forest, Sitka spruce grew better than Norway spruce on all the poorer and more exposed sites. So eventually it was chosen for higher altitudes, Norway spruce lower down, with Douglas fir on better soils and larches on drier ground.

Jim Shaw became Forester at Gwydyr in 1924 after a short stint at Tintern. By coincidence John McEwan moved there in 1928 to be District Officer – with Gwydyr and four other Forests in his charge. Sitka spruce was an important constituent of these forests:

> There were a number of plantations of sitka spruce, which was my speciality, all needing some attention. The management after planting had not been very carefully done in any of the forests, and as a working forester promoted to District Officer I was welcomed with open arms by every one of the foresters…
>
> McEwan, *ibid.*

Sir John Stirling-Maxwell was also influential in the choice of Sitka spruce. He published the results of his Corrour experiments to help other foresters choose their tree species.

Ground preparation

In the early nineteenth century, the land was often ploughed and cultivated using horses, drains were dug by hand and thick vegetation burned before trees were planted.

In the early twentieth century, ground preparation was often not considered necessary; occasionally the land was drained or ploughed first. Foresters who did prepare the land had no standardised methods to follow. Trees might be planted in a notch, slit or hole amongst vegetation. At Corrour, ground preparation methods were tested for three decades:

> Almost every acre needed some preparation before planting. Even on steep slopes, the peat was so tough and tight, that sooner or later it had to be cut through. And … the bulk of the plantations are included in a single deer fence which surrounds the east end of the loch.
>
> Stirling-Maxwell 1929/1951

Slabs of turf were hand-dug and turned over as a home for each tree and this 'turf planting' was copied at Inverliever Forest from 1910. Other tested methods included experimental ploughing with horses in Teidland Forest in 1927, to give each tree a small mound of soil.

Seeds, germination and planting out

Most nineteenth-century Sitka spruce seed came from the coasts of Washington and Oregon. It was fairly easy to obtain because when big Sitka spruces were logged, coning canopies lay on the ground. By the time of early air travel, considerable quantities of Sitka spruce were being felled from the rainforests and even more seed was available. The exact source of that seed did not concern foresters.

The Forestry Commission set up a tree nursery for all its forests and many private nurseries supplied seed and transplants. Sitka spruce seeds germinated easily and were grown for two years, then moved and grown on for two more years. As '2 + 2' transplants they were sent to their permanent forest homes.

Foresters who appreciated Sitka spruce planted it wherever they thought it might grow. They soon discovered that it 'checked' on heather-covered ground. But, in Inverliever Forest it recovered from even 20 years in check and grew again, especially if given phosphate fertiliser.

Sitka spruce in the Irish Free State and Northern Ireland

In 1922, the new Irish Free State formed its own forestry agency. Augustine Henry had evaluated many tree species for Ireland and prepared a useful treatise with Henry Elwes. As early as 1907, he had written:

> The results of my observations show that we may accept as a general law that all trees of the Pacific slope of North America from Alaska to Oregon are suitable for

> planting in Ireland, where they thrive amazingly... This [Sitka] spruce is much superior in Ireland, generally, to the common or European spruce... It appears to be perfectly healthy, is easily raised from seeds, and transplants well. It does well at high elevations.
>
> Joyce and OCarroll 2002

In 1928 John McEwan went to work for the forestry agency in Dublin. He visited forests from Cork to Donegal and decided that he must get the 40–70% death rate of new plantings reduced to less than 10%. At that time, the Irish Free State obtained its stock from Scotland, where conditions were similar. And John McEwan knew the nurserymen. He started a nursery for each Irish forest using Scottish stock but later decided to start with seed.

Table 6 shows that right from the start, the Irish Free State planted nearly one third of its new forests with Sitka spruce, but the seed was difficult to import in wartime.

Since 1922 the Northern Ireland Forest Service has run its state forests. Its first forest, Ballykelly had been acquired in 1910 as a mix of old oaks, beech and larches, but 11 acres (4.5 ha) of Douglas fir were soon planted. Castlecaldwell and Knockmany Forests were also early acquisitions. Sitka spruces planted in those early days still contribute to Knockmany Forest.

Sitka spruce in the 1930s

Difficult terrain and the requirement to plant quickly and prodigiously forced the state agencies to use land efficiently and grow one or a few species in blocks or 'compartments'. All the ground within a forest fence was used because space and time spent landscaping would delay the accumulation of a strategic reserve.

What researchers discovered

This is what Sir John Stirling-Maxwell wrote about Sitka spruce after nearly thirty years of trials:

> Of all the trees we have tried, this promises the best return. It is less subject to frost here than at lower elevations, as it seldom makes a second growth. In 1926, here as elsewhere, the autumn frosts destroyed a good many tops, including, in some cases the wood of the previous year. But this was quite exceptional. This tree is remarkably wind-fast. It grows faster than the Norway spruce, and is less intolerant of heather. Where the two species have been accidentally mixed on bad ground, it always leaves the Norway behind. Taken all round, its growth surpasses that of any other species with the possible exception of Douglas fir, and since it greatly outnumbers this and never succumbs to wind or snow, there can be little doubt that ten years hence it will dominate these plantations unless some unforeseen catastrophe occurs... The timber, even when grown on deep peat, appears to be much harder and tougher than that of Norway spruce.
>
> Stirling-Maxwell 1929/1951

Year	% Sitka spruce of total state afforestation area per decade
1925	27%
1935	22%
1945	11%

TABLE 6. Sitka spruce in early tree planting by the Irish Free State.

SOURCE: JOYCE AND OCARROLL (2002).

Year	Norway Spruce	Sitka Spruce	Total
1923	2	2	4
1925	5	3	8
1927	5	7	12
1929	4	9	13
1931	7	13	20

TABLE 7. Millions of spruces planted in early years by UK Forestry Commission.

SOURCE: MACDONALD (1931); DAVIES (1972).

The nineteenth-century Sitka spruce trees at Durris 'checked' for six years, but then grew better than all the other species until they blew over in 1927. During their last five years, they spurted 0.5 m (19.5 in) per year, meaning that they were still growing fast at age 48. They'd reached up to 30.4 m (100 ft) tall in that time. The only other trees still growing were a few, small Douglas fir and Norway spruce.

To evaluate Sitka spruce scientifically, foresters needed numerical data: how tall does it grow in different conditions? What volume of timber can it produce at age 30/40/50 on different ecological sites? At what age would it produce quality timber of the needed size? When would it be ready to harvest?

There were relatively few plantations to provide any data, so the answers were slow in coming. Preliminary 'yield tables' giving volumes of timber produced by Sitka spruce were published in 1931 using the few measurements available.

Roy Robinson had to tout the qualities of Sitka spruce. He emphasised that, on the type of site acquired, Sitka spruce would get away quicker than Norway spruce (still the foresters' favourite) and grow on sites where Norway spruce could not. After six to eight years it has grown to 1.5–3 m (5–10 ft) tall, compared with Norway spruce which has reached 1.2–1.8 (4–6 ft) tall. He was certain that Sitka would produce 50% more volume of timber than Norway spruce. But most foresters were not experienced with Sitka spruce and needed a nudge.

Table 7 shows that although foresters preferred Norway spruce in the early years, the number of Sitka spruce increased during the 1920s and by 1931 was double the Norway spruce.

Estate forestry

There was little financial support from the government for tree planting by private landowners. Thus Atholl Estate was typical in planting very few trees in the 1920s and 1930s for lack of finance.

Sitka spruce in Europe

In the first half of the twentieth century Icelandic foresters concentrated on protecting their native Downy birch (*Betula pubescens*) woodlands. From 1950

MAP 15. Commercial Sitka spruce in Europe, prepared by Susan Anderson.

DATA COURTESY OF BILL MASON

to 1990 they tested the potential of 28 species for afforestation. They eventually decided on native birch and Siberian larch (*Larix sibirica*) as the main trees for forestry. When its qualities in Icelandic conditions became apparent, Sitka spruce was planted for amenity and occasionally for farm shelter, but most often on hill land where farms had been deserted as being non-viable in modern conditions.

In the Faroe Islands native woodlands of *Betula pubescens*, rowan, aspen and juniper have been gone for more than a thousand years, helped by sheep, strong sea-winds and cool summers. Sitka spruce, Lodgepole pine and several Southern beech (*Nothofagus* species) have grown well in small, modern plantations.

FIGURE 11.9. Wartime fellings in Allerston Forest, North Yorkshire 1939.

PHOTO: COURTESY FORESTRY COMMISSION. © CROWN COPYRIGHT.

Scandinavia still has indigenous forests, but afforestation was important to supply its big timber industry during the twentieth century. In Norway and Denmark, with their westerly coasts, Sitka spruce has formed satisfactory plantations. In continental Germany and Holland, Sitka spruce was planted on a small scale as a timber tree and for ornament.

Map 15 shows the European countries which grow Sitka spruce commercially.

The Second World War

It has been calculated that annual imports of timber to Britain were reduced from 9.5 to 1.78 million tonnes (10.5–1.96 million tons) during the War, due to reduced purchases and sinking of transatlantic merchant shipping by submarines. A recent study of wartime records showed that: 'No other raw material of magnitude shows anything like the same fall in imports, nor indeed was any bulk material replaced by home production to the same degree as timber.' (Weir 2003)

There was no alternative but to use home-grown timber for pit props, bridges, military buildings and furniture, mulberry harbours and domestic needs.

By 1939, the Forestry Commission had acquired 584,367 ha (1,444,000 acres) of plantable land of which 175,634 ha (434,000 acres) had been afforested. However – the trees were not yet ready to harvest – the 'Strategic Reserve' had been growing for less than twenty years.

The volume of available timber nationally was *less* in 1939 than in 1914 because there had been so little incentive for private landowners to invest in forests.

What could be done? The same as before…

The owners of private forests were asked to sacrifice their woodlands once more especially in Scotland where cover of mature woodland was greatest:

FIGURE 11.10. Canadian Forestry Corps No. 5 District, Balblair House, Beauly, Inverness-shire 1940, Lord Lovat's estate.

PHOTO: COURTESY BOB BRIGGS, HTTP://FREEPAGES.GENEALOGY.ROOTSWEB.ANCESTRY.COM/~JMITCHELL/CFC.HTML.

> most of the plantations were too immature to make any really worth-while contribution to the timber needs of the 1939–1945 war, with the result that Scotland's older, privately owned woodlands were again subjected to heavy felling, amounting to 155,000 acres.
>
> Forestry Commission 1970

'Sawdust Fusiliers'

Outside help was again needed to fell trees because British men were at war or down the coal mines and other help was insufficient. To start with several thousand men from Newfoundland and 500 from Honduras were recruited to the Newfoundland Overseas Forestry Unit which worked in 71 camps.

Nearly 7000 'Sawdust Fusiliers' also came from Canada to fell Scottish trees. Loggers, sawmillers, blacksmiths, mechanics and cooks were sent to about 33 rural camps, each of about 200 men. Camps were mainly in eastern Scotland where mature woodlands predominated, with rail and sea transport close by. They brought their own equipment and machinery to build light railways and roads for extraction. Many workers on the Atholl Estate had joined the Forces, so Sawdust Fusiliers set up their camps and sawmills to cut down three woodlands the Estate had promised for the war effort.

In Inverliever Forest, Sawdust Fusiliers from the Cruachan camp felled the biggest trees (planted 1910, now almost 30 years old): in the emergency they were considered ready to use.

Tintern Forest lost one-fifth of its area, 370 ha (913 acres), to the war effort of which 121 ha (300 acres) were really too young, but they were needed. Corrour Estate produced 325,000 'lineal feet' of pit prop timber, for the healthy sum of £2822 (*c.*5845 CAD, $4065).

Wartime North America

Sitka spruce had been logged in North America since the late nineteenth century, and had a vital role during the Second World War. Large quantities were again cut from the coastal rainforests and Haida Gwaii, sent by rail to east coast ports and then shipped across the Atlantic Ocean.

Once in the UK the Sitka spruce was used to build the legendary aircraft 'Wooden Wonder' or 'Timber Terror', the de Havilland 'Mosquito' which was a fast, multi-use aircraft. More about the Mosquito in Chapter Fourteen.

Thousands of seedlings of Sitka spruce and other species were planted during the War onto the still almost-treeless Aleutian Islands to control erosion, beautify the landscape and provide future fuel. Sitka spruce transplanted successfully again onto Amaknak Island. The 1805 Sitka spruce trees started to produce seedlings on nearby disturbed ground soon after the War.

Post-War

What were the effects of the Second World War on UK forests and Sitka spruce? A large proportion of mature and nearly mature forests of many species was felled, but particularly Scots pine, spruces and larches. Fewer Douglas fir and broadleaved trees were felled, though at least one magnificent 200-year-old beech woodland disappeared. Afforestation declined during the War because management of existing plantations was paramount.

Between 1947 and 1949 the Forestry Commission carried out a census of its national estate. It noted a total of 225,000 ha (555,987 acres) of coniferous high forest: 27% of this was Sitka spruce.

Just after the war, Norway and Sitka spruces were the most important forest trees in Britain, making up more than half the state afforestation programme. By 1947, eight million Norway spruce trees and twenty million Sitka spruce had been planted in new forests. That was more than double the number of any other tree species. Foresters could still plant their beloved Scots pine on dry, infertile sites.

The 1947–1949 census showed a total of 67,597 ha (167,037 acres) of Sitka spruce in Britain. This was 9% of the total forest estate of 464,627 ha (1,148,119 acres) and 17% of the total area of conifers.

A century of experience with Sitka spruce had shown it would grow better than any other plantation tree in Britain and Ireland in exposed, high-rainfall areas. It grew on a variety of infertile soils where otherwise only low-yielding Lodgepole pine was possible. Phosphate fertiliser took it out of 'check' and stimulated growth. Sitka spruce was easy to transport and produced quality pulp and saw timber. It suited the degraded British and Irish landscapes beyond all expectations.

Chapter Twelve describes how Sitka spruce reached its zenith in even bigger forests on even worse land during the later twentieth century.

FIGURE 11.11 *(above left)*. Drains ploughed in parallel lines; rows of turves for planting trees will be dug between them, Minard Forest, Argyll, 1948.

PHOTO: M. LAURIE, FORESTRY COMMISSION. © CROWN COPYRIGHT.

FIGURE 11.12 *(above right)*. Taking out harvested spruce logs with a Clydesdale horse, west of Scotland *c*.1956.

PHOTO: GEORGE DEY, COURTESY OF THE FAMILY OF GEORGE DEY © ABERDEEN UNIVERSITY.

FIGURE 11.13 *(left)*. Sitka spruce planted in 1916 in Inverliever Forest being assessed in 1959.

PHOTO: GEORGE DEY, COURTESY OF THE FAMILY OF GEORGE DEY. © ABERDEEN UNIVERSITY.

FIGURE 11.14 *(above)*. The biggest Icelandic tree-nursery at Tumastadir, South Iceland in the 1950s; in the foreground are the tops of Sitka Spruce planted above the nursery in 1944.

PHOTO: HAKON BJARNASON, SKOGRAEKT RIKISINS.

FIGURE 11.15 *(right)*. A view of the nursery site 54 years later when the 60-year-old Sitka spruce has grown into a plantation, slowly on this site.

PHOTO: ÞRÖSTUR EYSTEINSSON, SKOGRAEKT RIKISINS.

CHAPTER TWELVE

Peat: The Final Frontier

> But the sheep were not so lucky. Many were lost each year: the very boggy ground was covered with bright, pale green vegetation which seemed to entice the sheep onto it to feed. But they sunk down into very soft bog and were lost forever – there were never even any bones left to find. Annually, about one-and-a-half-score sheep were lost in this way. This is why the soft, green, boggy ground was called 'The Black Death'!
>
> John Telfer of Eaglesham, talking about his farm, in Tittensor 2009

> In connection with the afforesting of waste lands I have travelled over the greater part of the Kingdom and examined much of the ground that could be set aside for this purpose, including the peat bogs of Ireland; while at altitudes up to 1,100 feet I have formed plantations on the bare and wind-swept hill-sides of Wales and Scotland, which today are not only a boon to the farmers in the way of shelter, but a considerable source of profit to the owners as well.
>
> Angus Duncan Webster 1905

Post-war expansion

A feeling of urgency to make the country a better place for its inhabitants triggered new towns, roads, and improvements in agriculture. Replanting war-cleared woodlands and afforesting new areas, particularly in Scotland which had borne the brunt of wartime felling, was part of this feeling. Landowners and MPs pressed for rapid development of state afforestation. The Forestry Commission was given the go-ahead to expand its land holdings:

> Faced with the need to replenish as quickly as possible the country's supplies of timber, the Forestry Commission embarked in 1946 on an extensive post-war programme, and now aims to plant in Scotland at a rate of 50,000 acres [20,235 ha] a year. The existing 875,000 acres [354,113 ha] of Commission plantations will be increased by some 250,000 acres [101,175 ha] by 1975, giving a total of about 1,125,000 acres [454,288 ha]. In addition, private woodland owners, who already hold about 950,000 acres [384,465 ha] of woodland well suited to economic management, are playing an active part in replanting and new afforestation, and are currently planting around 20,000 [8094 ha] acres each year.
>
> Edlin 1970

Getting private woodland owners to indulge in 'playing an active part in replanting and new afforestation' had actually been pretty hard! They had kept forestry going during the nineteenth and early twentieth centuries when the state had shown no interest in home forestry whatsoever. Between 1919 and 1946, small government grants of £2–4 per acre (*c*.£0.80–1.6 per ha) were intended to tempt landowners to engage in forestry. These finances were insufficient to clear their war-torn woodlands, plant and manage new woodlands for several decades until harvesting time came along.

Seeing the huge job ahead after the Second World War, governments finally realised that a reasonable financial incentive and free advice would encourage the private sector in woodland renewal and afforestation. One scheme, called 'Dedication' was initiated in 1947, but it required landowners to dedicate their woodlands to forestry in perpetuity. Not surprisingly, some were suspicious that this sudden interest in long-term dedication of woodlands was the back-door to land nationalisation.

However, the forestry department of the Atholl Estate was revitalised when it entered into the Dedication Scheme. By 1952, the Estate intended to afforest 81 ha (200 acres) and thin 304 ha (750 acres) annually, and employed 44 forestry workers for the purpose. Its period of stagnation was over. But it was not for another twenty years, when grants had been increased, that private estates participated fully in restocking and afforesting the UK landscape. The Dedication Schemes continued until 1981, when they were replaced by other financial incentives.

Planting and productivity

The Forestry Commission had planted a total of 404,685 ha (one million acres) of trees countrywide between 1919 and 1956. The West of Scotland Conservancy (about one-third of Scotland's forest area) planted 73,843 ha (180,000 acres) of new forests between 1919 and 1967. With an average annual rainfall of 1524–2286 mm (60–90 in), 57% of it was Sitka spruce, 15% Norway spruce, with a variety of other conifers and deciduous trees making up the other 28%.

A total of 2833 ha (7000 acres) of trees had been planted in Gwydyr Forest by 1954. Sitka spruce, planted extensively since 1921, was by now the principal and most valuable tree there, despite its susceptibility to check and spring frosts. One-quarter of the acreage was now producing thinnings: pit props went to the Wrecsam coalfield, poles to local farmers. By 1948, the earliest plantations in the Forest of Ae were also thinned, providing pit props for the Lanarkshire and Ayrshire coalfields.

Practical problems

The practical side of afforestation by landowners and state was beset by problems throughout the twentieth century. 'Vermin' which nibbled or gnawed trees (including young Sitka spruce) necessitated expensive, dug-in fencing around new plantations. Super-abundant rabbits still burrowed in. The viral disease myxomatosis brought into England in 1953, reduced rabbit populations by over 99%. Rabbits however, gradually developed immunity and resistance so that by

FIGURE 12.1. Typical upland moorland at 301 m (988 ft) of the sort used for tree-planting in the later twentieth century, Mean Muir, Ayrshire.

PHOTO: RUTH TITTENSOR, 2010.

the 1980s, when 'only' 40–60% of infected rabbits were killed by the disease, fencing them out of plantations was again necessary.

Two species of hare would nip off the tops of young trees along whole rows. Trappers dealt with them, feral goats and other animals which damaged small trees. Forest Rangers were trained to cull Red and Roe deer whose populations increased when new forests grew upwards and outwards. Red squirrels damaged conifers, particularly pines, by feeding on the sweet underbark and were controlled until the 1970s. They then colonised coning Sitka spruce plantations to feed on the seeds.

Effects of Sitka spruce

By 1954, Gwydyr Forest employed 180 full-time and 30 part-time people. Forests all over the UK provided rural employment but the work was not to the taste of all.

'Forest Villages' were needed for forest workers where new forests were in localities without settlements. The Forest of Ae in southern Scotland was the first to have a completely new village to house its workers, with ten houses built by 1948 and 80 the target. Inverliever Forest had a new village of wood houses by 1955: the population increased from 55 in 1907, to 87 by 1913, to 285 by 1952. During the 1960s, I enjoyed lodging in a wood-built forestry home in Rowardennan Forest, by Loch Lomond, while field-working on the steep slopes above. Forest villages of wooden homes were unusual in Britain where stone and brick are the usual building materials.

FIGURE 12.2. The scale of afforestation is shown by this landscape of ploughing and planting at Ravens Rock and Torrachilty Forest, Ross-shire, 1979.

PHOTO: GEORGE DEY, COURTESY OF THE FAMILY OF GEORGE DEY; © ABERDEEN UNIVERSITY.

Change of remit

In 1957, a major report proposed that a *strategic supply* of timber was no longer a suitable reason for the Forestry Commission's existence. A future war would use nuclear energy devices and would end quickly (I hope so). Timber would not be needed for similar purposes to recent wars. Forestry, its aims and practice needed to change.

Civil servants, landowners and MPs cogitated. They decided that forestry should henceforth be more about reducing the cost of timber imports, should have social aims and encourage employment in rural areas to halt population decline. It should now look mainly to the uplands of the north and west.

Uplands?

The Forestry Commission had been afforesting the uplands (in particular) since 1919. However, the memory of difficult wartime food supplies coloured government decisions. The marginal lands which had been ploughed up during the Second World War were to remain in cultivation to grow food for the increasing 'Baby Boomers'. This meant action to force even worse marginal land to be 'productive'.

Forestry was left with what civil servants saw as 'non agricultural', 'waste' and 'unused' – that is – poorer than marginal land. But this type of land had been classified previously by foresters as 'unplantable'. Was this now to be afforested?

Yes.

Afforestation could from now on take place on only the very worst 'non-agricultural' land, mainly moorlands at 244–305 m (800–1000 ft) altitude,

FIGURE 12.3. Panorama of recently-planted Sitka spruce, Whitelee Forest, Ayrshire.

PHOTO: BRIAN SPEIRS, 1980S.

with soils of acid peat, podsols and gleys, receiving high rainfall. Their farm carrying-capacity had been one sheep to 1.21–1.62 ha (3–4 acres). The choice of tree species for afforestation would be reduced even further. What on earth would grow *and* produce marketable timber?

Sitka spruce of course! And perhaps Lodgepole pine.

Foresters would now have to afforest poor mountain and plateau pastures, high moorland, rocks, blanket bogs, raised bogs, steep slopes and windswept places in the remotest parts of the country. Not the sort of workplace many people would consider. Map 16 shows the main upland areas of Britain and Ireland.

Whitelee Forest, in south-west Scotland, was an example of the type of dreadful land, bought in 1961 with hopes that trees might grow:

> At that time the Commission was desperate to acquire land to meet the large planting programme and it was not a question of what we will take, but what can we lay our hands on.
>
> Whitelee at high elevation on deep peat was in many ways a stab in the dark and a lot of people in the FC doubted if it would be successful. However, J. A. B. MacDonald, Conservator, South Scotland at the time was a forward looking man and decided to take it up.
>
> Peter Innes in Tittensor 2009

Afforesting peat landscapes

During the earlier twentieth century Sitka spruce grew better than other species on poor sites. But it was considered bad practice to plant it on deep, infertile peat because it went into check for lack of essential minerals.

After 1960, the Forestry Commission was faced with many deep peat sites, as well as sand dunes, rocks, scree and extremely windy places. There is plenty of hill peat on the English Pennine Hills and in Wales. But Scotland was desperate

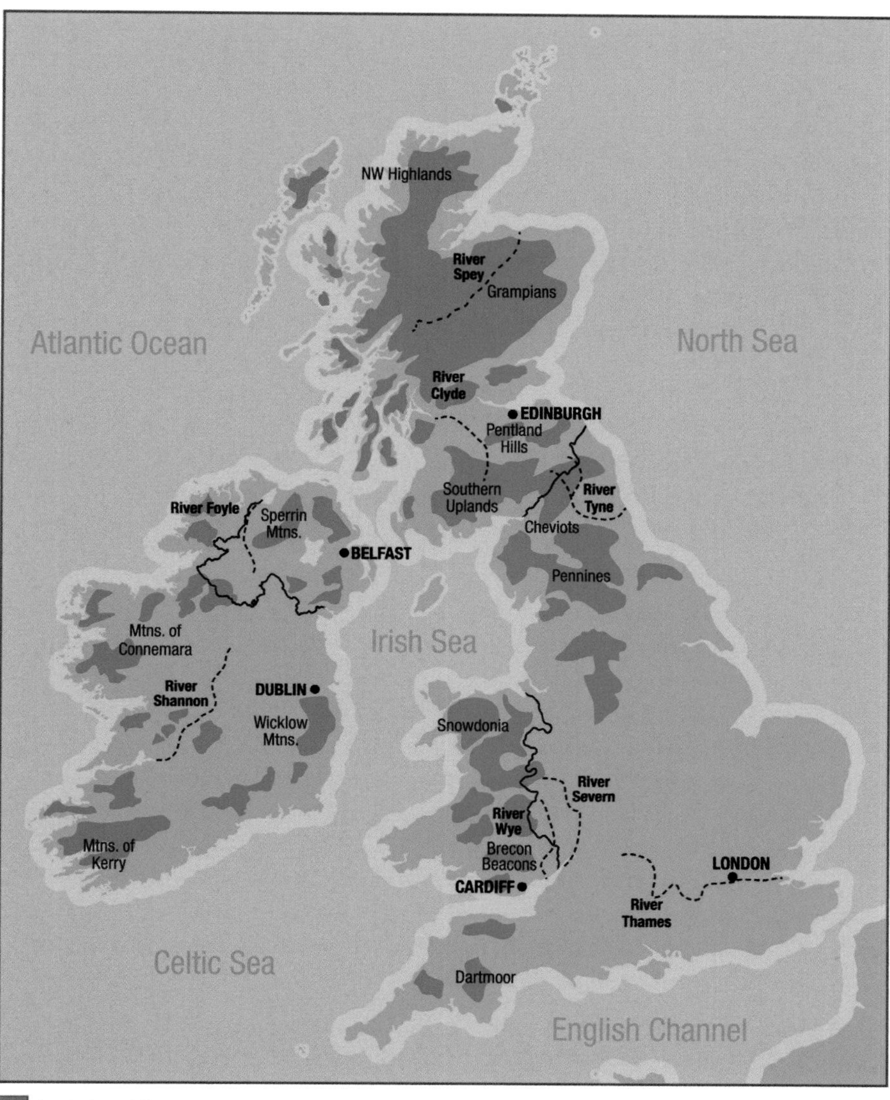

MAP 16. Uplands of Britain and Ireland, prepared by Susan Anderson.

SOURCE: ATLASES.

to have the extra afforestation land to assist the re-population of its uplands: there were still memories of earlier periods of depopulation. South-west Scotland and parts of Wales received most of the available finances for post-1960 afforestation.

The Republic of Ireland, where Sitka spruce trees grew very fast, also expanded afforestation on its peat lands.

The Sitka spruce peat challenge

When I started an oral history project about Whitelee Forest, Ayrshire in 2004, I was warned of tramping its peat landscape on my own, because the peat was deep, very soft and contained frequent pools of water covered in green moss called 'quaking bog'. My dangerous forays into Whitelee Forest in the early twenty-first century showed that Sitka spruce could indeed grow on this appalling ground.

It could grow, because between 1960 and 2000, many researchers, foresters, forest workers and engineering firms had taken up the challenge to grow productive forests on the dreadful ground they had been allocated.

Huge expansion of afforestation with Sitka spruce was needed to make good the losses of two world wars – and more. But foresters already knew that Sitka spruce grew better on soils other than deep peat where it often went into long-term check.

The challenge to grow Sitka spruce to marketable quality and size, forced development of better methods of afforestation and silviculture in Britain and Ireland. Foresters needed to know more about its ecological and physiological requirements. They did this by long-term observation and research.

Research

In the 1950s, the ecology of Sitka spruce in its native, North American ecosystems, was little known in Britain: would it actually *grow*, let alone grow well enough to produce a marketable harvest on deep, acid peat?

It was in 1952 that academic forester W. R. Day visited British Columbia, including Haida Gwaii. He wanted to study Sitka spruce ecology in its native ecosystems and to apply his findings back home. Douglas Malcolm continued this with research on its performance in different conditions in the UK and how this related to its native ecology.

The medium in which any tree grows or is grown, that is, the soil, was recognised as important. Gardeners work with the soil they have in their back gardens; foresters had to develop an important industry with soils they definitely would not have chosen:

> By choice we would not have wished to go into the peaty areas. The mineral soils give you far more flexibility in practising the type of forestry that professional foresters want to practice... And the forests that I enjoyed managing best were those forests... I had the slopes of the Great Glen, big Douglas fir areas as they were, but very productive. Lots of oak woods, lots of broadleaf within it. I love that.
>
> Gordon Cowie in Tittensor 2009

Foresters studied how effective was ploughing and breaking-up deep, impenetrable layers in certain soils. They found it increased the depth and spread of roots. John McEwan later noted how important this was in afforesting the worst lands:

> Pyatt's pioneering work is also important because forestry must not encroach on agricultural land, not even poor grazing. Up to 95% of our plantations had to be placed on land of second or even third grade quality, uplands and even some hill land. Pyatt's work has made it possible to grow economic forests on land we couldn't

FIGURE 12.4. Quaking Bog of Whitelee Forest, Ayrshire, a landscape type frequently afforested.

PHOTO: RUTH TITTENSOR, 2007.

> have used previously, up to seven or eight hundred feet, over eight hundred feet is not viable.
>
> McEwan 1998, but recorded in 1984

Richard Toleman was given the task of aerial surveying and ground mapping forest soils: they are not just 'peat' but peat of varied colours, constituents and depths, with different soil structure, horizons and acidity. His detailed maps of forest soils made it possible to carry out remedial work in each area before tree planting (as in gardening).

Researchers and foresters assessed the effects of draining and ploughing on subsequent tree growth; they tried planting little trees on the top or sides of

FIGURE 12.5 *(left).* Ploughing up high (*c.*360 m/1181 ft) near Loch Avich, Argyll, 1960s; the tractor is pulling a single furrow plough operated by cable from a winch on the back. The operator is ploughing alone hence the 'banksman' walking behind in case of accident.

PHOTO: DON MACCASKILL, IMAGE COURTESY OF BRIDGET MACCASKILL, STRATHYRE.

FIGURE 12.6 *(right).* Sinking BDT8 Tractor and Clark's plough, a frequent hazard for tractor drivers ploughing on peat; Whitelee Forest, Ayrshire.

PHOTO: ROY HARVEY, 1970.

ploughed ridges and checked their root growth. They analysed whether Sitka spruce grew better alone or in mixtures with 'nurses' of European larch and Lodgepole pine. They tried planting little trees at different distances apart.

Although Sitka spruce can exist on peat in some of its rainforest homes, it grows best on the moist, mineral-rich soils of river flood-plains. This suggested that peat soils in Britain and Ireland would produce better Sitka spruce trees if minerals such as nitrogen, phosphorus and potassium were added to the soil.

Research took time: years and decades. Mistakes were made meantime, and not all plantations grew as well as hoped, with 'check' a major problem, as in Rannoch Forest. Results from research forests such as Inchnacardoch and Borgie were available in due course.

New, large forests of mainly Sitka spruce continued to appear on open moorland and mountainsides in Scotland, Wales and northern England. They did what was required, producing harvestable and marketable products quickly: thinnings at 20–25 years and a clear-fell final crop at 40–80 years old.

The forestry industry could feel positive about its work.

However, the general public which was used to, and loved, unenclosed, treeless moorlands and mountainsides was not so happy. Sitka spruce-dominated plantations seemed uniform in appearance. Up-and-down ploughing, parallel rows of trees in rounded landscapes and sharp edges along land ownership boundaries vexed onlookers. A big green area would then quickly turn messy brown when felled: I have heard the word 'untidy' innumerable times describing felled forest, while forester Syd House recalls the more critical, 'resembles a WW1battlefield'!

Although Sir John Stirling-Maxwell had experimented with many broadleaf as well as conifer trees at Corrour, few broadleaves grew successfully enough for a timber crop. But during the 1960s and 1970s some broadleaves with attractive flowers or fruits, such as horse chestnut, rowan and wild cherry were planted around conifer blocks for amenity.

Foresters nevertheless 'ploughed on' with their main task of growing a timber resource on the worst land in the country.

Machinery for peat

Research and experience had shown that preparing soil for tree planting was as important as soil preparation for garden vegetables.

Ploughing peat down 0.9–1.2 m (3–4 ft) – a long way – was necessary to mix the soil layers, drain water and allow air into the soil for respiration of young tree roots. Ploughs could both drain and cultivate. Drainage channels first took away excess water from the planting area. Cultivation then mixed soil layers and provided ready-made, warm and dry, tree-planting ridges conducive to root growth.

In the nineteenth and earlier twentieth century, drains were dug by hand with huge spades. It was not until 1946 that a plough suitable for rough, afforestation land was developed. Engineering firms such as Clarks of Ae and Cuthbertsons of Biggar, designed strong ploughs which could work satisfactorily in peat and produce planting ridges – better than the hand turfing methods used at Corrour and Inverliever Forest.

Clarks developed the 'Parkgate ploughs' which could go down to 60 cm (24 in); the arched 'Humpy' plough brought up soil from 90 cm (3 ft) deep. Clark cultivation ploughs were pulled by a variety of tractors, for instance two BTD8 Crawler tractors in tandem or an International TD500 tractor with slatted wooden tracks.

Bowen 60 Crawler tractors with long and wide tracks, Priestman Cub Diggers on mats and with a bucket on a side dragline, or Ford Backacter Diggers with flotation tyres pumped up to only 0.413 bar (6 lb per sq in) pressure did much of the draining. These tractors could pass over soggy peat without sinking (usually!).

Begg, Clarks and Cuthbertsons, for example, designed double-mouldboard ploughs which produced a much wider ridge, but had to be pulled across the peat by two tractors in tandem. A wider ridge gave each small tree more space for its roots to grow out and so greater stability in wind.

FIGURE 12.7 *(below left)*. Filling trays with collected cones of Sitka spruce in British Columbia.

PHOTO: ALAN FLETCHER, 1970.

FIGURE 12.8 *(below right)*. Cone dryer and tumbler for extracting seeds; British Columbia.

PHOTO: ALAN FLETCHER, 1970.

FIGURE 12.9. The Whitelee Forest squad planted about 10 million trees in 30 years. In Scotland, forest workers increased tree-cover from 6% to 17% of its land area between 1960 and 2000.

CARTOON: JOHN MACKIE.

Tractors and ploughs could now move with so little ground pressure that they could plough peat which a person could not walk on without sinking. You would recognise this achievement if you tried to walk over soft peat!

Many technical developments were progressing unseen on the remote hills. They encouraged trees to grow into the first high-altitude forests since oak and pine woodlands were finally subsumed into peat in prehistory.

Adding fertiliser had not previously been an important part of forestry. Researchers in soil chemistry, however, found that nitrogen added to heathery ground encouraged Sitka spruce in check to start growing. It also became standard practice for phosphorus and potassium to be added to young plantations in less heathery areas, to encourage early growth. Hand fertilising individual saplings using buckets and spoons was replaced by helicopters with dangling crucibles, which could treat a wide area in one sweep.

Sitka spruce afforestation and growth on deep peat had become both possible and mechanised.

Developments were summarised by famous writer on woodlands Herbert Edlin:

> The Commission's forests are being established on the poorer type of land, most of which has to be ploughed. Giant ploughs pulled by powerful tractors make it possible to plant successfully on sour, peaty soil which not so long ago would have been regarded as useless. The young trees are planted in the over-turned ridge which is thrown up by the ploughs as they make the deep drains required to dry out the moorland. Thus the tree not only occupies a well aerated site, but its roots, in developing, have the benefit of the fresh humus formed by the vegetation which decays in the centre of the turf 'sandwich'. Phosphatic fertilisers are nearly always applied to the early growth.
>
> Edlin 1970

Head foresters tested the new machinery, tweaked ploughs to suit their particular ground and dealt with everyday practical difficulties. Sitka spruce was overwhelmingly the majority tree on peat; small amounts of Norway spruce

FIGURE 12.10. Thinned Sitka spruce plantation growing well in the Brecon Beacons, South Wales.

PHOTO: DAVID ELLERBY, 2005.

and Lodgepole pine were planted, but other conifers did not grow satisfactorily in acid peat.

An Aberdeenshire tractor driver told me how much he enjoyed ploughing steep hillsides or peaty plateaux throughout Scotland. He was proud to produce a clean pattern of parallel ridges and furrows. Sometimes he and other drivers had to jump out of their cabs as their tractors sank beneath them into the peat… Tractors with winches or buckets were used for 'de-bogging' machinery sunk down in the soft peat. Forestry jobs could be varied!

Squad workers had more mundane work; they spent their working hours wading through heather, bracken, or across drains, to where they planted millions of little Sitka spruce on ridges upturned by the tractor drivers:

> There was no shelter from existing woods or forests, or from any topographical feature. The squad just had to carry on whatever the weather… You walked down the line of the planting ridge with your spade and sprigs, planting little sprigs every six feet. When you got to the end of the line of the ridge you climbed down into and across the furrow, then on to the top of the next parallel ridge. Then you went back down the line, parallel to the previous ridge which you'd just planted. That went on all day for eight hours, except when it was time for your piece.
>
> George Young's comments on tree-planting in 1965, in Tittensor 2009

Forester Donald Fraser described life in 1965 on the remote Rannoch Moor, where 5400 ha (13,500 acres) of sporting estates bought by the Forestry Commission had been planted:

> Featureless moorland had to be surveyed and mapped. Fencelines planned considering such features as sheep passes and deer sanctuaries. A network of roads surveyed and tree species selected to suit the variable soils… This was also a time of great urgency when planting targets had to be achieved. It would be a few years later before we could afford to ease up, take stock and with the benefit of hindsight give more consideration to the multiplicity of important features our woodlands could offer; especially to the visiting public.
>
> Fraser 2004

FIGURE 12.11. Nowadays, felling and harvesting Sitka spruce on peat uses a track of brash to spread the weight of heavy forestry vehicles, East Burnhead, Ayrshire.

PHOTO: ANDY TITTENSOR, 2013.

Numbers planted

Between 1947 and 1949, the Forestry Commission carried out a census of woodlands in the UK. It had 225,000 ha (555,987 acres) of conifer high forest, of which 27% were Sitka spruce. The next census, 1965–1967, showed that this had increased to 36% of its total 500,000 ha (1,235,526 acres) of conifer high forest.

Sitka spruce made up 88% of new planting by the state between 1957 and 1988.

By 1967, in new compartments of five forests making up the Argyll Forest Park, the Forestry Commission had put in more Sitka spruce than any other tree. Sitka spruce made up 50%, Norway spruce 32%, pines 8%, larches 6%, others 4%.

During 1970 to 1971, Lodgepole pine at 17% followed Sitka spruce at 55% of all trees planted in state forests in the UK. Between 1981 and 1989, Sitka spruce contributed 65% of the area of state afforestation and 63% of private sector planting.

Difficulties

Fire is a forester's worst nightmare and a hazard particularly in dry spring weather. Gwydyr Forest suffered excessively, with several plantations burnt down. The hazard lessened when steam locomotives were replaced by diesel engines on the country's railways in the late 1960s.

The archipelagos of north west Europe frequently receive high winds off the Atlantic. Like peat, wind determined where trees were planted or not: it was difficult to determine how future winds would affect a plantation. Sitka spruce was particularly susceptible to wind in earlier forests because shallow ploughing produced superficial tree roots unable to hold a tree in wind.

Foresters working in the west of Britain and Ireland constantly feared gales. Inverliever Forest suffered from windblow. In Northern Ireland, three gales and a hurricane between 1953 and 1961 blew down 22.5% of its forest area. The higher the trees the more likely they were to topple or break in gales. Shorter rotations got round this problem by harvesting trees before they reached the vulnerable heights of 18–24 m (60–80 ft).

Taking an intermediate harvest of thinnings would also increase the likelihood of windblow. After much discussion, 'No-Thin' management regimes were started in those Sitka spruce forests liable to windblow. Closure of a pulp mill at Corpach, Fort William in 1980 coincidentally shut down the market for Sitka spruce thinnings in north-west Scotland.

Provenance: it matters where seeds come from

'You're not from round here are you? Where do you come from?' are questions I get asked often. People recognise that I am not of local provenance, not 'born and bred' in the region where I now live.

Provenance describes the appearance and adaptations of a plant or animal to life in its place of origin.

For Sitka spruce, its natural distribution and place of origin is within a narrow band 80–150 km (50–93 miles) wide, along a latitudinal corridor of 3600 km (2237 miles) adjoining the west coast of North America. The species has evolved in a variety of climates and soils along that coast. Its genetic variation has allowed it to evolve adaptations to local ecological regimes.

In the Alaskan north, it lives in a short growing-season, with late spring and early autumn frosts, cool summer temperatures but very long, summer day-length. In the Oregon and Californian south of its range, the summer day-length is shorter but warmer, while frosts are few; the atmosphere can be foggy. In the British Columbia middle of its range, its environment is in-between, but coastal rainfall and wind are high. It frequently grows right next to the sea, often on sand, but also on mineral-rich, flood-plain soils.

FIGURE 12.12. Timber lorry removing large-diameter logs from the stack at East Burnhead, Ayrshire.

PHOTO: ANDY TITTENSOR, 2013.

FIGURE 12.13. In 1922 American Forestry Association member Charles Lathrop (right) gave representatives of France and Britain a gift of 35 million conifer seeds to afforest their countries.

PHOTO RETRIEVED WITH ACKNOWLEDGEMENT FROM: HTTP://WWW.HISTORICALSTOCKPHOTOS.COM/DETAILS/PHOTO/999_JUSSERAND_JJ_BRODERICK_CHARLES_LATHROP.HTML.

For more than a century, the geographical and ecological source of seed did not particularly concern foresters in the UK. Sitka spruce was grown in many nurseries and plantations – a percentage of losses was considered normal and acceptable. Seed sources were sometimes known, but generally the specific site, or even region of their collection were not considered. Ecology and genetics, now vital to forestry, were then new-ish sciences.

Early seed: Washington and Oregon

We know from explorers' notebooks that nineteenth-century Sitka spruce grown in Scottish arboreta was mainly seed collected from the Washington coast. David Douglas's 1831 seeds came from the Columbia River area between Washington and Oregon; Jeffrey and Lobb's seeds came from further south, along the Californian coast. The pioneering nineteenth-century plantations of Sitka spruce in Scotland used seeds of unknown origin, except for Corrour. Luckily, Sir John Stirling-Maxwell used seeds from Haida Gwaii and Alaska. He discovered that they suited his high, snowy, peaty ground. But how did he acquire them? Seeds collected by Messrs. Douglas, Lobb and Jeffrey from Washington, Oregon and California would *not* have suited his conditions: we know that now from provenance experiments.

The new Forestry Commission of the 1920s really needed a large amount or regular supply of tree seed of the species it intended to plant in its new forests.

FIGURE 12.14. De-winged spruce seed of Haida Gwaii provenance.

PHOTO: FORESTRY COMMISSION. © CROWN COPYRIGHT.

Conifers of the North American coastal rainforests were high on its list, though Sitka spruce was probably not the first.

In 1921, it requested help from Canada in obtaining bulk seed. The Canadian government very helpfully sent 1361 kg (3000 lb) of Sitka spruce seed of crop year 1921 from Haida Gwaii, plus 1814 kg (4000 lb) of Douglas fir seed and 45 kg (100 lb) of Western hemlock seed, which was received by the Forestry Commission in 1922. Another 577 kg (1273 lb) of Sitka spruce seed and other North American trees arrived from the Canadian and USA governments in 1924.

That was a good start!

1361 kg (3000 lb) of Sitka spruce seeds *is quite a lot of seeds*.

Each female cone of Sitka spruce produces on average 80–100 seeds when cleaned; there are 500 or more cones per tree, so each tree could produce 50,000 or more seeds.

One thousand cleaned seeds weigh 2.08–2.51 g, so there are about 400,000 cleaned seeds per kilogram. Thus 1361 kg would consist of 400,000 × 1361 seeds, or 544,400,000 (over 544 million) seeds.

To acquire 544 million seeds, at 50,000 seeds per tree, the Canadians needed to collect seeds from about 10,881 trees, depending upon the seeds' weights.

How was this done? And over what time period?

When British Columbia became part of the Dominion of Canada in 1871, the Dominion undertook to construct a railway to connect the seaboard of BC with the railways system of Canada. In return, BC agreed to transfer to the Dominion a belt of land 32 km (20 miles) wide on each side of the new line. Canada's Forest Branch managed the land between 1884 and 1930 after which the BC Forest Service took control. Railroad companies taking lines to Vancouver agreed to log trees from a corridor of forested land 6 km (4 miles) wide, building the tracks along its centre. They were expected to log trees, clear the ground and collect the cones from the grounded canopies in autumn. Douglas fir cones, particularly, were collected from these railroad routes.

In 1922, Canada's Forest Branch built a seed extractory on the water front at New Westminster, Vancouver, to process tree seeds requested by the British Forestry Commission. It was replaced by a private seed plant near Vancouver in 1936.

With hundreds of millions of seeds being imported to the UK in the 1920s, an awful lot of afforestation was intended! If about 4940 Sitka spruce treelets per ha (2000 per acre) were to be dug in, then about 121,406 ha (300,000 acres) of afforestation were intended. Luckily, Sitka spruce seeds remain viable, and the 1938 kg (3000 + 1273 lb) supplied in 1922 and 1924 would have lasted ten years even though the germination percentage reduces over time.

Other seed sources: Haida Gwaii

It happened that in the 1920s, there was a major programme of commercial logging of rainforest and Sitka spruce on Haida Gwaii, including coastal areas of Moresby Island. This archipelago consists of six large and about 150 small islands, home of Haida people. The islands, which rise to 1219 m (4000 ft), are mid-way way along the latitudinal distribution of Sitka spruce at 52–54° N.

Early cone collections from Haida Gwaii were shipped by sea to Vancouver and processed there or at Washington seed extraction plants. But in 1954, a seed 'extractory' was built on Haida Gwaii which functioned until 1968. It was important to extract the seeds from their cones, so they could be dried (David Douglas had done it over an open fire each evening), cleaned of other bits and pieces, weighed, parcelled and transported afar.

A British forester, A. D. Hopkinson, visited Haida Gwaii in the 1930s. He was *very* impressed with the Sitka spruce trees he saw. *So* impressed, and so keen on the fact that the islands are at a similar latitude to Edinburgh (Haida Gwaii 53° N, Edinburgh 56° N), that from then on Sitka spruce seed for state and private forests in the UK and Irish Free State was collected from there.

In due course, the proportion of imported seed was gradually reduced until domestic plantations could provide some home-collected cones for seed. However, it was not until the 1950s that home-produced seed of Sitka spruce reached over 10% of what was used. About 20.5% of Sitka spruce seed in the UK is now home-produced; the Republic of Ireland prefers to import its seed.

Home-grown seed is 'son of Haida Gwaii': so is it still Haida Gwaii provenance or has a new provenance evolved in the UK?

Collecting cones and seeds

Seed buyers had little control over collection sites. Where logging took place to some extent determined its source. Early twentieth-century sites tended to be accessible to logging and extraction by boat, or near cities like Seattle or Vancouver with rail transport.

Mature, coning Sitka spruce in rainforests are so very tall and on Haida Gwaii, cones were collected from the grounded canopies of felled trees. Later cone collectors also paid Haida people to climb the tallest trees on Graham Island. They would climb into the canopy and cut through the narrowing trunk, dropping the whole top of the tree to the ground, where cones could be easily removed. This practice was later stopped because it left many 'topless' Sitka spruce trees – which are still there.

Cone collectors sometimes did the tree-climbing themselves: David Douglas climbed up trees. So did Alan Fletcher of the Forestry Commission who was sent on several occasions between 1970 and 1979 to collect cones for provenance studies of several tree species. He needed to collect from many places which he could easily locate on maps – this was before GPS was available. He and his colleagues used squirrels such as the American red squirrel (*Tamiasciurus hudsonicus*) and Douglas' squirrel (*T. douglasii*) to collect cones too.

No, they didn't train a 'Squirrel Nutkin' to rush up a tree and throw down cones at them! Rainforest squirrels make winter caches of cones under fallen logs, amongst tree roots or on the open ground. Alan Fletcher described how a squirrel collects one cone from a tree, strips off the scales and eats the seeds to test their ripeness. If it is satisfied that the seeds are ripe and edible, the squirrel re-climbs the tree and cuts through many cone stalks with its very sharp front incisors. It may do this to many or all the cones on a tree. When they have dropped to the ground, the squirrel returns downwards and collects them into a cache on the ground, usually using the same place as earlier years (like always putting your flour away in the same bin).

Alan Fletcher told me that humans start collecting Sitka Spruce cones when they see these 'Nosheries' on the ground, as they know the seeds are ripe. He said 1959, 1966 and 1970 (when he was there) were bumper years for cones and seeds. Just as well: *5500 lb (2495 kg)* of seeds were needed that time. I'll leave you to calculate how many trees and cones were collected in order to send home 5500 lb of Sitka spruce seeds!

Finally, David Douglas and modern collectors also used shot guns to bring down cone-bearing branches of Sitka spruce and other conifers.

Back home, estate foresters bought their Sitka spruce seed from private seed merchants, nurseries or the Forestry Commission. Some of the experimenting estates which had grown Sitka spruce early in the century, probably collected seed from their own mature trees in good cone years.

Sitka spruce grows in North America from latitude 39° 20' N to 61° 00' N. The latitude of the UK (including the Shetland Islands) is from 50° N to 61°

FIGURE 12.15. In North America, the red squirrel *Tamiasciurus hudsonicus* leaves a midden of cone fragments below Sitka spruce trees, indicating the seeds are ripe.

PHOTO: ANDY TITTENSOR, ALASKA, 2014.

N and would fit nicely into the natural geographic range of Sitka spruce. One might then expect this tree to suit all high rainfall regions of Britain and Ireland if its other ecological requirements are met.

Whether the Haida Gwaii were chosen because their latitude and climate are a good match for much of the UK, or whether it was for ease of cone collection from felled trees, is not known. Whichever ... it was very, very, very fortunate that the Forestry Commission bought its seed from that locality for many years. Although Sitka spruce had grown well in parts of the UK already, the use of Haida Gwaii seed during early Forestry Commission years was exceptionally good fortune!

This is because the adaptations of the Haida Gwaii provenance fit it to many localities in the UK.

FIGURE 12.16. A red squirrel at a seed plant in British Columbia knows which side its bread is buttered.

PHOTO: ALAN FLETCHER, 1970S.

But not everywhere.

It was not until foresters realised that seed from other parts of its range might fit parts of the UK better, that provenance of Sitka spruce was really considered. And that needed research.

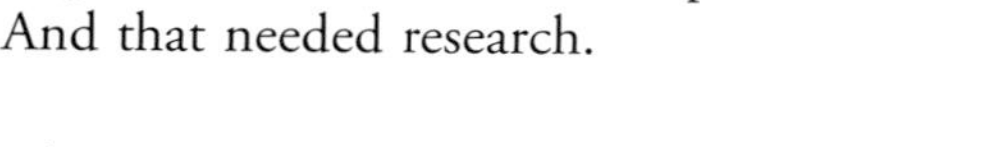

The need for good planting stock on peat

In 1920, forester John McEwan had to accept whatever plants and seeds he could obtain for starting off Monaughty Forest.

But in the later twentieth century when foresters had to grow Sitka spruce in extreme conditions, good planting stock, vigorous but not too large, was essential. On sites suited to Sitka spruce's ecology, genetic make-up is less important than on ecologically strenuous sites where provenance is particularly important.

'Good' seed of known origin was really needed for each and every forest. Luckily, a few people, who had been researching the subject, could provide advice.

What is 'good seed'?

In the wild, 'good' seed is that which is fertile, grows into a mature plant where it lands, reproduces and continues the species. It is adapted to the prevailing conditions, but with sufficient genetic variation to produce offspring which between them can grow in the range of environments where they happen to land.

For forestry, 'good' seed is that which germinates, produces a strong plant in a couple of years, transplants well and grows to maturity on the forest site.

It should be as required by the planter: plenty of foliage for photosynthesis, satisfactory growth, tall straight trunk without knots, timber of the required grade, out-competes weeds and re-grows after browsing … not much to ask for!

Research

The nursery of Johannes Rafn & Son, Copenhagen had been operating since 1887. Early on its Sitka spruce seed was imported from Oregon and Grays Harbor locality, Washington. It was an astute business and carried out its own research. The 1937–38 catalogue was keen to persuade its customers to buy Sitka spruce seed from Haida Gwaii, emphasising the firm's recent observations and highlighting the qualities of the seedlings:

> In 1921 we imported for the first time seed of Sitka Spruce from the Queen Charlotte Islands in Canada, which race seems noticeably suitable for the cold temperate climate of Northern Europe. From experiments and observations made in the Hjortsö and Paludans Nurseries here in Denmark, we have ascertained that the Sitka Spruce from the Queen Charlotte Islands begins to grow about a week or two earlier than the Sitka Spruce from the state of Wash. And without being damaged in any way by frost. In addition to this it also finishes its growth much earlier (about 20 to 30 days) than the Sitka Spruce from Wash.; a fact of great value and importance. Two years seedlings of the Canadian Sitka exhibit a much more vigorous growth than the Wash.-Sitka, and attain a height of 40 centimetres, and withstand the winter easily without shelter or covering and under conditions when the Wash,-Sitka would invariably be destroyed by frost… We highly recommend this Canadian Sitka spruce from the **Queen Charlotte Islands** for localities with insular climates like Great Britain and Denmark.
>
> Johannes Rafn & Son 1937–38

For a tree such as the Sitka spruce, the time when it starts to grow in spring and finishes growth in autumn is important: how much (height or volume) does it grow each year? And how much timber does it produces in its lifetime? J. Rafn & Son observed that growth varied according to the origin of a tree's seed. To a forester, useful provenance features might be: rate of growth, canopy shape, root spread, trunk height, quality of wood, frequency of coning, fertility of seeds and whether the tree is broken by heavy snow or falls in wind.

J. Rafn's work was invaluable and encouraged foresters on a journey to consider provenance, genetics and breeding. Research on provenance has been summarised in the excellent booklet by Messrs. Samuel, Fletcher and Lines (2007) following two research projects in the 1960s and 1970s.

To start with, there was considerable planning because Sitka spruce seed was to be collected from 67 places within its natural range and in Europe. Cones were collected in good coning years during a short period, to be certain seed was ripe and not blown away. Collections were carefully labelled to prevent muddling … 400 cones from about twenty trees at each site were collected.

Seeds were sent to countries which grow Sitka spruce commercially, so that each could carry out the same trials in its individual climate and environmental conditions. Experiments analysed, for instance, the start and finish dates of yearly growth, abundance of lammas shoots, vigour, wood density, susceptibility to frost in spring and autumn.

For the UK, researchers found two features which help decide the appropriate provenance of seed to use in different regions: 1) date when growth stops in autumn, and 2) rate of growth.

Seed from south of Puget Sound, Washington, produces trees which grow more vigorously and longer each summer. As growth is quick, their wood is less dense. Such trees are susceptible to autumn frosts in the UK, where frosts start sooner than where their ancestors evolved in North America.

Sitka spruce trees from further-north-seed that is, from the British Columbia coast north to the Skeena River and Alaska, grow more slowly and are less susceptible to autumn frosts because their ancestors evolved in shorter growing-seasons. Their slow growth means they produce dense, strong wood. However, foresters have to grow them for five to ten years longer to reach the same size as (younger) Washington Sitka spruce.

Seed from the Haida Gwaii, especially from Graham Island, produces trees with excellent growth in many UK conditions. So, depending where in the UK is one's forest, and what sort of spruce timber is required, the provenance is chosen from one of three regions within the tree's natural range: 1) California, Oregon and Washington; 2) British Columbia coast and Haida Gwaii; 3) Alaska.

It does not always work: Pat Armstrong, Head Forester at Whitelee, Ayrshire from 1975 to 1987, told me that he requested Haida Gwaii provenance, but was often sent transplants from Washington seed.

Researchers produced a map for practical foresters to show how different provenances of seed should be used: trees from Alaskan seed show superior growth in the extreme north and west; trees from Washington and Oregon seed do best in Wales, south-west and north-west England; trees from Haida Gwaii seed grow very well in most of Scotland.

Although Alaskan seed was planted early on in Kielder and Galloway Forests, very little is used nowadays in Britain and Ireland. Between 1921 and 1955, the only Alaska seed imported by the Forestry Commission was 28 g (1 oz) from Ketchikan and the same weight from Juneau, both Crop Year 1927; and 113 g (4 oz) from Sitka and 907 g (2 lb) from elsewhere in Alaska of Crop Year 1929.

But in 1951, Sir John Stirling-Maxwell commented favourably on his Alaska seed:

> We are beginning to find marked differences in frost resistance between various batches of Sitka. The best plants we have ever had were raised from Alaska seed, distributed by the Arnold Arboretum. They are planted out at the east end of Clach Mhor Plantation… We have in recent years only used seed from Queen Charlotte islands.

Stirling-Maxwell 1929/1951

FIGURE 12.17. Labelled bags of cones stored in a Washington warehouse ready for export to the UK.

PHOTO: ALAN FLETCHER, 1970.

Arnold Arboretum is the botanic garden of Harvard University. C. S. Sargent, its first director, corresponded with Sir John Stirling-Maxwell in 1906 and 1907, sending him seeds, but no written record of Sitka spruce survives.

Provenance Studies in the Republic of Ireland

In 1928 John McEwan worked in the forestry department of the Irish Free State but obtained his seed and plants from Scotland:

> When I arrived in Ireland the method was to get samples of plants from England and Scotland, particularly from Scotland, as they didn't grow their own plants in Ireland. We selected the best varieties and prices from these samples, and placed orders for tens of thousands of plants, big orders, because planting was going full out in Ireland with its twenty-seven forests. Most of our plants came from Scotland, where climate and conditions were similar, and I knew the Scots nurserymen very well and knew where the best plants could be got. The plants we bought in were lined out in the nurseries attached to each forest. In Scotland the Forestry Commission had its own distributing nurseries, and grew their own, and it paid dividends.
>
> McEwan 1998 but recorded in 1984

The Republic of Ireland participated in the same two post-war trials as the UK – but the results were different. It was the southerly, not the Haida Gwaii provenances that most often grew best. With an equable climate and few frosts in the Republic of Ireland, provenances from Washington and Oregon grew fastest in many areas and types of ground. British Columbia provenances grew least (Haida Gwaii were taken as the standard). The wood of the southern-seed-trees was 10% denser than the more northerly ones, so even though fast-growing, their timber could be used for construction.

Policy was changed to take note: 70% of imported Sitka spruce seed is now of Washington and Oregon provenance, 30% from Haida Gwaii – this is what is used above 300 m (984 ft) altitude (forests in Ireland are planted up to 450 m (1476 ft) altitude.

Seed for forests in the Republic of Ireland today comes mainly from the UK and Denmark. The state agency 'Coillte' set up its own seed-processing plant in 1988, so that some home-grown seed could be supplied.

Even while provenance trials were ongoing in Britain and Ireland, other researchers had started work on *tree-breeding* to select trees with the particular characteristics most desired by foresters and industry. More on tree-breeding in Chapter Seventeen.

Kielder: largest Sitka spruce forest in Europe

North from the Roman Hadrian's Wall to the border of England with Scotland is a recent landscape of contiguous spruce forests. Together, as 'Kielder Forest', they illustrate evolving management typical of spruce forests during the twentieth century.

The Cheviot Hills which cross the border are formed of hard, igneous rocks in which deep valleys carry tributaries of the River Coquet. Further south Kielder Burn and smaller waterways form shallow-valley tributaries of the North Tyne and River Rede. Underlying sedimentary rocks form rounded hills, but faults and igneous dykes stick up as high ridges or 'fells'. The ice ages left a thick layer of impervious Boulder Clays, so the soils are acid podsols, gleys, peaty gleys and peat, though basic soils developed on limestone outcrops.

With average annual rainfall of 1300 mm (51 in) many 'meres' and 'mires' (pools and peat) developed on the poorly drained long-treeless plateaux. Peat

FIGURE 12.18. The altitudinal limit of planted Sitka spruce in Coire Shalloch, Glen Clova, Angus at 610 m (1968 ft).

PHOTO: ANDY TITTENSOR, 2015.

started developing during prehistory into 'Border Mires' which are now over 5 m (16 ft) deep and considered to have high nature conservation value. The pastoral moorland vegetation of heather and sedge families, of *Molinia caerulea* or *Nardus stricta* grassland, or of bracken varies according to soil and terrain.

Peel Fell reaches to 602 m (1975 ft), Sighty Crag to 518 m (1699 ft) and Larriston Fells to 512 m (1680 ft), all remote from access or settlement. The landscape was described as: '...the vast, rolling, empty, desolate moorland of the early twentieth century...' (Wilson and Leathart 1982).

However it was not empty: it was a huge area of moorland and bog, grazed by thousands of Blackface, Swaledale and Cheviot sheep, black cattle and Red grouse. Coal, iron and limestone had been mined. The landscape had long been of interest to naturalists, ramblers and historians, quite apart from the farm families and shepherds who made a living there.

Civil servant Roy Robinson 'discovered' this high border country in 1910 and put it on his list of future acquisition sites. Groves of native deciduous trees hung onto inaccessible crags outwith the reach of ravenous sheep; Scots pine and Norway spruce had been planted around the eighteenth-century Kielder Castle; there was a small Sitka spruce plantation on the Scottish side. These were the only woodlands in 1910.

It was in 1925 that the Forestry Commission bought Scottish estate land for Newcastleton Forest. Struggling estates on the English side were bought in 1925 and 1932 to a total of 33,994 ha (84,000 acres); more was gradually added until the total acquisition reached 75,649 ha (186,932 acres) in 1962. The 1932

FIGURE 12.19. On the Corrour Estate, Inverness-shire, Sitka spruce has been grown as a commercial crop since the early twentieth century.

PHOTO: JOHN SUTHERLAND, 2014.

acquisition was Kielder Estate so this was taken as the overall name for the new forests of the Borders.

It is not recorded what happened to the sheep or to the families of farm tenants.

The first plantings in 1926 were with Norway spruce and Scots pine. As we now know, Norway spruce was not the most appropriate species for moorland, but it thrived on the better-drained, more fertile and sheltered areas. Scots pine was planted on drier, rocky places especially if heathery; Lodgepole pine was put on very deep peat with heather; Japanese larch, planted in small amounts on bracken areas, was not always successful. Sitka spruce, easy to establish, fast-growing and high-yielding was planted on peat plateaux, higher slopes and sites with infertile soils – and eventually over most of the annual planting area. Based on the experience of Corrour, 'turf planting' by hand was used.

During the 1930s, the Ministry of Labour persuaded unemployed people from declining mining villages and shipyards of north-east England to work in Kielder Forest. They lived in residential camps on Forestry Commission land, now under Kielder Water reservoir.

The early forest was predominantly Sitka spruce from Alaska and British Columbia seed. Small trees were grown in millions in nearby nurseries and planted at a high density of 10,000 per ha (4000 per acre) in even-aged compartments of 60–100 ha (148–247 acres). With all this work going on, several villages or part villages were built in the 1950s for forest employees and their families.

By 1962, 54, 633 ha (135,000 acres) of moorland had been planted at altitudes of 200–400 m (656–1312 ft) but up to 600 m (1969 ft) on Peel Fell. Even the border mires were planted with Sitka spruce, to the dismay of conservationists.

Kielder and Gwydyr Forests contrasted in the variety of trees planted. Kielder was intended to be a 'Sitka spruce tree factory' for producing cellulose so its amenity was not of prime importance. However, the terrain of Gwydyr Forest meant that production of large quantities of marketable timber was doubtful. Its cultural history, tourism, varied geology and scenery ensured that many species of conifer and broadleaf trees were planted from the start.

At Kielder it was intended to follow the usual pattern of Sitka spruce silviculture, with a rotation of 50–60 years. Intermediate yields from thinnings would be removed after about 25 years, providing income and encouragement for the remaining trees to grow more. The first thinnings of Sitka spruce were taken in 1948: men did it with handsaws and axes, using horses to pull the thinnings to the ride edge.

However, during the mid-1960s, it was found that the foresters' second-worst nightmare – wind blow – had affected many thinned compartments. Foresters eventually decided that 85% of Kielder Forest was not suited to thinning in the prevailing windy conditions. Clear-felling at an earlier age was the only option, which meant the harvest of (smaller) trees would be most suitable for the pulp market. It would also mean less work. The rotation length was of necessity reduced to about 30 years.

FIGURE 12.20. Harvesting close-packed Sitka spruce from Kielder Forest.

PHOTO: RUTH TITTENSOR, 2012.

FIGURE 12.21. Poles of unthinned Sitka spruce ready for collection from Kielder Forest.

PHOTO: RUTH TITTENSOR, 2012.

FIGURE 12.22. Sitka spruce plantation at Stálpastaðir, Skorradalur, West Iceland. The farm was deserted in 1946, planting trees started in 1952 and eventually the new forest held over 600,000 trees of many species.

PHOTO: EDDA SIGURDÍS ODDSDÓTTIR HTTPS://WWW.FLICKR.COM/PHOTOS/E_SIGURDIS/.

During the 1960s when early-planted compartments had reached the allotted age, the first ever timber crops from Kielder Forest were felled. Restocking followed, this time with more varied compartment sizes and a greater proportion of Sitka spruce: it had shown itself to be the tree which grew best at Kielder.

Between 1975 and 1981, the firm Northumbrian Water paid for the Kielder valley to be dammed for a reservoir and hydro-electric scheme to provide for the industrialised north-east of England. This drowned some forest, farms, a railway and a school.

By the 1980s, Kielder had grown to 20,234 ha (50,000 acres); 200 million trees had been planted of which 61% were Sitka spruce. Increased tourism suggested that the public appreciated the more varied landscape of interspersed moorland and forest.

The recreation potential of Kielder Forest was deliberately enhanced in conjunction with forestry. A Forest Park was designated and numerous activities became possible: camping, walking, mountain biking, orienteering, driving along scenic routes, watching wildlife, vintage car gatherings, even a marathon.

The reservoir's industrial use has latterly declined, but there are many springs which maintain the water level, making it a superb site for water recreation. 27.5 miles (44.3 km) of Kielder Water shoreline are available for quiet recreation. Some 300,000 people visit Kielder Forest and Water annually for day visits or longer holidays.

In 2011, the land-holding was 59,187 ha (146,254 acres), with 13% kept unplanted for nature conservation or amenity. Of the 87% forest, 75% of its area is coniferous, 4% broadleaves and 21% left open as fire breaks and internal access. Kielder Forest produces over 20% of England's annual timber harvest, bringing in £12 million.

Silvicultural management is changing from the earlier days of tree farming. Kielder Forest must still produce a big volume of timber for home use, but its remit now encompasses many uses in a varied landscape.

Chapter Thirteen will discuss how Sitka spruce fitted into changing forests at the end of the twentieth century.

CHAPTER THIRTEEN

Perceptions

> The Forestry Commission has been, and still is, subjected to severe criticism on account of its policy of growing huge areas of conifers in serried ranks of uniform species. To this complaint has been added that of the rambler, who finds it laborious, if not impossible, to walk across the deeply furrowed surface of a modern plantation.
>
> L. A. Harvey and D. St Leger-Gordon 1953

> The forests are dark places, dark in every sense of the word. They shouldn't be there of course; Sitka spruce is not indigenous to Wales. In the 1920s the land-purloining Forestry Commission cloaked the hills with a harvestable crop, a market force, that drove much of the native flora and fauna away. Nothing likes to live inside these forests. Pennant, the eighteenth-century antiquarian, wrote that the noblest oaks in all of Wales grew on the slopes above Gwydyr. But they were all chopped down and someone thought conifers would be a good idea; but they weren't, even if I am a little partial to the eerie light they cast.
>
> Judy Corbett 2004

> The Welsh are familiar with Sitka Spruce!
>
> David Ellerby, Planning Forester, Forestry Commission, Llandovery 2013

Awful conifers

As a young mother in the 1970s, I enjoyed a weekly morning as tutor for adults of varied age and background on non-vocational courses.

In our discussions on ecology and the local countryside, one of the frequent questions was 'Why do they plant those awful conifers?' I organised meetings with foresters on nearby estates so students could find out, 'Why?' Being in dry southern England the main commercial trees were Corsican pine, Hybrid larch and beech.

Forty years later, I am asked a similar question in Sitka spruce Scotland, something like 'They do leave a mess when they clear-fell; they're not going to replant with those awful conifers are they?'

How did these perceptions arise?

Changes in society from 1960

Forestry and farming contributed to rapid changes in UK society in the later twentieth century. The country became more affluent despite war debts: car ownership increased, the first motorway opened, people travelled more and further, took holidays and not just to the nearest seaside resort. More people could see what was going on in more of the countryside. Class structure, which had wobbled during the war, cracked further. It was the first time of mass-education of women to advanced levels (I was one of them).

Television ownership increased, contributing to greater breadth of knowledge. War had stimulated science and invention: my school was affected by aircraft development at nearby Filton, by Peter Scott at the Wildfowl Trust and by *Sputnik*.

During the Second World War, the country fed itself by rationing and enforced changes in farming. Afterwards, people were ready for more and varied food, which required reshaping the countryside again. More pasture was ploughed and cultivated, fields got larger, more artificial fertilisers, pesticides and herbicides were available and mechanisation increased. Export markets looked to the European Common Market rather than distant colonies.

During the 1970s the efficient farmers of Europe and the UK produced a food surplus: I remember 'butter mountains' and 'wine lakes'. A generation grew up which never knew hunger (people aged below 60 now). There was enough time to consider other features of the countryside than just necessities.

To have their cake and eat it

The population wanted plenty of products of farm and forest but did not care for the corresponding changes in appearance and ecological structure. There seemed a diminution and loss of something important to their felt British-ness, their 'traditional' countryside, their 'Eden', their 'green and pleasant land'. Bitter

FIGURE 13.1. A typical disliked landscape: looking over Harwood Forest, Northumberland, at Sitka spruce in its preferred, misty environment.

PHOTO: LUKAS SIEBICKE 2013.

FIGURE 13.2. Disliked straight boundary segregating Sitka spruce on Forrest Estate, Kirkcudbrightshire, from adjoining moorland.

PHOTO: RUTH TITTENSOR, 2011.

arguments developed between farming, forestry and their associated industries on one hand – and amenity, access and nature conservation lobbies, the general public and 'greens' on the other hand.

In a hard-hitting condemnation of nature conservationists, agriculturalist A. S. Thomas wrote:

> As mentioned before, it is only when men feel secure and well fed that they can philosophize about the importance of wild life; the sight of an uncommon animal or plant will thrill those who do not have to worry unduly about necessities.
>
> Thomas 1975

The uplands changed as significantly as the lowlands of Britain, but more imperceptibly. Moorlands and heathlands decreased when their lower bounds were enclosed into fields. Heather and rush vegetation gave way to grasslands when 'headage payments' raised stocking levels for sheep and cattle. Payments for hill drainage caused 'rough' vegetation to give way to fine grasses. Under severe pressure of hoofs and hiking-boots, peat eroded especially in upland England.

There was public disquiet at conifer plantations in the English Lake District as early as the 1930s. It was in the 1970s and 1980s that larger post-war forests had grown sufficiently to be visible as 'green blankets' on the hills. Taxpayers realised the size of their financial contribution to these disliked new landscapes, complained and criticised. Pre-war forests looked varied because foresters were trialling many tree species. Then, people had admired forests for their manifestation of renewal amongst First World War devastation. They now saw uniformity of colour and species on the hills.

The later twentieth century was also a time when the science of ecology gained public prominence. It took decades for ecologists to be accepted as professionals and not as eccentric bearded men in sandals (I was affected by that). London and Edinburgh Universities were the first to teach courses in ecology from the 1960s. At Edinburgh, ecology was taught as part of the forestry degree, which had positive ramifications in due course.

The BBC Natural History Unit opened in 1957, opening eyes to the ecological diversity of the world. Volunteers of many backgrounds now administered conservation organisations to 'protect' the countryside, its flora and fauna (but not necessarily its people).

The practice of 'nature conservation' was taken on board by an emotional public as an ethic and politic but not as a biological science. I tried to explain that nature conservation is a *method of applied ecology*, equivalent to farming, forestry or field sports, where ecological science is applied to the management of different groups of species according to peoples' wishes.

For about three centuries, observant naturalists had recorded wild flora and fauna so that considerable information about distribution and abundance of wild species had been amassed. After the Second World War, the Biological Records Centre was set up to co-ordinate species-mapping countrywide and digitally. Apart from the traditional tally of game bags and raptors, we don't have such detailed data from earlier centuries, but new research in environmental archaeology showed that plant and animal abundance had changed radically in previous millennia also.

Amenity and access groups became more vociferous about rural changes they disliked. Books, articles and celebrities discussed issues, added heat, argument, understanding or mis-understanding of ecology, nature conservation, farming and forestry.

Marion Shoard's 1980 book *The Theft of the Countryside*, Steve Tompkin's books about 'afforestation and the battle for the hills', as well as media apoplexy all added fuel to the fire. The frenzy against farming and forestry in the later twentieth century failed to consider the anthropogenic (human-induced) changes to the landscape of the previous *six millennia*. I reckoned that the countryside had been 'stolen' so many times before, that this particular 'theft' was merely the change we happened to be living through.

Few queried whether acceptable biological diversity and beauty might develop in the *longer term* in new landscapes. Onlookers continued to eat as much as they wished and buy new homes which covered yet more soil with concrete – while criticising the landscape. For an assessment of twentieth-century nature conservation try John Sheail's (1998) book *Nature Conservation in Britain: The Formative Years*.

Another profession with similar background expanded parallel with ecology. After concentrating on monuments, tools and pottery in the lowlands, archaeology gradually developed a whole-landscape approach. It was not until the later twentieth century, however, that the uplands received their share of research. Then archaeologists found that the uplands had been used as intensively as the lowlands even in prehistory.

Archaeology and ecology both developed from nineteenth-century cooperation between amateurs and professionals and both took time to be recognised as academic subjects. Cooperation between the two disciplines has increased our understanding of human interaction with early environments substantially.

Relevance to Sitka spruce

What has all this to do with Sitka spruce?

Lots!

The decades after 1960 were the years when afforestation was most rapid and extensive, when Sitka spruce was planted in hundreds of millions across the north and west of the UK and in the Republic of Ireland. Nature conservationists are still very unhappy about this enormous landscape change:

> Forest cover increased by a quarter between 1870 and 1947, but almost doubled between 1948 and 1995... The balance also shifted from native broad leaves to exotic conifers as expressed by the present dominance of Sitka Spruce.
>
> Bunce *et al.* 2012

Of all the afforestation carried out by the Forestry Commission between 1957 and 1988, 88% was Sitka spruce, a huge percentage. In Kielder Forest alone, 200 million Sitka spruce trees had been planted by 1982.

In the Irish Free State (later the Republic of Ireland), Sitka spruce contributed 22%, 11% and 33% to the *numbers* of trees planted by the state during the three decades to 1955; this increased to 46%, 58% and 60% in the following three decades ending in 1985. Sitka spruce contributed a total of 18% by *area* to all Irish state forests by 1948, but by 1970 its contribution had increased significantly to 70% by area.

A contribution above 60% was maintained in the Republic of Ireland until the end of the twentieth century, when the state held 346,535 ha (856,306 acres) of forest land. Private estates tended to plant a lower proportion of Sitka spruce than did Coillte, the Irish Forestry Board. In the UK, the rate of afforestation and of the contribution of Sitka spruce had, in contrast, declined by the century end.

By 2001, afforestation in the Republic of Ireland had increased the country's tree cover to 9% by area from 1% in 1900. UK afforestation had increased tree-cover from 5% in 1900 to 12% by century-end. Scotland had done most of this: its woodland area increased from 4% in 1900 to 6% in 1960, to an amazing 17% by 2000 – an increase of 11% in only 40 years!

Ireland and Britain experienced similar disquiet at conifer afforestation and at Sitka spruce in particular. In Britain, it was seen as 'blanket afforestation' of attractive, wild uplands. During the 1970s the main commercial species planted by the Forestry Commission were also conifers: Lodgepole pine, Norway spruce

FIGURE 13.3. Afforested moorland: Sitka spruce hillside with clear-fell and re-stocked areas, Kielder Forest.

PHOTO: GEORGE GATE, 1990, COURTESY FORESTRY COMMISSION.

and larches. On sand dunes it was Corsican pine, on Thetford Chase in East Anglia and in parts of the eastern Grampian Mountains it was mainly Scots pine. Conifer plantations are 'traditional' only in those places such as Atholl and Gwydyr estates which have planted them for nearly four centuries.

In Northern Ireland the situation was a bit different: public disquiet at conifer afforestation was more to do with the 'Irish land question' and the feeling of Anglo-dominance than with perceived changes in amenity and ecology (Noel Melanaphy pers. comm.).

Famous Welsh poet R. S. Thomas expressed unhappiness at afforestation with spruce as another aspect of decline in Welsh-ness:

Afforestation

It's a population of trees
Colonising the old
Haunts of men; I prefer,
Listening to their talk.
The bare language of grass
To what the woods say,
Standing in black crowds
Under the stars at night
Or in the sun's way.
The grass feeds the sheep;
The sheep give the wool
For warm clothing, but these – ?
I see the cheap times
Against which they grow:
Thin houses for dupes,
Pages of pale trash,
A world that has gone sour
With spruce. Cut them down,
They won't take the weight
Of any of the strong bodies
For which the wind sighs.

R. S. Thomas

Thomas, R. S. *The Bread of Truth*, 1963.

There was undoubtedly a certain amount of poor silvicultural management of Sitka spruce plantations. Poorly-planned ploughing produced stripes up-and-down hillsides; badly-designed drains which led directly into waterways took sediment with them; silvicultural methods tended to be similar countrywide; soils were ploughed whether necessary or not.

Relationships with nature and countryside

Upland landscapes are held in romantic affection especially by ramblers, climbers, holiday-makers and artists. In 2004, a Welsh slate miner wondered to me why people came to climb the mountains which had been his everyday, toiling, work-place. Northumbrian and Scottish hill-farmers have affection for their cold, wet uplands: is it romantic, poetic, practical, familiarity or because they were born there?

Keith Thomas (1983) and Chris Smout (2003) assessed perceptions of the natural world in England and Scotland. Ian Simmons (2003) analysed recent evidence from natural sciences, archaeology and history to present a record of moorlands in England and Wales since prehistory. He then told an alternative story through non-scientific perceptions of moorlands. It showed that human imagination has usually downplayed human influences, in poetry, paintings, photographs and writings, interpreting moorlands as 'open', 'natural' and 'wild', representing vegetation, fauna and landscapes descended from, or similar to, pre-human wilderness. Even ecologists and conservationists took this wilderness concept on board.

My own dawning realisation that 'natural' and 'semi-natural' explanations of ecosystems did not fit what I saw, is shrouded in memory's mists. But the oak woodlands clothing the mountainsides beside Loch Lomond developed my ideas. These woodlands had been described as 'natural' or 'native scrub'. This terminology was difficult to accommodate with archive information which showed they had

FIGURE 13.4. Bleak? Useless? Romantic? Wild? Moorland: looking north from Pugaven Brig, Ayrshire.

PHOTO: MORAG GILLETT, 1948.

been managed intensively for at least three centuries. Looking at the trees on the high-rainfall, high-midge, loch-side slopes gave clues too.

I then tramped around lowland landscapes, searched archives and worked with archaeologists. I gradually realised that heathland, coppiced-woodlands, hedges, flower-hay-meadows and chalk downland were prehistoric and historic remnants of what we humans do with and to the landscape. When we use the landscape – or just exist – we alter ecosystems. I could see new ecosystems developing before my eyes: airport grassland, silage fields, motorway-verge heathland, coal-tips, urban landscapes of trees favoured by councils – and – amazing gravel ecosystems in millions of gardens.

FIGURE 13.5. People's 'mind's eye view' of romantic, open moorland in Scotland: looking up the Gameshope Burn, Tweedsmuir, Scottish Borders.

PHOTO: ANDY TITTENSOR, 2012.

Fall-out from large-scale afforestation

For decades, public and nature conservationists were at constant loggerheads with foresters and farmers – indeed with anyone not concerned with 'protecting' the 'environment'. I was once asked by an adult student how I (an ecologist), could possibly be married to someone involved in farming! It showed the depth of feeling about countryside change. The fact that *he* was also an ecologist would have been too difficult to explain!

In the Republic of Ireland, state forests were legally barred to the public until Tolleymore Forest Park was inaugurated in 1955. In the UK, forests were free to public access, but the sheer prickliness of young Sitka spruce and new sheep-proof fencing meant people felt denied their use of what had been easily-accessible short-turf, sheep-grazed, tree-less land.

The Forestry Commissioners inaugurated the 24,281 ha (60,000 acre) Argyll National Forest Park in 1935 and the smaller Snowdonia National Forest Park in 1937. People used them for walking, climbing, pony-trekking and camping especially after the last war. The lowland New Forest has a variety of conifer and broadleaf plantations, and old trees (including conifers) planted for ornament, which, together with pony lawns, make a pleasing mosaic which hundreds of thousands of visitors actually enjoy.

New upland conifer forests were often perceived as un-landscaped by amenity and conservation organisations. This perception triggered groups to lobby for forests which felt more 'natural'. However, as an elderly forester recently explained, they were compelled to plant every bit of acquired land with trees, right up to the often-straight ownership boundaries (in the same way that farmers grow wheat or turnips right up to their field boundaries). Foresters' role, after all, was to produce as much timber as fast as possible.

In 1963, the Forestry Commission appointed a landscape architect to advise on improving the visual appeal of forests. Sylvia Crowe found that dealing with forests – which change seasonally and as they grow – was unlike her more usual static power stations and pylons.

Sitka spruce forests still produce comment by older people who remember the uplands of their youth. Famous biologist Derek Ratcliffe commented recently on southern Scotland:

> ***A Transformation of Scene***
> No one can travel through Galloway and the Borders without becoming aware of the huge extent of plantations of coniferous trees in the hill country and even, locally, in the lowlands. Most of them have been created since 1940, and have transformed the character of these previously almost treeless uplands.
>
> Ratcliffe 2007

The statutory Nature Conservancy Council became so concerned by 'coniferisation' of moorlands that, in 1986, it produced a solid book explaining how sites for nature conservation are evaluated, its own policies and practice. Although new conifer forests might develop their own wildlife potential, it was

FIGURE 13.6. Strangely-landscaped, hard-edged, Sitka spruce shelterbelts in the Ettrick Valley, Scottish Borders.

PHOTO: RUTH TITTENSOR, 2012.

the value of the *open ground ecosystems they replaced*, which was of concern for conservation. Leaving fragments of (for instance) peat bog near a new forest was not sufficient: a functioning bog needed its complete hydrological system undamaged by nearby ploughing and draining.

Users and onlookers were unhappy with inaccessible Sitka spruce forests yet were overwhelmingly certain that they barely supported wildlife. There were few ecological studies to support these assertions until ornithologists noted an increase in song birds, raptors and small mammals when small conifers grew into thickets on ploughed moorland. The forest squad of Whitelee, Ayrshire, observed that it took about ten years for moorland species to disappear and for forest-edge and forest plants and animals to colonise.

Public and conservationists were in a hurry. Conservationists wanted moorlands to be left treeless and 'natural'; public wanted bluebells, primroses and tawny owls in the new plantations right away. Woodland nature conservation is based on the dogma that 'ancient woodlands' are more valuable than new or young woodlands. Perhaps time is needed for flora and fauna to come together into (as yet unknown) Sitka spruce ecosystems?

Although practical foresters were delighted with the technologies which allowed them to afforest even raised bogs and quaking peat uplands, nature conservationists shuddered with horror when they found such 'special' sites ploughed and tree-planted. Planting forests over sites of potential nature conservation value, such as the Border Mires of Kielder, caused dismay.

Nature conservationists were frustrated by foresters' unwillingness to abandon the task given them by governments. Foresters thought they were doing as policy-makers had required of them, which they were, but not what the public wished for.

Coniferisation of existing deciduous woodlands, often privately-owned, was highlighted by Marion Shoard, a complainant of this era:

> But those that escape the bulldozer do not necessarily survive unscathed. Instead, many of them are being surreptitiously transformed into a dark and dismal travesty of their former selves. Leaving only a thin outer screen of deciduous trees to deceive the passer-by, landowners are scooping out the guts of our native woodlands. And our native broadleaved species such as oak, ash, beech and lime, are being replaced with serried ranks of conifers, whose permanent canopy stifles undergrowth and changes woodlands once alive with flowers and birds into gloomy timber factories.
>
> Shoard 1980

But she must have needed the products of those 'gloomy timber factories' to have her book published?

Her sentiments still echo through nature conservation today. In summer 2013, removal of 'non-native conifers' from the 'damaged ancient woodland' Fingals Wood on upland Dartmoor, Devon by two conservation organisations was heralded by the national press and TV as saving it from further 'destruction'.

How foresters felt

John McEwan's 1984 attitude to the role of Sitka spruce was typical of working foresters (Monaughty was afforested starting 1920):

> If I were planting today in Monaughty, I would put 90% of it under Sitka spruce, instead of the three or four species we planted then… The other 10% might be used for growing trees which might break up a little what people describe as blank stretches of dark-coloured packed conifer forests. I have no objection to them because I know the value of these forests and the financial returns which are made from these dense acres of dark green… If forestry is going to be a commercial enterprise and communities are going to be formed depending on forestry, on a forest industry, then sitka spruce has to be planted in bulk.
>
> McEwan 1998, recorded in 1984

Antagonism – by the media, statutory and voluntary nature conservation organisations and the public towards upland afforestation and coniferisation of lowland broadleaved woodlands – continued unabated. It affected the forestry profession. Senior forestry officer Gordon Cowie told me what it felt like to be constantly at the receiving end of these negative attitudes:

> My first awareness of anything other than isolated opposition to forestry was when the planting was taking place in the so-called Flow Country in Sutherland. Now we were encouraged initially by the Nature Conservancy people back in the Sixties. This was seen as something that would add to the diversity of the area. And then

> of course there was the realisation of what was happening to the site. I'm not apologising for it, but, we went into it without any negative connotations... It was seen as a good thing.
>
> And then there was the opposition, the initial reaction was 'Oh they don't know what they're talking about. These people from the cities'.
>
> But over a few years, we began to realise, no, we've got to be careful. But that was when the damage was done. And, there was a general perception that big, large-scale planting wasn't a good thing. But that would have been in the late Seventies, early Eighties.
>
> We began to develop a feeling of being under attack obviously. And, there was this, I suppose anybody in this situation, feeling from, the attitude that we were well-respected and what we were doing was appreciated, to suddenly a negative attitude, took some getting used to. And some, some ... we got over it. But it's taken a lot of adaptation. And a lot of necessary changes, changes that might not have come about if it hadn't been for these pressures.
>
> Cowie in Tittensor 2009

Political developments

After the Second World War, governments and their civil servants realised that a state enterprise alone could not fulfil the forestry expansion required and that the private sector should have more encouragement.

Although private landowners had contributed their assets to two world wars, it was not until 1967 that forestry grants and tax benefits were sufficient to persuade them to replant and expand their woodlands substantially. Research had shown that adding fertilisers to Sitka spruce plantations enhanced growth and marketable volume, so greater returns would accrue. Private tree planting increased.

Accountants spotted that the same tax benefits could be used by clients of a different kind. In 1979 the firm Fountain Forestry started to buy cheap land in the Caithness Flow Country. It felt that afforestation on the Flows would attract wealthy investors who could claim planting grants and tax relief against their other income. Fountain Forestry assisted pension firms, insurance companies and wealthy, previously non-landowning individuals to undertake financially-expedient afforestation in the Flow Country.

Grants for forestry on private land were controlled and overseen by the Forestry Commission to ensure government policy was fulfilled. The public could comment on grant-aid for private afforestation. However, tax benefits far outweighed the grants and were confidential, so government and public lost influence over the rate or methods of afforestation.

Amenity and nature conservation groups were furious that the uplands were being altered via the tax system, were lost to sheep-farming and lost to them and their own perceived needs.

'The Flows' were regarded by ecologists as natural, peat-mire ecosystems. They were breeding places for large populations of wildfowl and raptors as well as rare species like Black-throated diver (*Gavia arctica*). Historian John Sheail later wrote about the Flows:

FIGURE 13.7. Peat pools in the Flow Country, Sutherland.

PHOTO: ANDY TITTENSOR, 2015.

> This actively-growing blanket bog was the largest in the world. There was no larger primeval ecosystem in the UK. The pools and surface patterning (the 'flows') supported a specialised range of mosses and vascular plants. They accommodated some 66% of the breeding populations of greenshank, 35% of dunlin, and 17% of golden plover in the European Community. Afforestation already affected some 15 per cent of the Flow Country, which once covered some 401,000 hectares. Only eight of the 41 hydrological systems remained free of some form of planting. Most of the planting had occurred since the 1981 [*Wildlife and Countryside*] Act. If continued, it would represent 'the most massive loss of important wildlife habitat in Britain since the second world war'.
>
> Sheail 1998

During the 1980s, thousands of miles of drains were cut in the Flows' peat and 4000 sq km (1500 sq miles) were planted with Sitka spruce and Lodgepole pine, together or as monocultures.

The ability to plant and grow trees on this appalling, or wonderful, peat mire (depending on one's point of view) was possible because long-term experiments had been carried out since 1925 at Inchnacardoch Forest near Fort Augustus. Research was deliberately carried out on 'deep peat flats and basins interspersed between irregular rock covered ridge and hillock landforms; a matrix of generally low nutrient status peatland and infertile organo-mineral soils'.

Inchnacardoch researchers studied drainage, planting techniques and application of different N, P, and K fertilisers on the varied soils. They had

planted many conifer and a few broadleaf species but particularly Lodgepole pine and Sitka spruce. Researchers analysed trees' growth rate as single-species or in mixtures, their root development and the effects of exposure or shelter.

It was discovered that Sitka spruce would almost double in yield if large amounts of fertiliser (N, P, K) were added just as the canopy was closing, compared with no fertiliser. Another technique was to plant Lodgepole pine of Alaska provenance as a 'nurse' to Sitka spruce: it would grow more quickly to start with, uptake nutrients and dry out the peat somewhat. The Lodgepole pine was then shaded out by the Sitka spruce; when it died and decomposed, its nutrients were available for Sitka spruce to grow. For more detail on the Inchnacardoch experiments, consult Zehetmayr (1954).

There was no assessment of the long-term *ecological effects* of peat afforestation on flora and fauna at Inchnacardoch Forest. However, by the time its research was applied to afforesting the Flows, ecologists' attention was awakened to potential effects. Nature

FIGURE 13.8. Tree-planting on peat ridges in the Flow Country, Caithness.

PHOTO: FORESTRY COMMISSION. © CROWN COPYRIGHT.

FIGURE 13.9. Commercial plantations of Sitka spruce and Lodgepole pine in the Flow Country, with wettening moorland of the RSPB Forsinard Flows Nature Reserve in foreground.

PHOTO: ANDY TITTENSOR, 2015.

conservationists saw the technical achievements of peatland afforestation as irreversible damage to the peat mires and severe habitat loss for birds.

Residents knew that unemployment in Caithness and Sutherland was one of the highest in the country and that tree-planting had helped reduce it since the 1950s when the Forestry Commission started operating in the region. Infrastructure, sparse in many parts of rural Wales and Scotland, was sometimes built to accompany afforestation. Afforestation, then, could improve the social fabric and bring more cash into remote villages.

After a decade of tax-encouraged afforestation in this remote and poverty-stricken area, countrywide controversy about the organisation and practice of forestry came to a head.

The public and anti-afforestation organisations were infuriated that a forestry company and well-endowed individuals were increasing their wealth with assistance from taxes. There was bitter debate nationally until the then Chancellor of the Exchequer, Nigel Lawson, removed the offending or helpful tax system (depending, again, on one's point of view) in 1988. Afforestation of the Flow Country became an unattractive financial proposition from then on. David Foot's *Woods and People* (2010) gives a detailed analysis of the Flow Country saga.

Articulate romantics, visitors and new rural dwellers still complained that their favourite landscapes were being coniferised. The community of the town of Comrie, Perthshire felt threatened by afforestation proposed by Dunira Estate. In 1987, 1000 people signed a petition against it (the total population in Comrie recorded by the national census in 1991 was only 1439). Their local newsletter told them:

> The plan by Dunira Estate to plant Glen Lednock came to light in mid-February. Although the area proposed for planting *750 acres*, the boundary fence shown on the planting plan would enclose *over 1000 acres* of the glen in a high deer fence. By the time the District Council met in early March, a petition with 1,000 signatures opposing the forestry plan had been collected and opposition was also registered by a number of organisations including the British Council for Archaeology Scotland, Ramblers Association, Scottish Wild Land Group, Friends of the Earth, and the Association for the Protection of Rural Scotland. The Countryside Commission for Scotland have also expressed serious reservation…
>
> The whole future of our countryside is under intense debate. Forestry is a crucial part of that debate, and we must get it right: mistakes will stay with us for a very long time! It is estimated that forestry in Britain will be subsidised to the tune of about £50 million this year alone: a lot of money for an industry in which major developments are not subject to planning permission…
>
> Anon 1987

No forest was planted there in 1987.

And forestry was to change. The growing power of the so-called 'green' or 'ecology' lobby was feared by politicians. Government changed significantly the structure of British statutory forestry, nature conservation and amenity bodies. In Northern Ireland, sudden loss of tax benefits for conifer forestry had little effect because private investment in forestry was much less than that of the Forest Service, and private landowners planted twice as many broadleaves as conifers.

How was Sitka spruce affected by the tensions?

Although trees in pre-1988 plantations on the Flows continued to grow, Sitka spruce planting by the private sector suddenly declined there and countrywide. The government banned afforestation in those English uplands where only conifers would grow, places like the Lake District.

Large areas of the Flows were subsequently scheduled under nature conservation legislation. The Royal Society for the Protection of Birds bought 10,117 ha (25,000 acres) of both afforested and unplanted peatland. A long-term project to reverse afforestation is in progress: plantations on its Forsinard Flows reserve have been felled and drains blocked to raise the water-table and to encourage peat growth once more.

'Historic Living Landscapes' and forestry

From pre-war times, ecologists had thought of deciduous woodlands, heathlands, peat bogs, moorlands, mountain-tops and salt-marsh as 'natural' or 'semi-natural' ecosystems barely affected by humans. Cereal fields, urban parks and conifer forests were, in contrast, obviously artificial, being composed of introduced, planted species.

Working with archaeologists on lowland estates changed how I perceived the landscape. All the ecosystems – hedges, field margins, chalk heath and downland, coppices, wooded commons and commercial woodlands – had a human history as well as an ecological structure and dynamic. Human influences were often visible if you learned what to look for; archives usually confirmed them. The more I studied ecosystem history the more I found myself at odds with the paradigms of the past.

It was the same for every bit of countryside I analysed. Archaeologists and ecologists could see that upland ecosystems such as heather moorland were not as natural as assumed either.

Archaeologists specialising in plants and animals of past environments reinforced the message with substantial evidence of prehistoric interaction between humans and landscapes. They improved our understanding of how humans used or interacted with wild species. They found evidence of long-term tree management in peat mires like the Somerset Levels; they studied organic remains in estuary muds of the River Severn and Humber; from calcareous ground where mollusc shells survive; from plants used to construct early buildings; from the timbers of Iron Age crannogs; from shell middens; from pollen analysis of cores through peat bogs.

Archaeologists named their sites 'Historic Landscapes'; I named the living

FIGURE 13.10. An earth dam in an old forest drain holds back water and raises the water-table, Forsinard Flows Nature Reserve, Sutherland.

PHOTO: ANDY TITTENSOR, 2015.

FIGURE 13.11. A plastic dam also holds back water in a previous forest drain.

PHOTO: ANDY TITTENSOR, 2015.

mantle 'Historic Living Landscapes'. Putting the two together gives 'Cultural Landscapes', a good name for the environment we live in.

Evidence built up showing humans have interacted with a large proportion of the landscape since prehistory. Sites such as Star Carr and the island of Oronsay showed that even in the Middle Stone Age, humans altered ecology and landscape sufficiently for it to be visible in the archaeological record (see Chapter 6). Humans have also responded to ecological and landscape changes by changing their behaviour, their food and feeding, by migrating, by altering their surroundings – and still do.

Is anything sacred? How 'natural' are the Caithness Flows? This depends, of course, on what is meant by 'natural'.

Currently, evidence suggests that growth of blanket peat on upland Britain was a post-glacial process of soil development, of climate change and of human communities using woodlands.

Peat and lakes on the Thrumster Estate near Wick are gradually giving up information about long periods of prehistoric Caithness landscapes from 10,000 BP. At Oliclett, peat grew and covered a Mesolithic flint-working site. By the Bronze Age, about 3500 BP, people farmed near the upper peat margins. But they affected the landscape of the time by slowing down peat growth onto their ard-cultivated fields on adjoining mineral soils. Peat growth increased on the lower slopes and valleys when Bronze Age cultivation uphill caused water run-off downhill. Peat continued to increase on the lower ground in the next millennium but cereals were still cultivated on the mineral soils of the hills.

So is the landscape which has come down to us in such localities 'natural' or not?

It seems to me that 'natural', native' and 'wild' are not helpful when thinking about landscape change, ecological processes and nature conservation in my home country. Are they relevant to how we use places like the peat ecosystems of the Caithness Flows?

How do Sitka spruce forests differ from past landscape and ecological change? Can they fit into the picture?

We can view Sitka spruce plantations as the latest intervention in landscapes with which humans have had relationships.

Plantations of tree species from outwith the British Islands have been added by humans to the landscape for at least two millennia (see Chapter Seven).

A view of Sitka spruce forests as just the latest in a long line of 'historic living landscapes' or 'cultural landscapes' helps put them in perspective. Are they, then, recent equivalents of ancient cereal fields, heathlands and flower meadows? Are they comparable with contemporary new ecosystems such as motorway verges and vineyards?

This novel concept alters the existing, idealised, romantic notion of 'wild', 'natural' uplands like Dartmoor and the Lake District which are famous in British history and story-telling. A land with which humans have always related and participated tells a quite different story of the culture, dynamics and meaning of the uplands and their ecosystems.

If, as archaeologists tell us, woodland in the UK has been managed and planted for six millennia what is so different about modern conifer forests including Sitka spruce? What is this 'natural' phenomenon, this naturalness', 'natural woodland', which conservation takes as its base-line?

Perhaps our view of Sitka spruce afforestation of the uplands is coloured by its modern-ness. It was not, and is not, of the romantic past; it has happened before our very eyes, it was large-scale and rapid; people had little control over it and onlookers had no part in it. Its uses in our everyday lives are not obvious and it is looked after by people with whom we are unfamiliar as they are not local. It is

FIGURE 13.12. A variety of ages and sizes of Sitka spruce, but hard edges to this part of Ettrick Forest, Scottish Borders.

PHOTO: SYD HOUSE, 2015.

FIGURE 13.13. Attempts to improve amenity with edges of broadleaved trees are not always successful: Whitelee Forest, Renfrewshire.

PHOTO: RUTH TITTENSOR, 2005.

to phone numbers or email addresses that we have to turn for information, not to someone in a local pub or office.

False feeling of plenty

Years of complaints about Sitka spruce afforestation showed little understanding that it was an industry providing us with everyday items under extremely difficult ecological and working conditions. With full stomachs, the public could concern itself with more than just the food and timber needed from the countryside. Amenity, nature conservation and cultural values of farm and forestry land were of greater interest than production of necessities – a contrast to the time when Mosquito aircraft were being built!

The feeling of plenty was false: timber imports still provided over 70% of the country's needs. Despite more people visiting, and more urban dwellers 'escaping' to the countryside, the link between a working countryside and goods received appears not to be recognised: that their newspapers, kitchen units, house walls and roofs might be made of Sitka spruce.

Wider role for forestry

Government gave the Forestry Commission a wider role after 1988. Archaeology, education, landscape amenity, nature conservation, public health, recreation and waterway management were to be included along with timber production. Foresters in the private sector were given similar wide responsibilities enforced through the grant system.

Towards a more holistic forestry

Foresters with training in ecological science had first graduated in 1970, widening the professional expertise of the profession. Decades of scientific research on Sitka spruce were not wasted: research results were still required to grow it in forests which would be, in future, more complex in composition, structure, space and time. Sitka spruce was affected by two main changes, first its amount and second, its pattern of planting and management.

Between 1991 and 1999 it continued to be planted on both state and private land but the rate of planting declined. In state forests the 1980s high of 63% Sitka spruce by area reduced to 52% between 1991 and 1999: this included afforestation and restocking. On private land it contributed 63% in the 1980s but declined to 26% in the 1990s. In the Republic of Ireland, Sitka spruce maintained its high contribution to the century end.

Forests now grew a much greater variety of species. A Broadleaves Policy stopped further conversion of broadleaved to conifer woodland and encouraged foresters to plant them more frequently.

In the state forests of south-east Wales including the Brecon Beacons and the

Wye Valley, a dozen conifers and ten broadleaved species are now grown. But Corsican pine, larches and Ash are no longer planted in Wales due to diseases. Grey squirrels damage broadleaved trees so severely that their end use is uncertain. Forester David Ellerby made it clear that Sitka spruce is still grown on 'Sitka spruce land' in this part of Wales because it is the primary product.

The second change affecting Sitka spruce was more fundamental, affecting basic silviculture: the structure and dynamics of forests.

Foresters now assessed planting land taking into account *very small-scale* mosaics in soil, aspect and windiness. They prepared long-term plans for whole forests or estates, as tools to ensure continuity of management. Sitka spruce was positioned more carefully according to local conditions. Michael Caughlin, Head Forester of the Drumlanrig Estate, Ayrshire, showed me with what care he has done this, so as to plant a variety of trees in ecologically suitable places and reduce windblow in susceptible sites. With the Estate in a wide valley containing the beautiful Drumlanrig Castle, the forests must also look attractive from any position and at any time – and produce a timber crop.

At harvest time, smaller blocks of Sitka spruce are generally felled, with both windblow and amenity in mind. A more diverse age-structure and smaller-scale pattern develops. After 25 years of reconstruction, Sitka spruce forests have developed a quite different appearance. Working with smaller blocks also means regular or 'sustainable' harvests. 'Sustainable forestry' may be less efficient economically and ergonomically, but is more acceptable to the public. Private landowners have less choice in this because less efficiency reduces their income.

Suppose harvesting coups are reduced and reduced again, they would eventually become small glades or single-tree openings. There would never be a sudden, big opening in the tree canopy. This 'continuous cover' silviculture has long been a common technique in Europe where much greater forest cover ensures a constant supply of mature trees.

The Republic of Ireland has tended to continue with traditional methods of even-aged coups which are clear-felled and replanted; this is still extremely successful in rapid-growth, good-quality Sitka spruce production.

Taking account of nature conservation meant that foresters now included even more complexity into their management. Forestry regimes were expected to encourage 'biodiversity' (number of species per unit area) and 'iconic' species such as the European red squirrel, Black grouse and Water vole (*Arvicola terrestris*). Grants to private landowners were dependent on fulfilling this requirement.

They were also expected to increase the 'naturalness' of their commercial forests, a measure of nature conservation value. This was done by increasing forests' structural complexity further. Foresters grew more tree species, produced forests with more varied age and size of trees and left 5–10% of the land unplanted. They were also required to remove conifers from scheduled broadleaved sites within forests.

These big changes happened to coincide with a sudden availability of cheap conifer timber from Russia and east Europe in the late 1980s. The bottom dropped out of the home conifer market. With home-grown conifers more

FIGURE 13.14. Several tree species together with unplanted land produce an attractive forested landscape, Loch Ard Forest, Stirlingshire.

PHOTO: RUTH TITTENSOR, 2012.

difficult to sell, state foresters could accede to conservation and landscape demands for naturalness and restoration of nature conservation sites.

Was it disheartening when home-grown Sitka spruce forests, on which years of research and management had been lavished, metamorphosed from primary producers into venues for recreation, landscape and nature conservation?

Changing forests

Gwydyr

The original objectives 70 years previously required planting many tree-species to enhance tourists' appreciation of the landscape. Changes at the end of the twentieth century were a matter of degree. Trials of continuous cover management were started in some Sitka spruce compartments. Many older trees were retained for their great amenity value.

Kielder

Cellulose production has broadened to 'multi-purpose use'. Kielder's 59,000 ha (156,254 acres) developed into a favoured honey-pot for outdoor pursuits from (inactive) forest drives to (very active) mountain-biking. Archaeology, local history, ecology and nature conservation were taken up by amateurs, professionals and forest staff.

Nevertheless, a continuous supply of marketable timber is still fundamental. Foresters have cleverly maintained the Sitka spruce timber supply while re-structuring the Forest and enhancing the landscape. Forest Design Plans, first used in the early 1990s, guide management. Sitka spruce is planted in smaller blocks and for shorter rotations but must still be grown fast and straight to suit markets and our needs.

At least 13% of the landholding remains as moorland; of the 87% with trees, access tracks make up 20%; broadleaves replace conifers on ancient woodland sites and around tourist sites. Trees are felled from Border Mires. Conifers are cleared from a strip 50 m (164 ft) either side of waterways allowing grasses and shrubs to form a lighter riparian zone. Forest edges are graded in height and density, suitable for Black grouse. Kielder Forest holds the largest population of Red squirrels in England (70% of total); researchers carry out ecological studies of squirrels and other fauna.

Tintern

Sitka spruce is still planted on the poorer soils and wetter areas, Douglas fir on the fertile soils. As an important historical and tourist area, conifers are replaced by new broadleaved woodland on feasible sites such as Chepstow Park.

Continuous cover harvesting has replaced clear-fell, but instead of removing *smaller* trees, *large* Sitka spruce trees are taken out during thinning operations, allowing more light to enter the plantations. Foresters hope that more oak and birch will regenerate and form a future broadleaved forest. After about a century, all the Sitka spruce will be gone, but without any clear-felling (David Ellerby pers. comm.).

A second bridge across the River Severn allows more visitors from the English Midlands and West Country to reach Tintern Forest easily. They should approve of what they see.

Clocaenog

Continuous cover forestry primarily for nature conservation is under way. Big Sitka spruces planted during the 1950s and 1960s are thinned individually every five years, leaving 'frame' trees to form the forest canopy structure. Small trees regenerate in profusion and these will be gradually thinned to grow into the next canopy. All the big frame trees will by then have been harvested.

Compartments of younger Sitka spruce trees are more easily converted to continuous cover because they are less liable to windblow. In Norway spruce compartments, small groups of trees are taken out to encourage regeneration, but it is Sitka spruce which has taken advantage. This is a difficulty as Red squirrels prefer seeds of Norway to Sitka spruce, so under-planting it must provide their future sustenance.

The current 61% of Sitka spruce is to be reduced and the 19% of Norway spruce will be increased. No large-seeded broadleaved trees will planted as their seeds suit Grey alias Eastern Gray squirrels (*Sciurus carolinensis*). More small-seeded trees and age variety will give a constant supply of Eurasian red squirrel food. Nature conservation also takes account of other species and habitats. Visitors are encouraged to walk, cycle and horse-ride, observe animals and admire the landscape.

Inverliever

Sitka spruce is the predominant species of this greatly-enlarged forest, but Douglas fir, European, Japanese and hybrid larches also contribute to a forest which is now very varied in species and ages. Beech and oak remaining from previous estate days regenerate naturally. Most work is done by contractors and Canadians on short-term visits. Harvested timber goes by boat to Ardrishaig and then to Northern Ireland or local outlets.

Corrour

The plantations initiated on the Corrour Estate in 1892 form one of the oldest forests of Pacific coast conifers in the UK. So far, Sitka spruce is the most successful of all the conifers: its proportion has increased to 76% of all trees, it regenerates the most profusely in wind-blow gaps and it grows to good sizes. The owner intends to maintain these old plantations and to increase their structural and species diversity without losing trees from wind-blow. Corrour Estate also has strictly commercial Sitka spruce plantations for its everyday bread-and-butter. The potential for windblow means that thinning them part-way is risky. When restocking a harvested area, other tree species are planted along with Sitka spruce and more ground is free of trees.

Achievements of the private sector

The end of government tax incentives in 1988 reduced the number of investors who started afforestation in the Flow Country. Landowners more generally, their foresters and consultants still wanted – needed – to grow forests of predominantly Sitka spruce as much-needed income from the land and for the other benefits trees bring to the rural scene.

New financial incentives brought increasingly onerous regulations for private forestry. Compliance with conditions for grants necessitated sophisticated planning, adherence to European regulations and prior nature conservation surveys. The UK Forest Standard and Woodland Assurance Schemes (for certified woodlands) meant acceding to yet more regulations concerning variety of tree species, biodiversity, seed and plant provenance and silvicultural methods – and rigorously adhering to them.

Private landowners with over 100 ha (247 acres) of woodland must produce a long-term Forest Plan approved by the Forestry Commission. Those with smaller acreages must produce a short management plan, even for felling approval. Obtaining a grant requires attention to restocking and harvesting methods, waterways, archaeology and conservation sites, designated wildlife, public access and climate change. Plans for deer and grey squirrel management must be drawn up as part of the overall plan. Neighbourly-ness and compliance with local authority road regulations are also required when huge lorries need access to uplift harvested timber to market.

Estates in scheduled landscapes or scenic areas must demonstrate that their changing forests will *always* fit into the surroundings. Then, in this crowded country, there are everyday operational constraints (hazards!) such as phone masts, gas pipelines, non-mains water supplies, wayleaves, derelict walls, buildings and cattle troughs. The square miles of uninhabited Flow Country must have seemed like heaven in comparison!

Although upland landowners continue to grow Sitka spruce for commercial profit, they are forced to reduce its proportion despite its ease of establishment, quick growth and strong markets, its resistance to diseases and deer. However, foresters are testing several other tree species for a possible commercial future in changing climates. Sitka spruce is likely to still suit the north and west of Britain and Republic of Ireland; it has new and expanding markets which will contribute to the economy long-term if sufficient is planted now to give a continued supply of this valuable raw material.

Its contribution to our everyday lives is discussed in Chapter Fourteen.

CHAPTER FOURTEEN

Contribution to Modern Societies

> The material in World War I was high-grade lumber from the Sitka spruce – an obscure tree found in the coastal rainforests of the Pacific Northwest. Exactly how the European Allies stumbled upon Sitka spruce as the best lumber for aircraft remains a mystery, but they asked for 100 million board feet of lumber from this species for their war effort. In 1916 – when the Europeans began buying Sitka spruce from Northwest mills – the tree yielded only 10% aircraft-grade lumber. The quota of 100 million board feet set for 1918 would require that one billion board feet of Sitka spruce be cut, and approximately 900 million board feet be rejected. Logging a minor species on this scale could obviously not be sustained.
>
> Ward Tonsfeldt 2013

> Probably man's highest engineering achievement in timber.
>
> Ralph Hare, Mosquito Senior Designer, de Havilland Aircraft Company, 1940s

There was little demand for spruce timber in European North America until the early twentieth century: spruces were seen as nuisance side-products of logging for more desirable conifers. East coast spruces supplied 85% of needs, mainly for musical instruments. On the west coast, Sitka spruce was logged along with more valuable Douglas fir, Western hemlock and Western red cedar, but was difficult to market and sold at very low prices.

FIGURE 14.1. Natasha Paremski performs at the 2014 Summer Music Festival, Sitka; the Steinway piano has a Sitka spruce soundboard, while Sitka spruce trees grow on the hill behind.

PHOTO: CHRISTINE DAVENPORT, COURTESY KAYLA BOETTCHER, DIRECTOR OF SITKA SUMMER MUSIC FESTIVAL.

By 1911 the Wright Flyer and Curtiss Pusher aircraft had been designed, built and flown in the USA. Yet a significant role for Sitka spruce in aircraft manufacture was first recognised elsewhere. During 1916, Europeans were buying large quantities of it from Pacific coast lumber mills, much to the mills' surprise. The supply could not match the increasing demand because the amount of aircraft-grade timber per tree was only 5–10%, leaving most of the tree to be sold cheaply.

Early in 1917, the British and Allies refused to accept USA exports of conifer timber except aircraft-grade Sitka spruce.

That did it!

In order to survive financially, west coast saw mills were forced to organise more logging of Sitka spruce. In fact, two mills sensibly decided to concentrate on only aircraft-grade Sitka spruce after spring 1917.

Perceptions adjusted and a nuisance tree became a valuable timber.

Crucial attributes of Sitka spruce in North America

Size

The *enormous height and diameter* to which Sitka spruce grew was noted by early explorers and later, by the timber industry. It is the tallest spruce species in the world and size alone ensures its timber is valuable. In the 1950s, mills in British Columbia could saw out planks up to 9.1 m (30 ft long) and 0.46 m (30 in) wide or narrower planks up to 12.2 m (40 ft) long.

Anatomy

The woody cells with fibres form a cylinder within the stem of a young tree. The cylinder grows thicker outwards every spring and summer, forming a hard column of wood fibres, no longer living. A thin layer of living, non-woody cells and bark is maintained round the outside of the cylinder.

The *long fibres* of Sitka spruce wood are 1–3 mm (0.04–0.12 in) long so when the timber is pulped, they bind together, making good quality paper.

Pale colour and smoothness make the wood attractive for food utensils and small joinery items. Although it can be worked to a smooth finish, it is normally used rough-sawn for outdoor products. The sapwood is pale cream, the heartwood a little darker with a pinkish tinge.

The *regular shape*, especially when close-grown, produces uniform timber with few knots, so that long lengths without faults can be milled.

Chemistry

The *high cellulose* but *low resin and tannin* content make it satisfactory for breaking down into wood pulp; minor constituents are extracted for industrial use. Boiling shoots in water produces a liquid high in vitamin C.

FIGURE 14.2. Icy Straits Lumber yard at Hoonah, Southeast Alaska which specialises in using local trees including Sitka spruce.

PHOTO: WES TYLER, 2002.

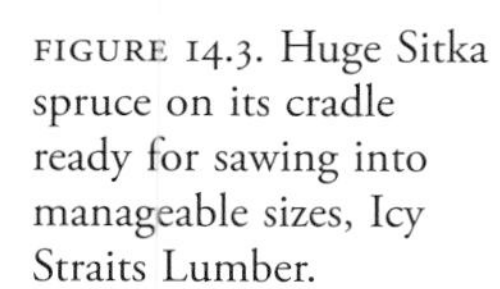

FIGURE 14.3. Huge Sitka spruce on its cradle ready for sawing into manageable sizes, Icy Straits Lumber.

PHOTO: WES TYLER, 2001.

Strength to weight ratio

Sitka spruce timber has the *highest strength to weight ratio* of any timber in the world.

It is usually straight-grained, resistant to mechanical pressure and has good shock-absorbing qualities. It is used to make ladders and other items which must be strong but light. Its strength and elasticity, as well as its availability in many grades and sizes mean Sitka spruce is an important construction timber.

Resonance

Slow-grown Sitka spruce from northern latitudes has close, regular, annual rings which make it an *excellent conductor of sound.* Its outstanding resonance, light weight and strength make a superior 'tonewood' for instrumental soundboards.

Prolific natural regeneration

Sitka spruce seeds germinate and grow easily onto open sites caused by windblow or landslip within the rainforest. Humans have provided more and larger open spaces (on logging sites) where it *regenerates naturally and prolifically.* So it is not usually necessary to collect seed and artificially restock logged areas. The resulting 'secondary' forests are rich in, if not dominated by, Sitka spruce of similar age so that they can be harvested efficiently.

Component of temperate rainforest

The ecological role of Sitka spruce in temperate rainforests has been discussed in Chapter Four. Its relative scarcity, unique role and traditional uses by First Nations give it immense cultural value.

Contributions to North American society

Driftwood

This is important to First Nations communities of the tree-less lengths of coastal Alaska. They have traditionally depended upon driftwood reaching them via five ocean currents or coming downstream in big rivers.

Red and White spruces are the main timbers arriving down-river onto the shores of Bristol Bay and northwards. The Aleutian Islands and south coast of the Alaska Peninsula, however, receive wood from the spruce-hemlock forests of Southeast Alaska. Many species arrive on their shores: Alaskan yellow cedar, cottonwoods (*Populus trichocarpa* and *P. balsamifera*), Sitka spruce, Western hemlock and Western red cedar.

In prehistory, driftwood was used for tools, utensils and weapons. Historically, people made snowshoes, dog sleds and house frames, as well as fuel. Sitka spruce driftwood was particularly important to Aleutian Islanders for the frames of their magnificent fast, light and seaworthy *kayaks* and for the *paddles,* as recorded by Russian explorers. Nowadays, less driftwood arrives than in living memory, but people still use it for drying racks and carving, fuel to smoke salmon and steam baths.

Ships and boats

Sailors in the past took conifers from coastal forests to replace their ships' *masts.* This involved felling a tall tree from near the shore, manoeuvring the trunk across the sea and then up onto the deck. Sitka spruce was also good for the *oars* and *paddles* of their small boats and for building *stairways* on the ships.

The framework of modern recreational boats is often constructed of Sitka spruce where strength and light weight are needed. The timber's light weight and ability to absorb shock endear it to builders of *racing boats*. *Spars* of sailing boats and *oars* of rowing boats are often made of Sitka spruce.

A *Polynesian boat* built of two big Sitka spruce was gifted by native Alaskans to the Hawaiians in 1913 to encourage traditional visits between Polynesian islands. Each Sitka spruce was more than 400 years old, 61 m (200 ft) high and 2.1 m (7 ft) basal diameter (Syd House pers. comm.).

Aircraft and gliders

Designers and builders of early aircraft looked for wood which would be lightweight but very strong. They started with ash and hickory (*Carya* sp.) but these were hard and rather heavy.

> There are two uses, however, for which Sitka spruce has no equal: the manufacture of wooden airplanes and stringed musical instruments. For early airplane fuselage and wing structures, the wood-of-choice was Sitka spruce because of its superior strength-to-weight ratio, its elasticity and ability to withstand sudden strain and shock, and the fact that it could be obtained in clear, straight-grained pieces of large size and uniform texture with few hidden defects.
>
> Mackovjak 2010

Sitka spruce was an important component of the frames of very early *American Aircraft* such as the Wright Flyer (1903), Curtiss Model D 'Pusher' (1911) and Boeing B&W Biplane (1915). Yet Sitka spruce grows neither in Ohio where the Wright Flyer was designed nor at Kitty Hawk, NC where it was first launched. The Wright Brothers developed their engineering techniques via bicycles and gliders, so they possibly learned about Sitka spruce wood that way.

Long lengths of straight-grained, defect-free timber were available from west coast rainforests. At that time, Sitka spruce had to be about 61 m (200 ft) tall and at least 1.5 m (5 ft) basal diameter to produce aircraft lumber. It had another enormous benefit for early war planes: the long, tough, fibres do not splinter when struck by bullets!

Although most planes are constructed of other materials nowadays, Sitka spruce is still used for *home-built, aerobatic planes* and modern *gliders*.

Pulp

Paper pulp was first made from Sitka spruce about 1907 by the 'sulphite process', where the wood is broken down chemically. The pulp bleaches easily and makes high-grade *newspaper, printing and bond papers, strong, 'kraft' paper* and *fibreboard*. High grade chemical pulp is used to manufacture *rayon* and some *plastics*. In the alternative 'mechanical process', the wood is ground down physically, giving low-grade newsprint and paper. Pulp is one of the most important uses of Sitka spruce in modern North America.

FIGURE 14.4 *(left)*. Sitka spruce log being sawn into slabs for large table tops; Icy Straits Lumber.

PHOTO: WES TYLER, 2001.

FIGURE 14.5 *(right)*. Framework of Sitka spruce for a dwelling constructed by a local doctor; Icy Straits Lumber.

PHOTO: WES TYLER, 2003.

Construction timber

During the second half of the eighteenth and first half of the nineteenth centuries, when Russia held Alaska, residents of Baranof Island built *ships* and constructed *homes*, *factories* and *St Michael's Cathedral* with Sitka spruce and other timber from nearby rainforest.

Sitka spruce is nowadays sawn into lumber of many sizes and grades and converted into construction timber: *roof and floor trusses*, *timber-framed homes*, *internal joinery* and *decorative woodwork*. After the Second World War, Japan urgently needed new housing and Alaska exported Sitka spruce sawlogs there. Japan and other Asian countries continue to buy most of Alaska's old-growth Sitka spruce.

Tonewood

This describes the timber used to make *soundboards* for stringed instruments such as guitars, pianos and harps. A soundboard is the wood against which the strings resonate via a bridge, increasing the volume and producing harmonic 'overtones'.

From the late eighteenth century, early keyboard instruments like harpsichords and spinets were made in eastern North America and their *soundboards* were usually cut from White pine (*Pinus strobus*) in New England. But in the following century, spruces were more commonly used for soundboards: Engelmann spruce gives a strong note with mellow overtones; Red spruce produces a clear note and a rich overtone; a Sitka spruce soundboard gives a very strong basic note with sweet and warm overtones.

Red and White spruces from the Adirondack Mountains and northern New England were the favoured species for piano soundboards in nineteenth-century eastern USA. However, during the twentieth century, Sitka spruce became the favoured tonewood even in New England – 4000 km (2486 miles) from its

FIGURE 14.6 *(above left)*. Brent Cole removing a round of wood from Sitka spruce timber for guitar tops.

PHOTO: ALASKA SPECIALTY WOODS, CRAIG, ALASKA, 2012.

FIGURE 14.7 *(above right)*. Annette Cole splitting the round into wedges for guitar soundboards.

PHOTO: ALASKA SPECIALTY WOODS, 2012.

FIGURE 14.8 *(below left)*. Splitting a Sitka spruce round for piano soundboards.

PHOTO: ALASKA SPECIALTY WOODS, 2012.

FIGURE 14.9 *(below right)*. Syd House, UK Sitka spruce specialist, with his 1960s Martin 6-string acoustic guitar with Sitka spruce soundboard.

PHOTO: COURTESY SYD HOUSE.

homeland! Perhaps large spruce trees in local forests had all been used. Harpsichord makers in Boston used Sitka spruce for soundboards after the Second World War. John Koster, a specialist in instrumental woods, remembers visiting a timber yard in the 1960s to choose Sitka spruce for his own harpsichord soundboards.

Sitka spruce is now losing favour amongst American harpsichord-makers, some of whom import Norway spruce from Europe. However, it is still used by American piano-makers and is the most favoured tonewood for soundboards of high-quality guitars. But heavy logging of Sitka spruce in Alaska has by now produced a situation where suitable old-growth Sitka spruce trees are difficult to locate and really require conservation.

Trees for tonewood are chosen carefully as large-girthed, slow-grown individuals with narrow, even rings but without knots and blemishes: 8–22 rings per inch (3–9 per cm) in a cross-section are favoured. Potential tonewood trees are harvested during winter, sawn into really thick, circular slabs which are weathered or artificially dried. Each round slab is then cut quarter-wise for piano soundboards.

Lutz spruce (*Picea* x *lutzii*), the hybrid between White and Sitka spruces, is now gaining favour for guitar soundboards in view of the scarcity of big, slow-grown Sitka spruce.

The Manhattan firm of Steinway & Sons uses only Sitka spruce for its piano soundboards. Its web site states:

> ***The Patented Steinway Diaphragmatic Soundboard***
> *Superior Materials*
> The materials used in the manufacturing of the piano are just as important as the design. Steinway pianos combine the resonance of Sitka spruce with the rigidity of hard rock maple to intensify the richness of the sound. All Steinway soundboards are made with Sitka spruce, the most resonant wood available. Soundboards in Steinway pianos are constructed from solid (never laminated) Sitka spruce with annual growth rings measuring 8–12 per inch. These close-grained lines enable the sound-producing energy to travel to the end of the board, which is custom-fit to the top of the inner rim. The energy travels more efficiently when the soundboard is close-grained.
>
> Steinway Pianos 2013

Lyon & Healy of Chicago also uses Sitka spruce for the soundboards of its magnificent hand-built harps. Taylors Guitars in San Diego makes high-quality instruments using Sitka spruce and:

> ...the company is equally committed to procuring wood that is responsibly harvested. In 2007, the company joined the Greenpeace *MusicWood Coalition* which is committed to FSC certification of Sitka spruce, an Alaskan forested wood that is becoming scarce due to heavy use in the construction industry.
>
> Taylor Guitars 2008

Arrows

Arrow shafts of Sitka spruce are made by the family firm of Hildebrand at Port Angeles, close to Washington's rainforest. Sitka spruce does not split or splinter,

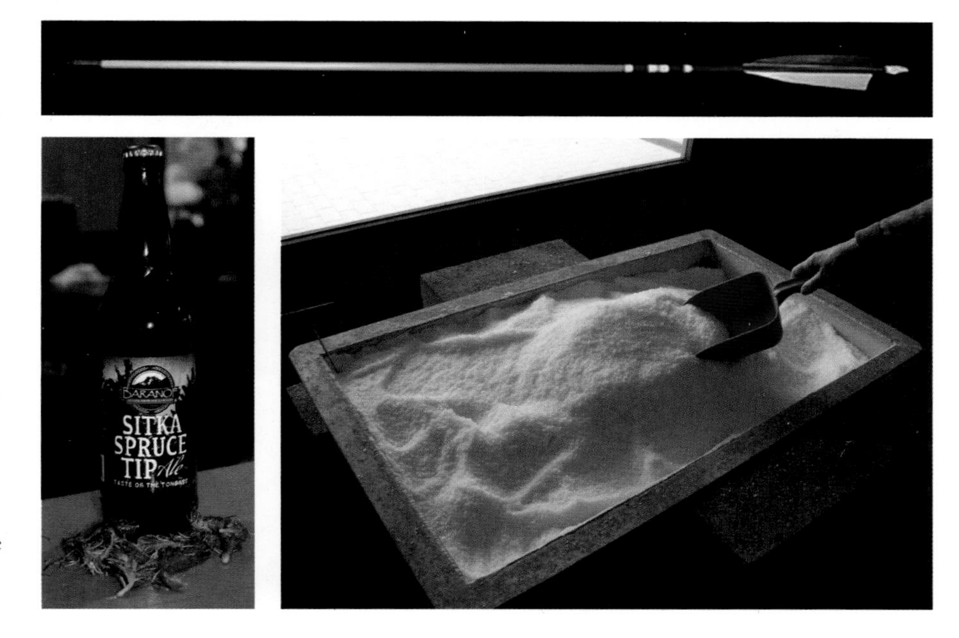

FIGURE 14.10. Sitka spruce arrow.

PHOTO: COURTESY NEIL HILDEBRAND, HILDEBRAND ARROW SHAFTS, PORT ANGELES, 2013.

FIGURE 14.11 *(left)*. Sitka Spruce Tip Ale manufactured at Baranof Island Brewery, Alaska.

PHOTO: ANDY TITTENSOR, 2014.

FIGURE 14.12 *(right)*. Large tray of Sitka spruce tip sea salt.

PHOTO: ANDY TITTENSOR, 2014.

absorbs shock and withstands sharp impacts, so produces excellent shafts for arrows – which fly faster than tournament-grade aluminium arrows.

Art

First Nations craftsmen carve Sitka spruce wood into beautiful *ornaments*, nowadays for sale to tourists. Beautiful modern versions of traditional art work on Sitka spruce wood are also popular amongst Tlingit people in Alaska (see Chapter Five).

Drinks

The Baranof Island Brewing Company produced its first *Baranof Island Sitka Spruce Tip Beer* in 2012. Young shoot tips are collected in May just as the brown scales are sliding off. They are harvested from trees away from municipal and First Nations lands; 1814–2268 kg (4000–5000 lb) of shoots are needed annually; those not used immediately are vacuum-packed and frozen for later. The beer, sold in Alaska, has been well-received by craft beer drinkers and I enjoyed some in the brewery's taproom.

The Alaska Brewing Company in Juneau makes a *Winter Ale* using glacier meltwater and Sitka spruce shoot tips collected from trees growing on the outwash plain of Glacier Bay. Unusual!

Foods

Shops in Southeast Alaska sell *sea salt*, *syrup* and *jelly* made locally using the springtime shoot tips of Sitka spruce.

FIGURE 14.13. Sitka spruce products from Alaska: Sitka Spruce Tip Sea Salt, Sitka Spruce Tip Jelly, Syrup, Herbal Body Lotion, Soap, Pitch Ointment.

PHOTO: ANDY TITTENSOR, 2014.

Cosmetics

Sitka spruce *body lotion*, *soap* and *ointment* are also made and sold in Alaska.

Cones

A resident of Sitka, Alaska, told me how he and his friends collected Sitka spruce cones for the Forest Service when they were lads in the 1960s. Perhaps some trees now growing in Britain came from their cones and seeds.

Bedding

Sitka spruce branches and foliage are said to be good for *bedding* on wilderness trips, but I have not tried them!

Construction of First World War aircraft

The US Army barracks at Fort Vancouver had a polo field which an unofficial mix of army and civilian enthusiasts used as a 'flying field' with planes like the Curtiss Pusher.

During the First World War, the European Allies (France, Italy and the UK) started to import large quantities of Sitka spruce to build many hundreds, even thousands, of war planes. They knew that Sitka spruce was the most suitable timber for aircraft frames, along with small amounts of other species. Geoffrey de Havilland's 1915 DH.3, for instance was made of Sitka spruce and ash.

During 1917, the Allies requested assistance from the USA in obtaining Sitka spruce for war planes in Europe and training planes in North America. The Allies owned probably 200–300 aircraft in 1914 but by the end of the war they had about 8,300 and production continued post-War.

They asked for *100 million board feet* of Sitka spruce by 1918! High-grade timber from big trees was the most valuable – and Sitka spruce became an important strategic material; 100 million board feet (235,974 m^3) represents all the timber from 590 ha (1458 acres) of modern UK forest at a harvesting time of 50 years old (using British yield tables of 1981).

A 'Spruce Production Division' of the US Army was set up. From headquarters in Portland, the Division was authorised to employ 28,825 soldiers and civilians to be trained at Vancouver Barracks. They then worked in 60 militarised logging camps and lumber mills along a coastal stretch of 484 km (300 miles) from the Canadian border south into Oregon. The Army also built an extremely large spruce mill on the polo-flying-field of Vancouver Barracks which was operated by 2000–5000 soldiers.

To connect forest logging sites with the mill, 13 railways, with over 210 km (130 miles) of track were built. During peak production in 1918, there were 10,000 soldiers in Oregon and Washington rainforests building railways. A recent study described them:

> The SPD's first strategy for getting spruce out of the woods during the spring of 1918 was to rive the huge logs into smaller flitches that could be loaded onto motor trucks and hauled to the mills. The second strategy was for the SPD to embark on a program of building logging railroads for the contract lumber companies to use. The SPD started construction of logging railroads in Washington and Oregon in the spring of 1918. These were numbered I through XIII. Most of these were relatively short logging lines reaching pure spruce stands or timber stands with a high percentage of spruce mixed with other softwoods.
>
> Tonsfeldt 2013

Logged trees were taken along the railways to local mills where the first cutting was done; branch lines then carried them to the Spruce Production Mill to saw the timber into smaller sizes. Aircraft-grade material was selected and then easily transported by main lines to shipping ports. To start with, two million board feet (4719.5 m^3) of Sitka spruce timber were produced per month at the mill, but that had to increase drastically if the Europeans were to build all the planes needed. Army personnel managed to increase production to 10 million board feet (23,597 m^3) per month by changing their sawing methods. When the mill started operating, only 10% of logged timber was considered aircraft grade. But the mill workers gradually found new ways of cutting it so as to extract 60% as aircraft-grade.

Production lasted just over a year, by which time the war had ended. The Spruce Production Division actually exported *150 million board feet* (353,960.6 m^3). There was sufficient Sitka spruce to supply only three-quarters of the total so other conifers made up the shortfall.

The post-war logging industry was able to expand quickly around and beyond the new roads and railways. Temperate rainforest was logged rapidly, leaving only conserved areas and occasional specimen trees for us today.

The Vancouver Barracks Spruce Production Mill closed after its strategic role in the First World War. Machinery was sold and men de-mobbed. The site became derelict and by 1923 aviation had started again; in 1925 the ex-polo-flying-field was renamed 'Pearson Field' after a young army test pilot.

The coastal forests of British Columbia, including Haida Gwaii, were also heavily logged during the First World War.

Crucial attributes of Sitka spruce in Europe

Versatility

Sitka spruce is the most versatile tree species in modern Britain and Ireland because it grows in a variety of conditions where other trees grow slowly or not at all. In oceanic Europe it is a very acceptable addition to the native tree flora for commercial and amenity use. In Iceland and the Faroe Islands it is one of the few possible commercial tree species.

Growth rate

It grows quickly for at least the first half-century, putting on height of 0.30–0.91 m (1–3 ft) annually. Sitka spruce is highly productive of cellulose volume because its wood is so fast-grown (in contrast to Alaska). The fast-grown wood suits pulp and paper manufacture, giving it huge commercial importance.

Size

Large Sitka spruce are only occasionally required because most of the current end-products can be easily milled from smaller, uniform-sized, close-grown trees, which produce a satisfactory volume of timber.

Anatomy

Long, strong fibres make Sitka spruce suitable for good-quality paper pulp and other cellulose products.

Chemistry

At least 33 different chemical compounds can be extracted with solvents from Sitka spruce timber. They include resins, gums, waxes and fats in tiny amounts but may become important products.

The chemical and physical properties of Sitka spruce timber in relation to its uses are described in detail in Moore's *Wood Properties and Uses of Sitka Spruce in Britain* (2011).

Strength to weight ratio

The light but strong wood is as important in Britain as in North America. It grows so fast, however, that British and Irish Sitka spruce cannot match slow-grown North American timber for strength. This means it is less suitable for construction or tonewood.

'Modulus of elasticity' describes its strength and toughness under deliberate bending stress; this measure is used to machine-grade its timber. UK Sitka spruce is usually Strength Class 16 which is insufficient for construction without a special approach to building design. Irish and other imported timber can be Strength Classes 18–24 so is more usually used in construction.

Knot-free

In Ireland and Britain, Sitka spruce trees are grown deliberately close. This means that they have few branches, so relatively few knots. This is useful when you are working its timber or producing poles for pulp.

Easily seasoned

Sitka spruce cells contain considerable moisture, so the timber is dried before using it in joinery or construction. When kiln dried, Sitka spruce wood is very useful to the sawmill industry – into which £300 million has been invested recently. However, if kiln-dried too fast, it splits and knots loosen. It takes several years to dry naturally: nevertheless this is done for some purposes. Chemical seasoning is less satisfactory because the cells are not sufficiently absorbent for uptake of liquid throughout a lump of timber.

Reproduction

Sitka spruce in Europe starts to produce fertile seed when it is 20–40 years old. It continues to 'cone' at regular intervals and abundantly. The seeds germinate and grow well in forest nurseries, are transplanted and then put in their final forest home. This procedure became so efficient that forests could be planted or restocked quickly with few establishment losses.

Vegetative propagation is nowadays the most common method of producing millions of small trees for afforestation and restocking, as discussed in Chapter Seventeen.

Creation of custom-assembled forests

Sitka spruce forests have added considerable tree cover to several poorly-wooded countries (Faroe Islands, Iceland, Republic of Ireland and the UK). They have benefits for recreation, public health, amenity and shelter for farmland; for wild animals requiring large, undisturbed territories; for colonisation and spread of wild flora and fauna.

Its pre-adaptive ability to grow in oceanic climates, on peat and close to coasts is an important feature of Sitka spruce in Ireland and Britain. Its anatomy and physiology suit the conditions of its new home: high growth rate, ecological plasticity, early and abundant reproduction, user-friendly timber with structural, chemical and cosmetic qualities.

Contributions to European society

It was in 1910 when W. Crozier, saw probably the first ever harvested UK Sitka spruce in use on Henry Baird's estate (see Chapter Ten). He noted the light-coloured, straight-grained, easily-dressed wood from trees then 28–37 years old.

Sitka spruce is nowadays harvested at 35–40, sometimes at 50–60 years old: pulp mills demand even-sized, small-diameter poles for their machinery. Felled Sitka spruce with a basal diameter of greater than 0.33 m (1 ft) has a limited market in the UK. However, in some places, plantations are allowed to carry on growing, so as to improve the scenery and in the expectation of changing markets.

Pit props

Forest thinnings were vital during most of the twentieth century for *pit props* which are posts that support the walls and roofs of mines. Coal was important to twentieth-century industry, shipping, schools, offices and homes, as fuel to heat premises and run machinery (including steam locomotives). The Loch Ossian forests, afforested by Sir John Stirling-Maxwell's Loch Ossian forests provided 19.8 km (12 miles) of pit props to the Second World War effort.

Pulp

The highest value product from UK Sitka spruce is *pulp*. It makes good-quality *paper* for magazines, photographs and computer use. The top half of a Sitka spruce tree, with the highest value, is used for paper pulp. Only the mechanical method is used in British pulp mills so pulp made by the sulphite process is imported.

There are two big integrated pulp-paper mills: the UPM Caledonian Mill at Irvine utilises 250,000 tonnes (275,578 tons) of Sitka spruce round wood from southern Scottish forests. The mill produces 20% of the UK's *lightweight coated paper*. The Iggersund Paperboard Ltd pulp mill at Workington, Cumbria receives Sitka spruce by sea from Scottish west coast forests and makes *paperboard* for *packaging*, *food cartons* and *graphics*. There are many paper-only mills.

Paper use is declining world-wide, so Sitka spruce growers and pulp makers are out and about looking for new markets. Timber research teams, as at Napier University in Edinburgh, are testing possible ways of breaking down a Sitka spruce tree into a variety of *chemical components* for new products. One potential use is for new designs of *aeroplane*.

FIGURE 14.14 *(above left)*. From this … Sitka spruce logs at UPM Caledonian Paper, Irvine, Ayrshire.

PHOTO: SYD HOUSE, 2009.

FIGURE 14.15 *(above right)*. To this … high quality glossy paper produced by UPM Caledonian Paper.

PHOTO: SYD HOUSE, 2009.

FIGURE 14.16. Everyday paper products of Sitka spruce used in the UK.

PHOTO: ANDY TITTENSOR, 2015.

Rayons are fabrics which were first made in France during the nineteenth century when the silk-worm industry was affected by disease. They are produced from high-quality cellulose derived from the wood pulp of spruces, pine or hemlock by a series of chemical and physical processes. Discovered by a Swiss chemist, *cellophane* was also made from spruce wood fibres which were pulped and transformed into a fine, transparent film of cellulose. Since the 1960s cellophane has been made from oil.

Fencing and pallets

Approximately 3.2 million m^3 (1,356,083,202 board ft) of sawn Sitka spruce timber is produced annually in the UK and used by 186 or so sawmills. *Fence posts, rails* and *planks* are in huge demand for surrounding new properties because farmers and householders now prefer fences to hedges. The demand is partially fulfilled by Sitka spruce which requires drying and treating with preservative.

FIGURE 14.17 *(left)*. Wood strands of Sitka spruce, the basis of strand-boards.

PHOTO: COURTESY OF NORBORD EUROPE LTD.

FIGURE 14.18 *(right)*. The Kronospan factory, Chirk which makes composite boards using Sitka spruce from North Welsh forests.

PHOTO: ANDY TITTENSOR, 2013.

Sitka spruce is one of the most commonly-used UK timbers for *pallets* because it does not split easily and can be nailed satisfactorily. Pallets are used in very large numbers for transporting goods in same-unit loads. They are double-layered beds of parallel wooden planks onto which goods are secured and moved by a fork-lift or similar truck.

Composites

Composites make the most of home-grown Sitka spruce by manufacturing products stronger than the fast-grown timber. Tree trunks arrive at a mill, are sawn into shorter 'sawlogs' and then fragmented into small strands, chips or fibres. A chemical to bind the fragments is added and the whole mass is pressed or moulded to produce board or spars which resemble wood but are stronger than their parent logs. They can be cut into the size and shape needed and used in *construction*.

Particleboards: particles are produced by grinding rough-sawn logs into small pieces. They are then glued together and forced into a mould of the required shape. The resulting *particleboard* is used for *panels* and *insulation board*, including *kitchen units*. Clocaenog and Gwydyr Forests supply timber to the firm Kronospan near Chirk, one of seven major particle-board manufacturers in the country.

Oriented strandboard: is a fairly recent product from the Republic of Ireland. Coillte, the Irish forestry agency, owns an OSB factory at Waterford Harbour. Sitka spruce timber is broken down into its constituent wood fibres, then wafers of fibres are laid in layers alternating at 90°. Like other manufactured Sitka spruce boards, it is strong enough for construction.

Fibreboard: *medium density fibreboard* is made by breaking down logs into bundles of fibres and adding resin. Pressure treatment forms this into fibre mats or *hardboard*. Ireland's high quality hardboard is used, for example, for *wall linings* of prestigious internal spaces.

FIGURE 14.19 *(above left)*. Dried wood strands blended with resin and wax are laid in cross directional layers to form a mat which is pressed in heat and pressure, cooled and cut to size, giving a quality Oriented Strand Board (OSB).

PHOTO COURTESY OF NORBORD EUROPE LTD.

FIGURE 14.20 *(above right)*. Close-up of OSB panel outer surface.

PHOTO COURTESY OF NORBORD EUROPE LTD.

FIGURE 14.21 *(below left)*. End-uses of OSB include packaging and crates.

PHOTO COURTESY OF NORBORD EUROPE LTD.

FIGURE 14.22 *(below right)*. JJI-Joists showing how parallel lengths of Sitka spruce timber are held together by strand board to give strong spars.

PHOTO: WHITE HOUSE STUDIO, 2013, COURTESY JAMES JONES & SONS LTD, FORRES.

Building construction

Sitka spruce can be used for roof and floor trusses if it reaches Strength Class 16, but imported Sitka spruce, being Class 18–24, is preferred.

FIGURE 14.23. Long JJI-joists used to construct the roof of Lakeside Primary School, Lincolnshire.

PHOTO: COURTESY JAMES JONES & SONS LTD, 2012.

The firm James Jones, in its Forres factory, manufactures joists from Sitka spruce composites specifically for construction. These *'JJI-joists'* are made from long lengths of composites with both sides embedded into solid wood lengths of imported Sitka spruce of Strength Class 24. The joists are produced to each customer's specifications, and enable deep, long spans for walls, floors and roofs to be constructed.

In the Republic of Ireland, *structural timber*, not pulp, is the present and likely future, most valuable Sitka spruce product. The usually-higher strength class of Irish Sitka spruce means that the country can often use its own timber for buildings. A study of stress graded Sitka spruce samples in 2000, showed that 95% of it was Strength Class 16 or more and therefore suitable for construction.

Transmission poles

In Ireland, *transmission poles* are sometimes made of Sitka spruce because there are insufficient home-grown, tall Douglas fir with the correct degree of taper. Sitka spruce poles need preservative treatment if they are to last the necessary 40 years.

Joinery

Sitka spruce has an attractive pale colour and easy to work, so is common in indoor *joinery;* it can be glued and holds nails well so *agricultural boxes* and *packing cases* are an important product.

Biomass chips

Smaller logs, and brash left on the ground after harvesting, are easily reduced to chips, good for *biomass kilns* which run *boilers* for heating. For instance, Kielder Forest village and Floors Castle, Roxburghshire use Sitka spruce chips from their own forests. The Duchess of Roxburgh supposedly remarked that Floors Castle is now warm for the first time in its three-century history!

At Steven's Croft near Lockerbie, the energy firm Eon is producing *electricity* for the UK's National Grid, using Sitka spruce logs transported and then chipped on-site for the kiln.

Tonewood

In sixteenth- to nineteenth-century Europe, European silver fir and Norway spruce were the most common timber for *soundboards* of keyboard instruments such as the spinet and clavichord. Scots pine from the Black Forest was used for

FIGURE 14.24 *(above left)*. Bunkers for Sitka spruce chips and bark with planks air-drying in front, James Kingan & Son Ltd.

FIGURE 14.25 *(above right)*. Ejection of sawdust, sold to equestrian enterprises, James Kingan & Son.

FIGURE 14.26 *(below left)*. Chips of Sitka spruce piled up ready to go into the Roxburghe Estate's kiln.

FIGURE 14.27 *(below right)*. The Estate's boiler which heats Floors Castle and estate dwellings.

PHOTOS: RUTH TITTENSOR, 2012.

FIGURE 14.28 *(left)*. Steinway piano in the Usher Hall, Edinburgh opened to show the strings and Sitka spruce soundboard below them.

PHOTO: RUTH TITTENSOR, 2014, COURTESY NORMAN MOTION AND THE USHER HALL.

FIGURE 14.29 *(right)*. The same Steinway piano from underneath showing the Sitka spruce soundboard close-up.

PHOTO: RUTH TITTENSOR, COURTESY NORMAN MOTION AND THE USHER HALL.

soundboards by nineteenth-century German piano makers. Steinway opened its piano factory in Hamburg in 1880. However, after the Second World War, the factory was forced to import Sitka spruce for its soundboards because the pine trees had been damaged by shrapnel.

Sitka spruce grown in Alaska now supplies soundboard wood for American and German Steinway pianos.

Other European tonewood is mainly from old-growth Norway spruce from the Carpathians and Alps. However, some Sitka spruce grown in the Jura and Vosges is also suitable. Soundboards of modern Gaelic harps are made from Sitka spruce but on the cross-grain, unlike most other soundboards which are cut along the grain.

Cross-laminated timber

Cross-laminated products have been developed and manufactured with Sitka and Norway spruces. An unequal number of panels – on top of but at right-angles to each other – are glued together under pressure. Panels are usually up to 3 m (10 ft) wide and 30 m (98 ft) long; the multi-layered board is usually 20–40 mm (0.79– 1.58 in) thick. It is strong but light and eminently suited to construction. Walls can be prefabricated from the panels, making on-site construction quick and easy.

Thurso College, Caithness and Murray Grove flats, London have been built of this new and beautiful material. It has many 'sustainable' features including long-term sequestration of carbon dioxide. Of course sustainable wooden buildings are *not* new. Some medieval homes and barns were timber-framed with an in-fill of wattle panels. Some have lasted more than 600 years, showing us what sustainability really means.

Sawmills

Master Carpenter Christian ap Iago of Newtown, Powys describes Sitka spruce as 'the work horse of timbers in our region'. He kiln-dries Sitka spruce, cuts out knots and weaknesses and produces 'glulam' which consists of small timbers bonded together to make a bigger stronger material suitable for construction. Christian specialises in outdoor buildings such as stand-alone offices, summerhouses and garages.

Sitka spruce is also the work horse of family firm, James Kingan & Son in Galloway. Ian Kingan, born in 1926, remembers the large water-wheel which powered the saw mill. He worked in the mill during the Second World War. Afterwards, the hills from which the water flowed were ploughed and planted with conifers… This caused erratic water-flows, an erratic water-wheel, erratic energy pulses and inefficient and hazardous sawing! So Kingan's Mill got itself another source of power, a steam-engine.

Prisoners of War helped to thin the Mill's Sitka spruce plantations post-War and luckily the material was bought by Blackpool and Manchester parks departments for *rustic materials*.

Pit props were a major market for bigger Sitka spruce from the nineteenth century onwards, but when coal mines started to close after 1965, Ian found a replacement product in *pallets*. By this time, of course, Kingan's had replaced the steam-engine with electricity!

Large-diameter trees, 'the finest timber', were always Kingan's speciality, but large Sitka spruce timber is now more difficult to market. So the mill concentrates on what the big sawmills eschew as too troublesome: *sheds*, pointed *fence posts*, *feather-edge fencing* and *specialist orders for planking*. Horse premises buy its *bark*, *chips* and *sawdust*. *Pallets* made from Sitka spruce are still in great demand.

James Jones, originally a family firm, is now one of the biggest sawmill and timber treatment enterprises supplying British timber. It runs five sawmills in Scotland and two pallet-making sites in Wales and England.

FIGURE 14.30 *(left)*. James Kingan & Son Ltd. Dumfries-shire, stackyard with Sitka spruce logs, large by UK standards.

PHOTO: RUTH TITTENSOR, 2012.

FIGURE 14.31 *(right)*. Sitka spruce and larch pallets going into the drier, James Kingan and Son Ltd.

PHOTO: KIT ALLEN, 2012.

Production manager Rob MacKenna showed me round its gigantic Lockerbie site where all operations are computer controlled. One hundred lorries of Sitka spruce logs from private and state forests within a radius of 161 km (100 miles) deposit their loads in the stacking yard daily. From the weekly intake of 12,500 tonnes (13,779 tons) the site produces over 7000 m^3 (2,966,432 board feet) of sawn timber. One-third is sold to each of *pallet*, *fencing* and *construction* firms.

Carbon dioxide uptake and fixing

The role of Sitka spruce forests in taking up and holding carbon dioxide is discussed in Chapter 17.

Shelterbelts

In the exposed regions of Britain and Ireland, stock and crops need shelter from wind, rain and snow. During the earlier twentieth century, Sitka spruce was planted as shelterbelts particularly in Wales and Ireland; government grants now encourage farmers to use native trees.

Other uses

At Crookedstane Rig near Moffat in southern Scotland a Sitka spruce plantation is the base for a family firm which ferments *Sitka spruce beer* for local events and distils *essential spruce oil* for general sale.

Some Scottish towns such as Biggar, Lanarkshire, use Sitka spruce for their public *Christmas tree* because the prickly foliage means they are less likely than Norway spruce or fir trees to be stolen.

The de Havilland Mosquito aircraft

First World War British and Allied aircraft were manufactured from Sitka spruce imported from North America; thousands of wood-framed aircraft

FIGURE 14.32 *(left)*. The huge stackyard at James Jones' Lockerbie mill, with Steven's Croft wood-based power station behind.

PHOTO: ROB MCKENNA, 2014, COURTESY JAMES JONES LTD, LOCKERBIE.

FIGURE 14.33 *(right)*. The complex operations of the mill are controlled by one man and a computer at James Jones, Lockerbie.

PHOTO: ROB MCKENNA, 2014, COURTESY JAMES JONES LTD, LOCKERBIE.

FIGURE 14.34. Zacharry's non-alcoholic Sitka spruce drink.

PHOTO: RUTH TITTENSOR, 2011.

FIGURE 14.35. Zacharry's Sitka spruce oil from the Southern Uplands.

PHOTO: RUTH TITTENSOR, 2011.

continued in production. But by the Second World War most military aircraft were constructed of metal. De Havilland had continued with wood for its 1934 DH.88 'Comet' and DH.91 1937 'Albatross' and in 1940, its ingenious inventors designed and built the prototype of a *light* and *very fast wooden* plane at Salisbury Hall, Hatfield.

The wood was Sitka spruce and the plane was the DH.98 Mosquito.

This wooden plane, in the era of metal or metal and wood planes, was scoffed at, but de Havilland insisted that metal industries would be overworked in war time while wood-working industries would be underemployed. And when the Mosquito's light weight, speed and capabilities were realised by decision-makers, the firm took orders for an initial 50 and eventually for thousands.

The Mosquito started as a fast, unarmed bomber, as a pathfinder for heavy bombers and in photo-reconnaissance; it was adapted as a long-range fighter, for mine laying and for high-speed military transport. Pilots liked the Mosquito because they were likely to return home if the plane were damaged, due to its strength: it had the highest survival rate of all Second World War planes. Maintenance staff liked it because it was simple to replace any damaged wooden parts. Officials liked its speed of over 644 kph (400 mph); after 1942, electronics experts liked its extremely small radar signature.

Sitka spruce became, once again, of strategic and military significance.

Furniture factories like Parker-Knoll were requisitioned and workers learned how to make the necessary spars, ribs, bulkheads and smaller components from Sitka spruce wood: in large numbers and accurately. Best quality Sitka spruce was required. Even when timber had been milled and graded for aircraft use, much was discarded. Every piece had to conform to strict standards: grain was to be straight not spiral; growth rings were to be close and evenly-spaced; resin pockets, shakes and splits were not allowed. All wood was dried down to 14–16% moisture and every batch was tested for its strength.

Sitka spruce formed the wooden framework and shape of Mosquito aircraft. Wing spars were made of half-inch (12.7 mm) laminates of Sitka spruce bonded

FIGURE 14.36. Mosquito aeroplane PR.IX LR432 during the Second World War.

OFFICIAL WARTIME PHOTO VIA DE HAVILLAND AIRCRAFT MUSEUM, COURTESY IAN THIRSK.

together, the front spar laminated horizontally and the rear spar vertically, each lamination running the whole spar length of 16.5 m (54 ft): this added to strength. The wing ribs used Sitka spruce and birch. Hard pieces of ash and walnut attached them to major structures.

The tail-plane was constructed similarly to the wing, with a frame of Sitka spruce spars and ribs. The seven bulkheads inside the fuselage were of Sitka spruce blocks separating two plywood skins; those nearer the back were thicker, up to 1.5 in (38 mm), to support the rudder.

The surfaces of wings and fuselage were two layers of birch plywood with other species forming longerons or infilling between the layers. Irish linen and eight layers of waterproof paint then covered the plane.

At Hatfield, the centre of Mosquito-manufacture in England, *3054* of the total *6710* DH.98 Mosquitos were constructed and *1505* damaged Mosquitoes were repaired. 1032 Mosquito bombers were built at de Havilland's Canadian factory, Downsview, Toronto and 108 in Sydney, Australia.

Apart from Sitka spruce, guns, bombs and rockets, navigators and pilots were needed … and many men were killed in action. Sitka spruce-built Mosquito aircraft were a major factor in the Allies winning the Second World War. It justified its nicknames of 'Wooden Wonder' and 'Timber Terror'.

Several Mosquitos, including the prototype, can be seen at The Mosquito Aircraft Museum, part of the de Havilland Aircraft Heritage Centre not far from St Albans, England. Canada Aviation and Space Museum at Rockcliffe Airport, Ottawa has a Mk. 10 Mosquito on display. Alberta Aviation Museum in Edmonton holds a Mk. 6 Mosquito B.35. The one flying Mosquito (Mk. 26) is held at the Military Aviation Museum, Virginia Beach, USA.

FIGURE 14.37 *(left)*. The fuselage of a crashed Mosquito aircraft showing longitudinal bands of Sitka spruce enveloped by balsa wood.

PHOTO: ANDY TITTENSOR, COURTESY MOSQUITO AIRCRAFT MUSEUM, ST ALBANS, 2013.

FIGURE 14.38 *(right)*. The flap of a Mosquito aircraft undergoing construction with Sitka spruce ribs.

PHOTO: ANDY TITTENSOR, COURTESY DE HAVILLAND AIRCRAFT MUSEUM, ST ALBANS. 2013.

Transporting Sitka spruce

The twentieth-century Sitka spruce forests of north and west Scotland have reached the age, size and volume for harvesting. The landscape where those forests were planted is mountainous, steep, rugged and intersected by long sea-lochs; roads are few, single-track and meandering; railways are almost non-existent.

Planting millions of seedling trees in such places was difficult; harvesting and extracting them half-a-century later is a significant problem. Sea is a traditional transport method in the UK. However, during the twentieth century, the coastal merchant marine was replaced by thousands of miles of new, wide, well-surfaced roads, in England at least. Wales, Scotland and Northern Ireland were not so lucky (or unfortunate, depending upon one's point of view).

Scottish west-coast communities depend upon the sea for transport of people and goods. During 50 years in the merchant marine, Captain David Neill became familiar with the indented west coast from the bridge of ferries and cargo boats. Between 2003 and 2005, his ship, Whitaker's *Bowcliffe*, transported felled timber from small Scottish ports to Ireland. He told me that *Bowcliffe* was loaded with 1300–1400 tonnes (1433–1543 tons) of 102–305 mm (4–12 in) diameter Sitka spruce, from quays at Ardrishaig, Campeltown, Corpach, Portavadie and Sandbank. Lorries brought the felled logs from forest roadsides to these quays along narrow, winding roads. *Bowcliffe* had no crane, so could work only at a quay with a permanent crane or large enough for the timber to be stacked. A mobile crane, which could travel between quays, then loaded timber into *Bowcliffe*'s hold.

Sitka spruce was a regular and frequently-available cargo; with a loading or discharge time of eight hours, the *Bowcliffe* could make between two and four timber trips a week, depending on the voyage distance. Captain Neill frequently shipped the timber to Londonderry in Northern Ireland or to Youghal, Dundalk and Passage West in the Republic of Ireland. Occasionally he took Sitka spruce to the major port of Troon to be put on board larger ships for onward transport.

Scottish Sitka spruce is now taken in boats with twice the capacity of *Bowcliffe* and equipped with their own cranes. And at Fishnish on the Isle of Mull, a new 'timber pier' built by the Forestry Commission allows large volumes of Sitka spruce to be loaded and transported from island forests using minimum mainland road mileage. It will save 800,000 lorry miles (1,287,475 km) annually. A consortium of local firms has inaugurated an even more flexible system, using a large landing craft which can be loaded with 600 tonnes (661 tons) of timber from any coastal site with a suitable support for the craft's ramp.

Chapter Fifteen will return to the forests themselves, the plants and animals which live with them.

FIGURE 14.39 *(above).* Reducing heavy use of rural roads by transporting timber by boat from the new 'Timber Pier' at Fishnish, Isle of Mull, Argyll, 2013.

PHOTO: J. B. DAVIES PHOTOGRAPHY, COURTESY TSL CONTRACTORS LTD, ISLE OF MULL.

FIGURE 14.40 *(below left).* Harvesting typical tall, close-grown Sitka spruce on the Corrour Estate, Inverness-shire: destined for the sawmill.

PHOTO: JOHN SUTHERLAND, 2014.

FIGURE 14.41 *(below right).* Oars for the royal barge *Gloriana* 2012, made by J. Sutton of Windsor with Sitka spruce supplied by Stones Marine Timber Ltd of Salcombe, Devon.

PHOTO: SARAH WOOLLEY.

CHAPTER FIFTEEN

Plantation Ecology: Plants and Animals Re-assemble

> To an increasing degree the modern woodlands of Britain are coming to be composed of planted conifers of foreign origin. Scientific study of their ecology is naturally of the utmost importance to successful forestry, but it lies outside the scope of this book, and indeed is still in its earliest infancy.
>
> Arthur Tansley 1939

> Unintentionally they are a remarkable ecological experiment. For the first time for many hundreds of years, maybe two thousand years in some places, trees have been restored to the moorland. Mainly not native trees it is true, but trees nevertheless. And an end at last to ages of intensive grazing. So naturalists, however much they may abhor foreign introductions, especially on so massive a scale, cannot simply ignore them.
>
> William Condry 1981

Early observations

The aim of nineteenth-century landowners, explorers and their sponsors was to enhance the amenity and commercial productivity of their landscapes. Potential changes in native flora and fauna were probably not considered. They were involved in testing the basics of *what would grow* and not how pre-existing plants and animals would be affected by their activities. However, foresters on-site, like any countryperson, were probably familiar with the vagaries of weather, soil, drainage, plant and animal populations of their particular domain.

Scottish forester Angus Duncan Webster was also a good naturalist. He not only produced a book on British orchids, but recorded the common flora of the woodlands he had planted on the Penrhyn Estate, Caernarfonshire. He described the mild but stormy climate, variety of geology and soils and altitude to which trees could be grown. His first example was a several-species plantation at 229–305 m (750–1020 ft) on a north-facing hillside:

> The natural vegetation of the woodland consists of *Empetrum nigrum*, *Erica vulgaris*, *Vaccinium myrtillus*, *Oxycoccus palustris*, the latter on damp ground, *Pteris aquilina*, *Polypodium vulgare*, *Allosorus crispus*, *Athyrium filix-fœmina*, *Lastraea filix-mas*, *Lycopodium selago*, and various species of grasses, these generally occurring in the more open parts of the wood and amongst the rocks which crop out here and there over its surface.
>
> Webster 1890

Later opinions

Despite a long tradition of natural history in Britain and Ireland, new forests were ignored by naturalists and academics who preferred to study what they thought was 'natural' vegetation and fauna.

During the twentieth century, even serious natural history books and specialist journals ignored 'conifer plantations'. Natural history journals published studies of broadleaf and native conifer woodlands but not woodlands of introduced conifers.

Scottish ecologist Frank Fraser Darling included one paragraph on 'planted conifer forests' in his early book: '...he will find the floor a dead, quiet dark place in which there is little joy but that of shelter and stillness' (Fraser Darling 1947).

He later retracted and suggested more tree planting to encourage rural renewal.

Robert Lloyd Praeger in his 1950 *Natural History of Ireland* described the current distribution of native juniper and yew in his homeland, but not other conifers. McVean and Ratcliffe in their chapter on woodlands in *Plant Communities of the Scottish Highlands* of 1962 concentrated wholly on what they considered native forest types; Burnett's 1964 tome *The Vegetation of Scotland* ignored conifers apart from native Scots pine woodlands.

In his 1975 *The History of the Countryside*, Oliver Rackham distinguished 'plantation' from 'woodland' and wrote two pages about plantations in contrast to 57 pages on woodland. In his view, plantations consist of planted trees usually of just one or two species which do not maintain themselves. Woodlands, he suggested, arise naturally and can maintain themselves even if felled. However, after 40 years of environmental-archaeology research the difference between woodland and plantations is less clear and perhaps not useful.

By 2013, Michael Proctor's *Vegetation of Britain and Ireland* gave only one page to 'conifer plantations' in a whole chapter on woodlands on acid soils. He noted, however, that conifer plantations can be surprising. His surprises included the liverwort *Colura calyptrifolia* on spruce trunks in Welsh plantations; the moss *Daltonia splachnoides* unexpectedly colonising Sitka spruce in south-west Ireland; and wreaths of woodland mosses hanging from Sitka spruce branches in west Ireland. He wondered: '...is it a wake-up call to look harder at a habitat that many of us dismiss as "dull"?' (Proctor 2013).

Assumptions

Ecologists, conservationists and foresters assumed that because conifer plantations were new, looked uniform, were difficult to walk through when young, seemed dark and forbidding when older – they supported few flora and fauna (see Kimmins 1999 for an assessment of this attitude) – plants and animals could not be easily seen, so were absent.

They also assumed that conifer plantations needed 'improving' for biodiversity and looked for ways of 'improving' Sitka spruce forests for nature conservation. But they provided no data to show that they actually *needed* improving. Foresters responded to the widespread criticism and took pains to change how they managed Sitka spruce – but without having a base-line against which to check their 'improvements'.

Writing later about the Republic of Ireland, two eminent foresters wrote: 'Unfortunately the very words "Sitka spruce", as already mentioned, have come almost to have acquired a pejorative quality in the uninformed public mind', and quoted an item in the Irish Times of 2002: 'Endless rows of low-quality conifer like Sitka spruce are ugly, damaging, to rivers and lakes, and an arid habitat for wildlife' (Joyce and OCarroll 2002).

Did the critics go look? If Sitka spruce forests were so prickly and uninviting, how did people enter them to check for (absent) flora and fauna? Compared with ecological studies of broadleaved woodlands in Britain and Ireland, projects and publications about plantations of introduced conifers were negligible. Yet Sitka spruce soon became the commonest tree in Ireland and Britain.

What species did and do critics *expect* or *want* to be present in a Sitka spruce forest planted on upland moorland or peat bog? Did they expect typical spring flowers of deciduous woodlands, such as Early-purple orchid (*Orchis mascula*) and Yellow archangel (*Lamium galeobdolon*) to grow in new forests in quite different climates and soils?

Sitka spruce forests planted on poor acid soils which had no woodlands for four or more millennia were being compared with woodland sites where brown earth soils, flora and fauna had more than six millennia to accumulate.

FIGURE 15.1. After moorland was fenced for afforestation and to exclude grazing stock, 'hummock and hollow' started developing; Whitelee Forest, Renfrewshire.

PHOTO: RUTH TITTENSOR, 2005

Criticism seemed to be based more on emotionally-felt loss of beloved treeless landscapes than on scientific appraisal of new woodlands.

What species might an ecologist predict when moorland is afforested, based on knowledge of moorlands, of other conifer forests growing on once-open ground (such as yew) and of temperate rainforests with Sitka spruce? They were confronted with a new situation. The expected response would be to study the situation and find out – or wait and see.

Welsh ornithologist William Condry set the ball rolling in his book on British woodlands – by writing a *whole chapter* on 'Conifer Plantations' based on many years' observations. He made the poignant remark: 'Obviously if you plant a moorland with trees you are going to alter its ecology and therefore the wildlife almost completely. To see what actually happens let us look at afforestation stage by stage' (Condry 1974).

And he used his own observations to describe changes in bird life – which started when Whinchats (*Saxicola rubetra*) used newly dug-in fence posts as perches!

Our knowledge of Sitka spruce plantation ecology

This section outlines our knowledge of the ecological composition of Sitka spruce plantations, with some examples. Our understanding is still rudimentary and lacks an overall picture. The bibliography lists a selection of research results.

Herbaceous flora

Plants other than Sitka spruce may grow within the soil (for instance algae and lichens), rooted in the soil (herbaceous flora and shrubs), on the tree trunk and branches (for instance mosses) or where branches meet the trunk (for instance ferns).

The 'herbaceous' or 'ground' flora is what is most visible. In 1986, M. O. Hill listed four post-war studies of the status and changes in ground flora of commercial forests in the UK, but they were in lowland countryside and not Sitka spruce.

He suggested that without long-term observations of upland forests, it would be sensible to analyse a series of similar forests of differing ages. From these, he hoped to collate data which could be put together into a 'coherent story'. What he found was that although wild flowers behaved independently according to their 'personal ecological needs' there is an overall sequence of change.

After sheep flocks had been removed and excluded from moorland by new fencing prior to afforestation, lichens and mosses colonised the ploughed soil. Tussock grasses and shrubs increased in size and abundance, for instance Purple moor grass (*Molinia caerulea*) and heather. As Sitka spruce trees began to form a closed canopy, wild flowers might decline according to their need for light. Many had disappeared visually by the time the trees were clear-felled, but mosses

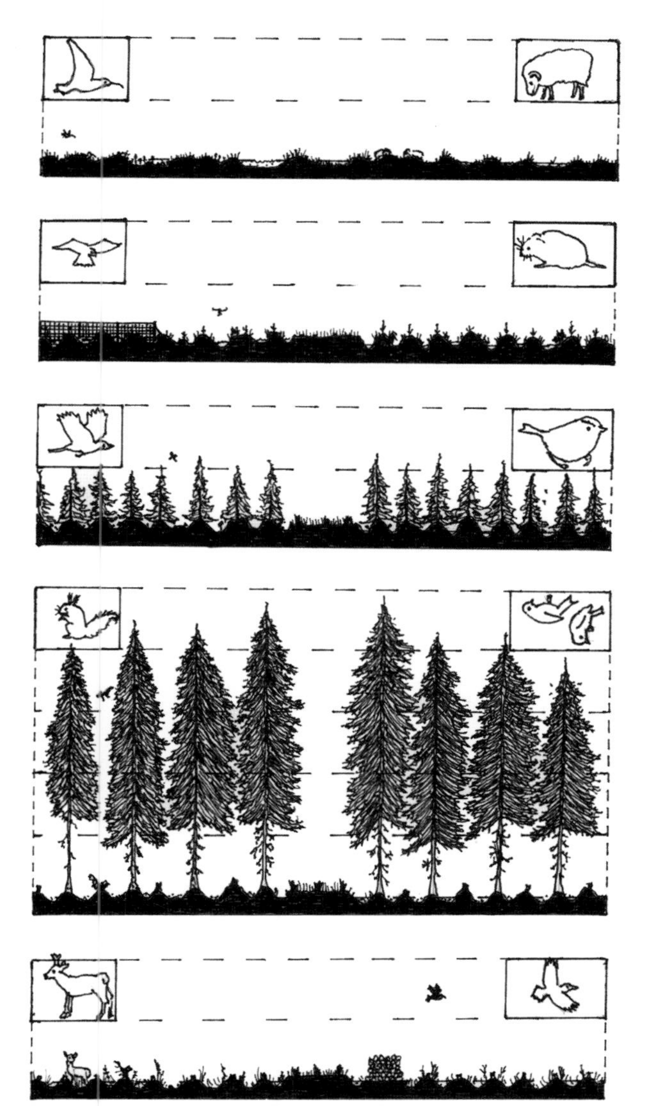

FIGURE 15.2. Ecological changes after afforestation of moorland in the UK.

PREPARED BY ANDY TITTENSOR, 2015.

were visible in all the years of a rotation. Another selection of plants grew from buried and blown seed when an area was clear-felled: heather, gorse (*Ulex europeaus*), foxgloves (*Digitalis purpurea*) and sedges germinated and grew in the opened areas. Plants which could not survive in deep shade and whose seeds do not survive in soil as long as the forestry rotation, disappeared, for instance Cotton grasses (*Eriophorum* spp.). Plants whose seeds could germinate in a clear-felled site grew, flowered and set seed again before the next tree canopy formed, surviving from one rotation to another. Examples were Heath bedstraw (*Galium saxatile*) and bramble (*Rubus fruticosus* agg.).

Berried plants such as bramble, bilberry (*Vaccinium myrtillus*) and Wild raspberry (*Rubus idaeus*) were dispersed right into forests by birds which defaecated their seeds from branches above; they grew but had to await the next lot of light before they flowered and fruited again. Land which was too high, rocky or boggy to afforest allowed existing plants to continue; lakes and watercourses provided habitats for aquatic and marsh plants.

Mosses and Liverworts

While recording his memories of its natural history before, during and after the afforestation of Whitelee Plateau, Bryan Simpson described to me the lichens and phenomenal moss hummocks:

RUTH So, what, what do you feel about the forest nowadays? And what natural history do you see when you're up there?

BRYAN Well, things have changed from the point of view that, I think probably one of the, the biggest gains is the fact that, the forest being there, you have an increase in the number of lichens. You don't have an increase in the number of species of mosses and liverworts and bryophytes and what, but you have an increase in the number of lichens, because you've got the trees there, and they accumulate. Oh they do accumulate a few bryophytes as well. But, the fact that the grazing animals were being kept out of there is one of the most important things.

RUTH Right.

BRYAN What has happened over twenty-odd years is, you've got these wonderful cushions of common mosses … growing, sometimes three feet in height ... golden, golden green cushions. And there are several different species that form these cushions, your common *Polytrichum* form big deep green masses there. It's wonderful. And when they flower, which is not really a flower, it's a sporophyte, they call a sporophyte, it's a little capsule where the spores are in, when they come up with all their different colours you know, it's quite a sight, it's quite an interesting sight.

RUTH Yes, it is, yes.

BRYAN Along the edge of the forest, you also get quite a variety of fungi there … associated with the trees I suppose. And, also, quite a variety of different types of lichens like to grow on the forest edge. Especially on the edge of the track where it's all gravelly and rotten. You get this wonderful mix of lichens and bryophytes … because, they've had over twenty years to establish themselves and they've done a really good job of it you know. And, it's, it's really attractive and very interesting to study these things at the side of the tracks on the edge of the forest. My friend John, the lichenologist, has found some quite unusual species of lichens on the forest edge, reckons in Whitelees, where we haven't had before. And, over on the, the north side of the forest, where we have these things called boulder fields, which I think have already been spoken about, John has ... spoken about these, yes. They've got an amazing variety of stuff over there, on the boulders. So, yes, the forest's not just a, a mass of trees, like people tend to think.

Simpson 2006

The moss cushions had reached up to 0.89 m (35 in) high sometimes topped with lichens and heather; they grew *en masse* along peaty road verges and over unplanted rides. They consisted mainly of *Poytrichum commune*, *Hylocomium splendens*, *Hypnum cupressiforme* and *Sphagnum* species, with wet hollows between. After centuries retarded by stock grazing followed by 40 un-grazed years, peat development was producing 'hummock and hollow'. Nine species of *Sphagnum*, the gloriously-coloured bog mosses, were recorded at Whitelee Forest.

What lichens get up to in Sitka spruce forest

Bryan's field colleague John Douglass surveyed the lichens of Whitelee Forest and adjoining moorland. I asked him to describe their ecological role:

RUTH So, what eats lichens? Do they, do they start food chains?

JOHN Oh absolutely. Well, because they're at the bottom of the food chain, they are producers, they'll, you know, they … they do have a lot of things associated with them, and maybe to start out with, you've got things like bark lice, which are things called psocids ... pronounced like 'sockid'... So these are minute little creatures about a millimetre in length. And you've got other things called oribatid mites, and they're, they will also live on and feed on the lichens, they'll feed on the reproductive structures that are called soredia, these are very powdery structures.

RUTH Right.

JOHN So, they just munch away on that sort of stuff. Or they'll munch around on the surface of the lichen. And they use it as an environment, as a habitat, they'll grow, they'll live underneath them. Quite often if you take a lichen off a tree or peel a bit back or peel some of the lobes back, you'll find hundreds of these little mites just living and mooching around under there. So it's an entire little world, a little ecosystem.

So I mean, there's actually, there's probably, well, dozens and dozens of invertebrates which will either eat lichens or eat things that eat lichens.

RUTH Yes. Yes.

JOHN There's various spiders associated with them, there's moths and caterpillars. Some caterpillars actually eat lichens, others, the moths are very good at using them as camouflage or ... or... So, you know, there's various ways in which animals are associated with it. Other animals, mammals and things, and, well, if you go

FIGURE 15.3. Beautiful bog moss hummock under Sitka spruce canopy; Whitelee Forest, Lanarkshire.

PHOTO: ANDY TITTENSOR, 2006.

> to the birds, birds use them in their nests, you know, use them as nesting material, which helps to camouflage the nest.
>
> Douglass 2002–6

John listed the many habitats used by lichens within Whitelee Forest: Sitka spruce trunks and branches, track verges, furrows and drains, peat cuttings, wooden fences, concrete and wooden fence posts, rock exposures, boulders, underhangs of gulleys, burns (streams), drystone dykes and stells (sheep folds), blanket bog. Some habitats were present before the Forest, some came with it. A few lichens which had somehow survived stunted on moorland then proliferated in the new shelter of the forest and grew more normally. In four years John identified 105 species of moss, 34 liverworts, nearly 200 lichens and 14 ferns on the Forest.

Major role of fungi

The function of fungi in woodlands is to recycle nutrients and energy. Some break down dead organisms into soluble mineral nutrients by decomposition, others assist trees take up nutrients from the soil water into their roots: these are *mycorrhizae*.

Mycorrhizae are associated with the roots of green plants in soil. They use plant sugars made by photosynthesis in the green plant; the green plants take up mineral nutrients and water from soil via the fungi. Mycorrhizal fungi are necessary to tree growth in nutrient-poor soils, because they allow a tree to obtain sufficient nutrients to grow.

Sitka spruce mycorrhizae are fungal threads forming sheaths around its roots. Between trees, in the soil and leaf litter, the threads form thick mats, 'mycelia', sometimes linked with those of other Sitka spruce trees. Mycelia produce familiar toadstools above the ground.

In poor soils, Sitka spruce seedlings with mycorrhizae have more extensive root systems and take up potassium and phosphorus better than those without. In North America, Sitka spruce has relationships with native mycorrhizae. In Britain and Ireland, it needs to form relationships with our mycorrhizae to grow well. Its seedlings can form relationships with a variety of fungi, such as those connected to birch and Scots pine, but as the trees get older, they are more selective.

There is a rich mycorrhizal flora in upland spruce forests, increasing with the age of the trees until the canopy is complete, when the tree needs most nutrients and grows fastest.

When and wherever seeds of Sitka spruce germinate they need threads or spores of suitable mycorrhizae to be already present in the soil. That is fine in native forests but in clear-felled plantations the fungal spores and mycelia may disappear if replanting is delayed, because, similarly, *they* need Sitka spruce! However, small mammals like voles, which eat fungi, disperse the spores in their droppings.

FIGURE 15.4. High rainfall and humidity within the 6070 ha (15,000 acre) Whitelee Forest encourage Witch's Hair lichen (*Alectoria fuscescens*) on a derelict farm fence-post.

PHOTO: ANDY TITTENSOR, 2006.

FIGURE 15.5. Lichens and algae on the rugged bark of a Sitka spruce in Benmore Arboretum, Argyll.

PHOTO: ANDY TITTENSOR, 2012.

Research at Inchnacardoch Forest showed that Sitka spruce grew better in plantations with other conifers than alone, because they 'nursed' it via their mycorrhizae. It also grows better on sites with leguminous plants such as broom which increase the soil nitrogen. In Ireland, the two native gorse species, *Ulex gallii* and *U. europeaus* have the same positive effect on Sitka spruce and are deliberately retained.

There is a succession of mycorrhizae associated with Sitka spruce and Lodgepole pine roots. *Laccaria* species are followed by *Inocybe longicystis*, by *Lactarius rufus* then *Cortinarius croceus*. The dark red toadstools of *Russula emetica* often finish the sequence, as under the canopy of Whitelee Forest. All are common woodland fungi.

Mycorrhizae of Sitka spruce forests in the Republic of Ireland are similarly diverse, with 144 species in sample plots compared with 269 in the UK, and 127 in Canadian Sitka spruce plots.

Butt rot fungus (*Heterobasidium annosum*) is a decomposing fungus which colonises recently-cut conifer stumps, grows down into the stump's roots and into roots of nearby living trees. It decays the lower trunk and weakens or kills them. Prophylactic (*before* infection) control of Butt rot fungus is being tested by researchers.

We give fungi in Sitka spruce forests a value according to whether they have positive, negative or no effects upon this commercially valuable tree species. However we label them though, they contribute to the web of ecosystem functioning and show that fungi in Sitka spruce plantations are integral to the energy and nutrient cycles.

Changes in bird life

Declines of favourite moorland species such as curlew (*Numenius arquata*) and lapwing (*Vanellus vanellus*) motivated bird watchers to observe what was happening when Sitka spruce forests were planted on moorland. Three well-known ornithologists stated in 1949: 'The bird life of uniform conifer forest is recognised as one of the weakest in Europe' (North *et al.* 1949).

Foresters usually erected perimeter fences round new forest on moorland, to prevent ingress by sheep and cattle. In the absence of nibbling teeth, grasses, plants of the heather family, gorse, broom (*Cytisus scoparius*) and herbaceous plants grew to new heights and density. Previously-absent flowers and insects attracted small birds such as Tree pipits (*Anthus trivialis*) and Grasshopper warblers (*Locustella naevia*).

Long-term observer of Welsh wildlife, William Condry, knew that sheep-grazed moorlands were actually rather poor in wildlife species and suggested that replacement conifer forests could be better! He noticed that the new Welsh forests suited some species, for instance: European red squirrel, its predator Pine marten, Common crossbill (*Loxia curvirostra*), siskin (*Carduelis spinus*) and Roe deer. His early, 1950s research in upland Wales showed how the bird and mammal life alters after afforestation.

He observed that moorland birds, for instance curlew and lapwing, declined first. Within a decade after fencing, vole populations reached exceedingly high levels amongst the grasses and flowers surrounding the small trees. They ate seeds and grasses, tunnelling and nesting in them. Short-eared owls (*Asio flameus*) came and fed on the voles, produced bigger clutches of nestlings which attracted Red foxes. When vole numbers eventually crashed, owls and foxes reduced or disappeared.

The habitat gradually changed to bushy grassland. Pipits disappeared from the much denser vegetation, but small birds like Willow warblers (*Phylloscopus trochilus*), dunnocks (*Prunella modularis*) and yellowhammers (*Emberiza citrinella*) ensconced themselves. When the spruces were recognisable trees, finches such as bullfinch (*Pyrrhula pyrrhula*), chaffinch and redpoll (*Carduelis flammea*), and thrushes such as blackbirds (*Turdus merula*), Song thrush (*T. philomelos*) and Ring ouzel (*T. torquatus*) colonised the young woodland. When tree canopies met, Wood pigeons (*Columba palumbus*) and Garden warblers (*Sylvia borin*) started to nest in them, but the smaller yellowhammers and Willow warblers moved away.

Small birds of undergrowth relocated and the trunked trees now supported canopy-nesters like jays (*Garrulus glandarus*) and crows. Small birds which eat conifer seeds then colonised the forests: siskins, goldcrest (*Regulus regulus*) and Coal tits (*Parus ater*). When the Sitka spruce reached 20–30 years old, obvious undergrowth became sparse. Along fire-brakes and rides, vegetation was descended directly from the previous moorland but could grow unhindered. Kestrels (*Falco tinnunculus*), sparrowhawks (*Accipiter nisus*), Tawny owls (*Strix

FIGURE 15.6. Common crossbill close to Whitelee Forest, Ayrshire.

PHOTO: ROGER PERRY, 2011.

aluco), Long-eared owls (*Asio otus*) and buzzards (*Buteo buteo*), which are tree-nesting raptors colonised the trees as they approached harvesting age. A similar succession of birds was noted by ornithologists in other regions, for instance the Caithness Flows.

Crossbills are adapted to feeding on conifer seeds. The British Trust for Ornithology (BTO) followed the distribution and abundance of crossbills in Ireland and Britain from 1968. The endemic Scottish crossbill (*Loxia scotica*) is confined to conifer forests in parts of the Scottish Highlands while the Common crossbill occurs throughout Britain in both lowland and upland conifer forests. In Ireland, crossbills live mainly in the afforested hills.

There has been an overall upwards trend in numbers and range since 1968 which is associated with the post-war forests of Lodgepole pine, larches and Sitka spruce reaching maturity and seeding. In Wales, crossbills are particularly abundant in years when Sitka spruce trees cone abundantly. Since 1984, their *winter* range increased by 177% in Britain and 11,150% in Ireland; since 1972 their *breeding* range increased 253% in Britain and 8133 % in Ireland. Increases were less after 1991. Figures, distribution and abundance maps are presented in the *Bird Atlas* published by Balmer *et al.* in 2013.

In Kielder Forest, newly-felled areas attracted fauna of the early stages of afforestation. Long lengths of 'forest edge' attracted Black grouse and pheasants again. Older plantations attracted yet more bird species: Tree creepers (*Certhia familiaris*), Green and Great spotted woodpeckers (*Picus viridis* and *Dendrocopus major*), Tree pipits and Spotted flycatchers (*Muscicapa striata*).

The usual favoured habitat of capercaillie (a very big grouse) in our lifetimes has been native Scots pine forest in Scotland, where they eat bilberries in summer and pine needles at other times. But Nicholas Picozzi discovered capercaillie living in the Sitka spruce-dominated Keillour Forest, Perthshire.

Here, Sitka spruce needles provided their main food in autumn, winter and spring while sedge fruiting heads were their summer food.

The 875 km^2 Isle of Mull off the west coast of Scotland has 4192 ha (10,358 acres) of state-owned conifer plantations, of which Sitka spruce makes up 61.5% or 2557 ha (6318 acres). These maturing and restructured forests provide a great variety of nesting sites for Hen harriers (*Circus cyaneus*) for instance: in second rotation forest, failed plantations, beneath trees up to 5 m (16 ft). Open moorland within the forest boundary is no longer grazed by sheep and cattle. Its grass and dwarf shrub vegetation is therefore tall and thick, making suitable habitat for small mammals, the prey of Hen harriers. They also nest in this vegetation. The combination of conifer forests and ungrazed, unburned moorland provides suitable ecological conditions for this raptor.

The effects of afforestation on Golden eagles (*Aquila chrysaetos*) in three regions of Scotland were also analysed. Raptor enthusiasts will be pleased that in most cases the amount and distribution of forest has not caused golden eagles to cease occupation of their home-ranges nor has it reduced their nesting success. The amount of high, open, ridge habitat within their territory is more important.

The 'loss' of moorland birds, compared with the 'gain' of woodland species is frequently lamented. There was particular concern at losses of birds when 140,000 ha (345,947 acres) were rapidly afforested with Sitka spruce in southern Scotland: 'The breeding population losses of some birds are substantial. The curlew has suffered a massive decline across the region since 1950 and, while not all of this is the result of afforestation, a large part of it undoubtedly is' (Ratcliffe 2007).

Contributors to the Whitelee Forest oral history project had seen curlews, snipe, lapwing and other moorland birds decline *before* afforestation, possibly due to drainage. Farmers had paid Irish workers to dig parallel drains over vast areas of moorland in an attempt to dry the dangerous bogs and increase the amount of heather and grassland for stock grazing.

When favourite species decline, it is helpful to see it as *change* rather than *decline* or *loss*. Woodlands covered 60–90% of the land surface for at least four millennia during post glacial times. A lesser proportion of the overall Scottish landscape was thus available to curlew and lapwings than today where only 17% is tree-covered.

Decline, *loss* or *gain* (alias ecological change) occurred throughout the post-glacial period and current population changes are the latest – but not the last.

Animals we call 'vermin', 'pests' or 'resources'

Animas classed as 'vermin' or 'pests' by foresters are controlled because their numbers and behaviour cause unacceptable damage to trees. Vermin are members of the forest ecosystem which happen to conflict with our wishes or needs. Rabbits, Brown and Blue hares, feral sheep and goats, Black grouse,

FIGURE 15.7. Even when felled, dead, and placed in old drainage furrows, Sitka spruce provides a habitat: song birds and hen harriers nest in those branches. Forsinard Flows Nature Reserve.

PHOTO: ANDY TITTENSOR, 2015.

FIGURE 15.8. Male Hen harrier in flight, Outer Hebrides.

PHOTO: PAUL HAWORTH, 2006.

FIGURE 15.9. Hen harrier chicks in their nest in long, grassy vegetation, Outer Hebrides.

PHOTO: PAUL HAWORTH, 2006.

Grey and European red squirrels, Field (*Microtus agrestis*) and Bank (*Clethrionomys glareolus*) voles have been vermin at times during the last century. They nip off terminal shoots of small trees, browse foliage, strip bark or girdle trunks at various heights. All except the voles and European red squirrels are eaten by humans and are therefore also resources.

Rabbits were super-abundant and damaged young forests in the decades before 1953 when myxomatosis was introduced to the UK. In King's Forest, Thetford, rabbits damaged 100% of the trees in 2428 ha (6000 acres) of young pine plantations and it took the forester and twelve trappers from 1947 to 1951 to clear – 77,079 – of them.

Nowadays it is more usual to see a plantation surrounded by a high deer-proof fence than a dug-in rabbit-proof fence round a plantation. Several species are now treated as pests but are also resources because their carcases are edible and saleable. There are six species of deer in the UK: Chinese water (*Hydropotes inermis*), Fallow, Muntjac (*Muntiacus reevesi*), Red, Sika (*Cervus nippon*) and Roe deer. Red and Roe deer are native species. Deer eat shoot tips of small trees and foliage of large trees, they strip bark from older trees and males fray saplings while rubbing velvet from their antlers.

FIGURE 15.10. A fallow deer has frayed this young Sitka spruce with its antlers, Margam Forest, Glamorgan.

PHOTO: FORESTRY COMMISSION.

After low wartime numbers, Roe deer quickly colonised the growing Sitka spruce plantations and reached population levels where bark stripping caused significant damage. Red and Sika deer have also increased in Sitka spruce forests, especially where fences need repairing. Muntjac deer browse heavily on smaller trees in forests but Sitka spruce are affected the least. The other deer are more common in forests of other tree species.

Fallow deer prefer deciduous or mixed woodlands with a good understorey of shrubs, grasses and herbs, but they browse young conifer shoots in winter and in very wooded Perthshire they damage Sitka spruce. When they strip bark, the bole tears, resulting in deformity or it breaks and the tree dies. Broadleaves are not now planted where deer populations are high so Sitka spruce is then the main timber tree.

During the past 50 years, there has been considerable research on deer in the UK. This means that legislative framework for their management has a strong scientific basis. Government agencies oversee deer management on both state and private land. Deer stalkers are obliged to take advanced training, to report details and send specific parts of all their victims to central laboratories. Forest-living deer are culled to a level which maintains them at densities acceptable to timber growers. Broadleaved trees can regenerate where numbers

FIGURE 15.11. The effects of Field voles (*Microtus agrestis*) gnawing the stem of a young Sitka spruce from Tywi Forest, Powys.

are 4–5 per km^2 (12 per square mile), but Sitka spruce is the only tree which regenerates if deer numbers are greater than 15 per km^2 (40 per square mile).

Four species of deer live in the Republic of Ireland (Fallow, Red, Sika and Muntjac deer) but none is native. They are over-abundant in Sitka spruce and other forests possibly because deer culling is less intensive and not overseen by an agency. Broadleaf woodland, biodiversity and economic returns are all suffering:

> Species Selection: Most tree species (in particular broadleaves) are vulnerable to significant deer damage. However, Sitka spruce is the least susceptible to damage and is frequently planted as the main species in areas with high deer populations.
> Coillte 2005

Another edible pest is the Feral goat which herds in mountainous areas; goats are culled and eaten if they damage forest trees. A self-perpetuating flock of Feral sheep lived within remote parts of Whitelee Forest for several decades but little is known of their habits.

Coning plantations of Sitka spruce are colonised by squirrels. There are native and re-introduced populations of European red squirrels, which, until about 1970 were controlled as vermin for girdling bark on several tree species (A. Tittensor pers. comm.; discussed by Shorten in *Squirrels and their Control*). This century they are believed to need conservation. Clocaenog and Kielder Forests hold the largest European red squirrel populations in their respective localities.

Although introduced Grey squirrels live at low densities in Sitka spruce forests, conifer seeds are not their primary food and bark stripping is rare: they are perceived more as competitors to European red squirrels.

The rare Black grouse were shot and eaten in the past, but that is now frowned on and they are considered a conservation resource. Their preferred habitat consists of a mosaic of woodland, moorland and grassland, but they sometimes live in at their edge of young Sitka spruce plantations and those which have succumbed to fire. Closed-canopy plantations are used by males during severe winters. In their new Sitka spruce habitats Black grouse eat spruce shoots instead of their more usual heather and bilberry.

'Disease' organisms

Sitka spruce is remarkable in its resistance to tree diseases caused by fungi and invertebrates.

The Green spruce aphid (*Elatobium abietinum*) sucks juices from the needles but does not kill a tree; the Great spruce bark-beetle (*Dendroctonous micans*) can kill Sitka spruce but in the UK it is kept at a low level by biological control and has little effect.

Other beetles are more serious pests of dead trunks and stumps, so that clear-felling can cause an outbreak of disease. The Large pine weevil (*Hylobius abietis*) lays its eggs in, or under, cut-stumps and the larvae make galleries within the wood as they feed. Mature weevils emerge from the stumps in late summer and feed on the inner bark of newly-planted young trees.

Populations of these invertebrates can be reduced by clear-felling much smaller areas and delaying replanting for several years to deny them food from the small trees.

By making value judgements of species as pests or food we perhaps overlook the fact that they are species which have adapted successfully to living as members of Sitka spruce ecosystems.

FIGURE 15.12. Painted Lady butterfly.

PHOTO: ANDY TITTENSOR 2015.

Mammals, reptiles and amphibia

In Whitelee Forest local people have seen badgers (*Meles meles*), stoats (*Mustela erminea*), weasels, otters (*Lutra lutra*), American mink (*Neovison vison*), Red foxes, Roe deer, Grey squirrels, bats, Water voles, Brown and Blue hares, rabbits and evidence of Pine martens. Grey squirrels, bats, Roe deer and Pine martens colonised the new Forest, the rest lived on the moorland previously and adapted to the new habitats. There has not yet been a study of how these species use the Forest habitats seasonally for food or shelter.

Common frogs (*Rana temporaria*), Common toads (*Bufo bufo*), adders (*Viper berus*) and two lizard species frequented the previous moorland and have also adapted to Forest conditions. Local people state that the 90 waterways were acidified by afforestation, that this and growth of long vegetation across them caused a drop in fish populations.

Small animals

Invertebrates are often brushed off or crushed by us. In forest ecosystems they function as pollinators, decomposers, herbivores, predators and prey, so are vital participants in Sitka spruce forest ecology, as described by Oxbrough *et al.* (2010).

Beetles in Kielder Forest were analysed in young, old and felled Sitka spruce compartments and then compared with those in adjoining moorland.

Similarly to birds, the beetle species altered when the trees grew and were felled. Surface-living, quickly-running species colonised the felled areas, which otherwise held few species. After re-planting with small trees however, beetles lived in their highest numbers and variety, because prey was abundant. Closed canopy plantations held the least beetles, but these are only a proportion of the complete rotation.

Mabie Forest, Dumfries-shire is part of the 97,000 ha (239,692 acres) Galloway Forest Park and consists mainly of Sitka spruce. In an open, grassy area with some oaks, there is a butterfly reserve which holds 23 of Scotland's resident 32 butterfly species (Table 8). Butterflies also use the sheltered forest rides and feed from 'new' flowers.

The functional role of spiders in forest ecosystems is in nutrient cycling and as pollinators. In Ireland, ground-living spiders within and adjoining Sitka spruce plantations one and eight years old were surveyed – they are easy to catch in pitfall traps. Researchers caught and identified 16,741 spiders of 141 species: large numbers and high diversity. There was no difference within and next to plantations. A Scottish study found that snails, however, are few in species but in great abundance within Sitka spruce plantations (R. Marriott pers. comm.).

TABLE 8. Butterflies recorded in Mabie Forest Butterfly Conservation Reserve near Dumfries.

SOURCE: UK BUTTERFLY MONITORING SCHEME, 1995–2013; SEE ALSO BUTTERFLY CONSERVATION, MABIE FOREST NATURE RESERVE LEAFLET (2014).

English Name	Latin Name
Clouded Yellow	*Colias croceus*
Common Blue	*Polyommatus icarus*
Dark Green Fritillary	*Argynnis aglaja*
Dingy Skipper	*Erynnis tages*
Green-veined White	*Pieris napi*
Large Skipper	*Ochlodes sylvanus*
Large White	*Pieris brassicae*
Meadow Brown	*Maniola jurtina*
Northern Brown Argus	*Aricia artaxerxes*
Orange-tip	*Anthocharis cardamines*
Painted Lady	*Vanessa cardui*
Peacock	*Aglais io*
Pearl-bordered Fritillary	*Boloria euphrosyne*
Purple Hairstreak	*Favonius quercus*
Red Admiral	*Vanessa atalanta*
Ringlet	*Aphantopus hyperantus*
Small Copper	*Lycaena phlaeas*
Small Heath	*Coenonympha pamphilus*
Small Pearl-bordered Fritillary	*Boloria selene*
Small Tortoiseshell	*Aglais urticae*
Small White	*Pieris rapae*
Wall Brown	*Lasiommata megera*

Kielder Forest ecology

Kielder Forest attracted wide interest as it grew into the largest man-made forest in England. Seventy years after its start in 1926, Kielder had changed and been changed. A symposium gathered together ecologists for an up-to-date picture.

Pairs of the small raptor, the Merlin (*Falco columbarius*) increased from ten in 1982 to 29 in 1991, because they found more nesting sites in the longer forest edges. Fledged broods were as successful in the forest edge as in other nesting sites. Field voles in long grassland amongst young trees were studied for 23 years, during which their populations went up and down in waves of three to four years. Their abundance was not coincident with numbers of predatory Short-eared owls or kestrels, whose numbers and density declined steadily. Although Tawny owls ate a large proportion of the voles, places where owls were absent had the same density of voles as where owls foraged: there seemed no connection. It was suggested that clear-fell openings might prevent voles expanding their ranges and populations, but other explanations are possible.

Goshawks *(Accipiter gentilis)*, large birds of prey, were re-established at Kielder Forest in the 1970s and have increased successfully without the expected persecution. Although they feed mainly on birds during their breeding season, goshawks took between eight and 261 European red squirrels per year for food (average 79) over a 23-year period. This is a very small proportion of the squirrel population (estimated at 1294 to 5556) which are thought to be limited by cone crops, not predators.

As Kielder Forest holds 70% of England's European red squirrels it is an important stronghold. The deliberate decrease in Sitka spruce and increase in broadleaved trees may favour Grey squirrels, which consume bigger seeds. A list of flora and fauna recorded in Kielder Forests is available from the Forestry Commission.

'Dirt' (soil) and water

Plant roots, fungi, earthworms, bacteria and voles (for instance) carry on their lives in the 'dirt' under Sitka spruce forests. 'Dirt' consists of ecosystems where some organisms decompose other, dead ones. Alchemists (fungi) change organic remains into mineral nutrients in solution ready for re-use by growing plants. Microbes such as protozoa and bacteria have a functional role as alchemists too. They change living and dead soil organisms into reserves of inorganic nutrients, into compounds which help hold soils together – and carbon dioxide. Their diversity rises to the middle of a rotation period and then declines.

Sitka spruce roots spread shallowly in upland UK soils, but if treelets are planted on mounds instead of ridges they grow deeper. Humus accumulates on the soil surface during afforestation as new trees and other plants arrive and grow, and when animal carcases decompose. Nutrients usually build up so that fertilising the second tree crop is not always necessary.

Water processes are closely related to soil, topography and rainfall. Breaking the surface of peat by ploughing alters its hydrology. Water feeds more quickly into local waterways via new deep, forestry drains and people familiar with nearby stream and rivers notice that they rise and fall more quickly and carry heavy sediment loads after forestry work. Furrows hold water initially, but as the years pass, sediment and vegetation infill them.

Beddgelert Forest, Caernarfonshire, was an extremely difficult area to afforest, with an average annual rainfall of 2540 mm (100 in) and ground almost entirely of peat bogs. In 1926, a huge hollow containing a raised bog was ploughed and planted with Sitka spruce. As they grew, the bog lost its water content and by 1954 had almost completely dried-up.

In the Republic of Ireland, as in the UK, new forests were restricted to poor soils:

> In fact the unstated land-use policy up to relatively recent times might be articulated as only the worst is bad enough for forestry… The result was a situation where for many years only the worst acid and infertile land was acquired and planted. It is therefore not surprising that the related streams are likewise acid.
>
> Joyce and OCarroll 2002

Sitka spruce grows naturally on soils with some saline content, in river flood plains and on mineral soils. Water from soil or atmosphere or both is a particularly important ecological requirements throughout its life, whether as rain, fog, mist, dew, snow or in the soil. The drier the soil, the more important is summer mist or fog.

Biodiversity research

In 1995 a Biodiversity Assessment Project was set up in response to the UNCED Rio Summit. Its aim was to provide a base-line for biodiversity in planted forests, initially by survey. Its aims were conservation of genetic resources, of species, of 'special' habitats and 'improvement' of plantation forests.

To assess the biodiversity of a forest is very time-consuming. So ecologists sometimes assess only 'indicator species' which reflect the presence and abundance of other species in the locality. In this biodiversity project, Forestry Commission ecologists studied *composition* (species present), *structure* (how forests are put together in 3-D space) and *functions* (processes such as natural regeneration).

Numbers of indicator species in Sitka spruce (introduced) and Scots pine (native) forests were analysed. Overall species richness showed no difference between the two, with 623 species in Sitka spruce and 627 in Scots pine forests. Half of the groups were more diverse in Scots pine, the other half in Sitka spruce. For instance, there were fewer lichen species in the spruce than the pine forests but mosses and liverworts occurred in similar numbers, and fungi were more diverse in spruce habitats.

The ecologists were unable to predict how indicator species groups might behave in other 'native' or 'introduced' woodlands as there was no pattern to species distribution according to the forest type in which they grow. It was not the tree species which primarily determined what lived where, but other factors such as latitude, topography, amount and type of dead wood.

Natural regeneration

In response to criticism, foresters have sought to produce Sitka spruce plantations with structures and processes akin to 'natural' forests. Natural regeneration is an example. Sitka spruce plantations nowadays produce abundant young trees in canopy gaps of many sizes. Upturned root plates provide similar conditions to the fallen trunks of its native forests where young trees often grow. However, a mass of seedlings and saplings is more complex to manage for timber than straight rows of planted seedlings with rows and plants equidistant!

The Forest Service of the Republic of Ireland (Coillte) states on its web site:

> Sitka spruce grows well in Ireland because it is suited to our soils and climate. As proof of its 'ecological fit' the species flowers, produces seed and is able to regenerate naturally. The species has thus adapted to the Irish environment rapidly and many native animals, insects and birds now inhabit Sitka spruce woodlands.
>
> Coillte 2015

Glasfynydd Forest is situated at 305–546 m (1000–1793 ft) on bleak, exposed mountains of west Wales, with 2.0–2.3 m (80–90 in) annual average rainfall. By 1952, a compartment of Sitka spruce only 23 years old was already regenerating, with many seedlings growing in peaty drains and on mineral soils in gaps produced by thinning. This was probably helped by the unusual absence of rabbits, although mountain hares were there.

In Galloway Forest Park, south-west Scotland, up to 20% of restocking is now from prolific natural regeneration, not planting. In fact, it is difficult to maintain the 10% of open space normally left treeless – it gets filled-in by young Sitka spruce!

In Clocaenog Forest, Denbighshire, there is a glut of thickly-packed, young Sitka spruce trees on open ground produced by wind-blow or clearfell. Red squirrel conservation is a main aim in Clocaenog Forest, yet they prefer to eat seeds of *Norway spruce* – which barely produces any offspring there. This makes difficult choices for the foresters who would prefer the more desirable seed-tree to regenerate like the Sitka spruce.

FIGURE 15.13. Wood wasp or Horntail (family Siricidae, Hymenoptera) on spruce, Torrachilty Forest, 1980.

PHOTO: GEORGE DEY, COURTESY OF THE FAMILY OF GEORGE DEY. © ABERDEEN UNIVERSITY.

Conclusions

Plants and animals have individually assembled into Sitka spruce plantations and are carrying on their lives. Our knowledge is patchy, we can see few overall ecological patterns and we are a century behind our understanding of deciduous forests.

However, there is no longer any need (was there ever?) to compare Sitka spruce plantations unfavourably with other woodlands. I suggest we put away our nostalgic, rose-tinted daydreams of desired Sitka spruce forests. We can then look forward and discover, without prejudice, just how Sitka spruce forests do evolve.

I suggest we also finish with the dogma of 'naturalness' which has forced organisations and individuals to try merging (native, natural woodlands) into one identity with (Sitka spruce look-alike native natural woodlands) – instead we might let Sitka spruce forests develop ecologically without pre-conceptions. This will increase ecological variety in the form of *new* ecosystems and perpetuate post-glacial ecological change into the future.

CHAPTER SIXTEEN

Sustainability in North America

> Throughout the Panhandle, the largest forest stands and individual trees grow on river floodplains, sloping alluvial fans, and uplifted deltas. These stands, where the volume of wood exceeds 30,000 board feet per acre, are the most coveted by timber companies and the most valuable for wildlife as well... At the end of the twentieth century, industrial logging had erased nearly one million acres of this productive forest ecosystem.
>
> Kathie Durbin 2005

> The long-held idea of a pristine forest, a blanket of green stretching from sea to sea ... and unsullied by human intervention is not supported by historic or scientific evidence.
>
> Ken Drushka 2003

> We love Sitka spruce here!
>
> Andrew Thoms, Sitka Conservation Society, verbal comment 2014

Although immense trees were used to build canoes and homes, there is little evidence that First Nations people clear-felled trees on a large-scale comparable with modern logging and extraction. But from California north to Seattle and on Vancouver Island, people set fire to some rainforests, thinning them out in favour of 'parkland' so that more shrubby food plants would grow. The maritime climate of coastal BC and Alaska precluded using fire to change forest structure there.

First Nations people harvested from forests rather than deliberately changing them. Their effects upon coastal rainforests were subtle and fluctuating; archaeologists and ecologists are learning to recognise the signs, but there is scope for more research enquiry.

Interaction between humans and rainforests increased in variety and amount during recent, European centuries: first the fur trade, ships' masts and spruce beer, then fort and farm construction, harbours, boats, fuel for cooking and warmth and export to a forest-less United Kingdom. European settlements and their stump-pocked farmland involved laborious felling of huge trees in chosen enclaves in the benign, southern rainforest terrain; in Alaska, a shorter growing season, less flat land, inclement weather and huge distance from markets saved rainforests from axe, saw and harvesting machine until later.

Logging by incomers

Seventeenth–early nineteenth centuries

Europeans felled and used trees in eastern North America intermittetently since Leif Eririksson took away a boat-load of timber in 1000 AD. During the sixteenth century large fishing fleets needed timber for harbours, buildings, fish-drying racks, fuel and mending boats. At the same time, the significant trade in Beaver pelts affected the forests vicariously: when Beavers (*Castor canadensis*) became scarce in coastal areas, Europeans travelled far inland up river valleys. Their fur trade travels stimulated early European exploration of inland North America.

When Russian naval expeditions explored coastal Alaska in the 1740s, they harvested pelts of Sea otters (*Enhydra lutris*), a trade which lasted about 60 years until near extinction of the otters. The fortress town of New Archangel (now Sitka) was built on (what is now) Baranof Island in 1804, as Russia's Alaskan capital. Russian settlers needed timber and wood for harbours, forts, warehouses, homes, a cathedral and fuel. A boat-building industry developed at New Archangel.

Naval surveyors and commercial explorers from Spain and Britain visited the northern Pacific coast too. When James Cook sailed along the British Columbia and Alaska coasts in 1778, he anchored frequently and his sailors took tree trunks for new masts as well as other plants and animals for food. The next decade, a 60-year export trade in mast spars started. A Spanish military fort was built on Nootka Island in 1789. The forts built by the Hudson Bay Company were timber, for instance Fort Vancouver, finished in 1825; the Company's workers opened transport routes through forest to link the forts.

People walked and canoed to the west coast from the newly-independent eastern country of America after 1776 to settle along the coasts of Oregon

FIGURE 16.1. Stand of Sitka spruce and Western hemlock saw timber near Ketchikan, Alaska, 1940.

PHOTO: US FOREST SERVICE, COURTESY WEYERHAEUSER USA, WEYERHAEUSER ARCHIVES T20-0051.

and Washington. The still-British coast north of Puget Sound also attracted incomers who explored, collected natural history specimens or attempted to farm and settle down.

Hopeful immigrants came from China, India, France and Quebec in the later nineteenth century to work for logging and mill companies. They were ill-treated, suffered discrimination, and diseases like cholera were common. The new nation of America and province of Canada needed large quantities of timber and wood; so did the old home country, the tree-less UK. The logging and milling industry developed, moved and expanded to wherever big timber could be obtained.

Industrial logging in the later nineteenth century

Cutting down and removing huge Western red cedar, Douglas fir, Western hemlock, Sitka spruce and other trees was a monumental task needing tough, hard-working, resourceful teams of men. They worked long hours, felled trees up to 5 m (16 ft) in diameter from high springboards using axes and long, two-man saws. They cut off side branches; men called 'buckers' then cross-cut some of the fallen trunks into shorter lengths, others removed bark. With a horse, ginpole or donkey-engine, each felled timber was dragged and lifted into position for the ox-train. Long teams of oxen then dragged or 'skidded' the felled trunks to the collecting yard, river or lake – whatever the state or slope of the ground. If possible, skid roads were made of logs placed transversely across the track. To aid the oxen, a lad with an oil-can sprayed the skid road to encourage the trunks to slide along it.

There were water flumes, skylines and inclines to build. In due course, industrial-sized mills, freighters and river rafts had to be fed with timber. Machinery was developed for some of these jobs: petrol-driven saws, locomotives, tractors and then railway networks.

Railways too, opened up forest access for logging: long, wide swathes of trees were cleared for railways, both cross-continent and local routes. Branch lines to logging sites and mills were a feature of the early twentieth century. Railways themselves needed huge amounts of timber for cross-ties (sleepers), for their wooden wagons and carriages, for building bridges across deep ravines and for snow tunnels.

During the nineteenth century, the logging industry was mostly small-scale; selected species and sizes were felled, leaving small trees and brash. In the early twentieth century it became large-scale and industrial. Export increased after the Panama Canal was opened in 1914 and European markets became very accessible to North American west coast industry.

From 1890 to 1945, the coastal logging industry south from Puget Sound to the Columbia River was photographically recorded by the brothers Darius and Charles Kinsey. Their photographs show many logging processes in detail, highlight workers carrying out their varied tasks, the animals at work, the changing machinery.

FIGURE 16.2 *(right)*. A logger stands on a spring board, preparing to 'fall' a Sitka spruce, Edna Bay, Alaska, 1940s/50s.

PHOTO: COURTESY ALASKA STATE LIBRARY, DORA M. SWEENEY PHOTO COLLECTION P421-291.

FIGURE 16.3 *(below)*. Former abundance of timber: a very large Sitka spruce logged by timber firm Bloedel Donovan on the Olympic Peninsula in 1925; note the lumbermen and their tools.

PHOTO BY DARIUS KINSEY, COURTESY WHATCOM MUSEUM BELLINGHAM, USA REF. 2004.50.15.

They even photographed trees as beautiful objects and the forests as scenic or destroyed landscapes. The very tall, straight, close-grown trees of the rainforest are recognisable from their individually-patterned bark. Sitka spruce is not the most common tree in the photographs, probably because the photographs were taken in the southern, more-mixed-tree-species rainforests.

The First World War Spruce Production Division sites and their soldier workers were included in the Kinsey portfolios. They show us the size, not just of trees, but of the industry; they demonstrate the pioneer spirit and tough life, but also off-duty times and the all-important, well-provisioned canteens. In more than a half-century of high-quality photography, the Kinsey brothers produced probably 50,000–60,000 negatives of the Pacific Coast temperate rainforest logging industry. Photographs are held in museums and university archives in the Pacific North West and a selection can be viewed online.

Official protection of temperate rainforest

Enclaves of Pacific rainforest were protected and scheduled as wilderness, for nature conservation, or for multipurpose, sustainable land use from the late nineteenth century.

After half a century of over-harvesting, the Sitka spruce-containing, temperate rainforested Olympic Peninsula was designated as the Olympic Forest Reserve in 1897 and Olympic National Forest in 1907 as a means of conserving the old-growth forest. It became the Olympic National Park in 1938 and a World Heritage Site in 1981. It is the largest temperate rainforest reserve in the USA; although regarded as wilderness, outdoor recreation is allowed but there are few roads and developments. A peripheral belt of commercial rainforest around the National Park is run by the Forest Service as the Olympic National Forest. It has sufficient roads to harvest and extract timber, oversees wildlife conservation and encourages visitors.

FIGURE 16.4. *We have Timber in Canada.* Empire Marketing Board poster: 1935.632. The poster is nowadays an uncomfortable view of timber importing. Poster Design: Keith Henderson, 1935.

BY PERMISSION OF PICTURE GALLERY, MANCHESTER ART GALLERY, MANCHESTER CITY GALLERIES.

Sitka spruce logging

By the time William Boeing bought rainforest land near Grays Harbour in 1903 and started to build aeroplanes, the logging industry had been operating in the Puget Sound-Olympic Peninsula area for about half a century. There were about 250 big logging firms and, when new settlements put down roots, each built its own water-powered sawmill.

FIGURE 16.5. Tree trunks of many species piled into a 'cold deck' containing over 2,360 m³ (1 million fbm) of timber, *c.*1930; the loggers perching on it show the size of the deck.

PHOTO BY DARIUS KINSEY, COURTESY WHATCOM MUSEUM, BELLINGHAM, USA, REF. 1978.84.696.

Figures from the United States Department of Agriculture (USDA) Forest Service show that annual production of Sitka spruce had reached 165 million board feet (389,357 m³) in 1899, with a maximum production of 508 million board feet (1,198,746 m³) during the First World War. The logging industry had reached Haida Gwaii (Canada) and Tongass Forest in Southeast Alaska by then: 10 million board feet (23,597 m³) of best Sitka spruce came from a mill at Wrangell in 1919.

Although fewer aeroplanes were required at home and in Europe during the 1920s, top quality Sitka spruce was still sold for that purpose from Haida Gwaii and Alaska, reaching a high price of $350 (£70.28 at the representative exchange rate of 0.2008 in 1930) per 10 million board feet in 1927.

At that time, foresters noted that all the top grade Sitka spruce had been removed from littoral sites – where 'stumpers' could be felled straight into the sea and transported as rafts. The long coastline of the Southeast Alaskan archipelagos made it possible. This realisation prompted forest infrastructure to be built across less rugged islands to reach the best inland timber trees too.

Such selective logging and removal of just the best timber while leaving the rest of the trees (a practice called 'highgrading'), was a deplorable and deleterious practice common to both Canadian and American forestry. It considered only short-term gain, failed to heed the future of any forest or of forestry and altered the character of ecosystems and future succession.

When, during the inter-war period, metals rather than wood became the predominant structural materials for manufacturing aeroplanes, annual production of Sitka spruce in the USA gradually reduced to 80 million board feet (188,778 m³). Despite that, and because some American and British planes

FIGURE 16.6. Raft of Sitka spruce logs harvested from Prince of Wales Island, for Ketchikan sawmill, 1924.

ALASKA STATE LIBRARY, U.S. FOREST SERVICE PHOTO COLLECTION P207-31-05.

were wood-built during the Second World War, production increased to 400 million board feet (943,895 m^3) annually during the War. An Alaska Spruce Log Program was created and run by the Forest Service of the US Department of Agriculture: no other species than Sitka spruce was felled and then only high grade spruce. Most of the spruce came from the very south of Alaska where roads, nine camps, harbours and boomed marine pools (where log rafts could be put together) were built. The timber was hauled and slid into the boomed pool, built into large rafts and towed south for 700 nautical miles (1296 km) along the coast to Puget Sound where it was milled. Alaska contributed 35 million board feet (82,590 m^3) of Sitka spruce to the war effort between 1942 and 1944.

After two world wars and their thousands of aeroplanes, Sitka spruce had become scarce in California, Oregon and Washington. When, in the mid-nineteenth century, serious logging of Pacific Coast rainforest had got under way, the old-growth forests and their trees probably seemed inexhaustible!

Before the Second World War, Japanese business had cast envious eyes towards Alaska as a source of good quality Sitka spruce pulp to make rayon, an artificial fibre; the best trees were logged, and shipped to Japan.

After Second World War bombing, Japan had to be rebuilt. As a start, Japanese Prisoners of War in Southeast Alaska felled Sitka spruce which was exported to Japan for construction.

Sitka spruce from Alaska and British Columbia

In Haida Gwaii and particularly in Alaska, the growing season is short and cool, Sitka spruce grows more slowly and lives longer than further south in its range; the timber has narrower growth rings and greater strength. Its retreat caused by logging south of Vancouver Island meant eyes turned northwards in anticipation of economic riches there.

Political and security considerations also promoted expansion of forestry in Alaska in the 1940s and 1950s: a larger population, more infrastructure, better communication, industrial development and statehood were considered vital by federal government and citizens of Alaska.

The medley of tree species in Alaskan rainforests is different from further south, consisting of Western and Mountain hemlock, Sitka spruce and Alaskan Yellow cedar alias Nootka cypress and Red alder. Western red cedar is restricted to southern Alaska. Sitka spruce is less abundant than Western hemlock in Alaska but is now more desired by mills and Asian markets.

Old-growth forest at lower altitudes currently consists of about 75% Western hemlock and up to 25% Sitka spruce. In the cool climate and short growing-season, Sitka spruce grows slower than further south, so its annual growth rings are close together and the wood is denser; it also lives longer – to a millenium occasionally. On alluvial and riparian sites, Sitka spruce is currently protected by stream buffer regulations.

The British Columbia Forest Service grew from the jurisdiction over lands, timber and wood granted by the British to the new Dominion in 1867, from the need to control forest fires inland and from excessive clearance of timber on public land.

Public ownership of forest land in British Columbia and the USA Pacific Coast made it relatively easy to single-mindedly fell temperate rainforests without reference to either reforestation or to First Nations' domains. To keep a steady timber industry, statutory forest agencies gave logging or mill companies 50-year contracts and therefore very large areas of land to work on.

Pulp

The home of top quality Sitka spruce, Kosciusko Island, was the site of the first big sawmill in Southeast Alaska, which opened in 1879. It was followed by large and small sawmills which came and went across the region. They were joined by pulp mills requiring vast volumes of ready timber after the Second World War.

It was post-war Japanese involvement with Alaska forestry which stimulated the pulp industry during the 1950s. On Baranof Island, the Alaska Pulp Corporation mill opened in 1959; it was important to the economy of the small city of Sitka, because 400 of its 8800 population worked there.

The pulp mill took its desired Sitka spruce logs from commercially attractive old growth on valley bottoms, including alluvial fans and floodplains. Secondary forest of Red alder with scattered Sitka spruce regenerated naturally from seed on these low-lying sites. The upland forests, with more hemlock and Yellow cedar, were not so commercially attractive at the time.

Another mill opened near Ketchikan on Prince of Wales Island in 1954. On this less mountainous landscape, a network of roads could be built and much of the old-growth forest was highgraded time and again during the next 40 years. Salmon populations in logged river catchments reduced noticeably at the same time.

By 1984, 89% of the United States' commercial Sitka spruce grew in Alaska. The other 11% grew in Oregon, Washington and California. At that time, Alaska

FIGURE 16.7 *(left)*. Modern forestry: Heli-logging at Shelikof Island, Craig, Alaska. Taking timber from logging site to heli-site.

PHOTO: BRENT COLE, ALASKA SPECIALTY WOODS, 2009.

FIGURE 16.8 *(right)*. Collecting logs at the heli-site; the logs contain 5.9–9.4 m^3 (2500–4000 board feet) of timber.

PHOTO: BRENT COLE, ALASKA SPECIALTY WOODS, 2009.

sawmills still produced some small items for local markets, but 98% of the sawn, squared timber (cants) was exported to Japan.

Like many forest industries the pulp industry was short-lived. For 40 years the industry provided considerable employment but left behind derelict buildings, unemployment, and damaged old-growth forests.

The Alaska Pulp Mill at Sitka closed in 1993, the pulp mill at Ketchikan in 1997. Replacements for the timber industry had to be found so that people still had work. Tourism, supplying summertime cruise ships and health services helped fill the employment gap.

Sitka spruce in small-scale wood industries

New small-scale forest-based industries have gradually developed in Southeast Alaska, supported by the USDA Forest Service. For example, Brent Cole of Alaska Specialty Woods makes bespoke Sitka spruce guitar soundboards for a worldwide clientele; he obtains much of his slow-grown Sitka spruce from blown and dead trees, salvaged timber and occasionally from selected live trees. He uses 'heli-logging' to get the big logs to his workshop.

Icy Straits Lumber and Milling of Hoonah specialises in providing bespoke (custom made) construction timber for a variety of beautiful, local buildings (for instance, a clinic, scout cabins and a Forest Service laboratory), builds furniture, small boats and decking. Locally-sourced Yellow cedar, Western hemlock and Sitka spruce are its main resources.

Learning to manage secondary Sitka spruce forest

Forest management for the pulp sector in Alaska followed the usual pattern: select a harvesting area from the old-growth apportionment; fell the best trees

FIGURE 16.9 *(left)*. A log attached to the helicopter cable; each log weighs up to 1134 kg (2500 lb).

PHOTO: BRENT COLE, ALASKA SPECIALTY WOODS, 2009.

FIGURE 16.10 *(right)*. Helicopter above the Sitka spruce forest, ready to transport a log to the mill.

PHOTO: BRENT COLE, ALASKA SPECIALTY WOODS, 2009.

and leave the rest; remove and transport timber; move on to another old-growth forest. At the original site allow for secondary forest to grow.

Natural regeneration after logging was usually prolific, but was less satisfactory on coastal and fluvial sites where Salmonberry and abundant Red alder were competitive. It is now known that the Red alder ecosystem gives way after about 250 years and is replaced by Sitka spruce and the other conifers as a part of natural succession.

Young Sitka spruce trees in profusion are sometimes thinned out to a wide spacing of 4–5 m (14–16 ft) between individuals so that a more 'natural' forest stand grows up. Foresters and ecologists believe that after clear-felling, a tree-species combination which is similar to the previous old-growth forest of the site eventually assembles – as if the clear-cut were just an extra-large windblow area. Studies of regenerated forests of increasing age, has shown that there are recognisable phases of development. A new version of old-growth type forest containing trees of many ages and stages of decay with their dependent flora and fauna takes 250–300 years.

However, the economy, and timber markets cannot wait that long! In recent years it has been decided that there should be a transition to young-growth, short-rotation commercial silviculture in Tongass Forest – a system hardly used in the region yet. Studies of possible methods of thinning, managing and marketing 'young-growth' forest are ongoing. Sitka spruce is suited to this type of management, but Yellow cedar grows so slowly that what looks like 'young growth' could be one century old!

Foresters in the UK and Republic of Ireland, on the other hand, have had over a century in which to perfect short-rotation Sitka spruce silviculture, harvesting and landscaping – and may have some helpful tips.

The market value of Western hemlock and Sitka spruce in Southeast Alaska is being overtaken by (old growth!) Yellow cedar which is popular for construction and decking because it is decay-resistant. Unfortunately Yellow cedars are succumbing to the reduced snow depth of climate change, which is nowadays insufficient to keep their superficial roots alive in winter.

Sitka spruce is currently exported as round logs, to China, Korea and Japan. This hinders development of more and bigger, local, wood-using industries. Export is controversial, but is allowed because the home market for sawn timber is rather small. Only when the Alaskan economy is in recession are restrictions placed on timber exports; this also reduces further loss of old-growth rainforests.

First Nations and rainforests

Native peoples in North America claim to be as worthy of land and other rights as later immigrants, based on their prior occupation of the land.

British Columbia

In total, 57 million ha (141 million acres) of British Columbia's 95 million ha (235 million acres) of land are forested, with 22 million ha (54 million acres) detailed for potential logging. Since 2003, the province has been developing an overall vision for economic and cultural progress among its First Nations people.

Income from forests on traditional First Nation lands was initially shared with those groups affected by logging. Now, a type of direct agreement with First Nations, similar to those with logging companies, is being developed. It recognises their unique cultures and provides for traditional harvesting and marketing of non-timber plants or products, for protection of cultural-historic sites and for protection of freshwaters. Timber harvesting is included. This approach of 'Ecosystem-based Management' is being received with interest. Grants assist First Nations to engage in wood-based business projects and forestry training.

Alaska Native Claims Settlement Act (ANCSA)

After Europeans explored and settled in North America, they disputed the rights of the then inhabitants to the lands on and from which they lived and with which they had strong spiritual attachment. Haida and Tlingit people in Alaska gradually obtained a degree of influence and justice in relation to their land claims and human rights. However, their generally low standard of living, unrest and the discovery of oil under Alaska in the 1960s, forced negotiators to settle Natives' land and financial claims.

In 1971, the USA legislature passed the *Alaska Native Claims Settlement Act* (ANCSA) in an attempt to end further claims, encourage economic enterprise and provide Native groups with specific lands. The ANCSA instigated twelve development corporations, covering geographical regions of Alaska, to be owned and run by the groups who traditionally lived in each region. Each corporation was set up and financed, with indigenous shareholders as the beneficiaries. The aim was to educate more Native people in modern economic practice, maintain native settlements and encourage their culture.

Sealaska Corporation

Sealaska Corporation is the regional corporation for Southeast Alaska. It runs a variety of economic enterprises of which forestry was the first. Its forestry enterprise, Sealaska Timber Corporation, owns 117,359 ha (290,000 acres) of Tongass Forest lands. The rest of the Forest is managed by the Forest Service and private landowners.

The proportion of Tongass Forest allotted to Sealaska Corporation by the 1971 Act means that its managers are of necessity resourceful in using the area to obtain income for its shareholders. At the same time the intention is to work sustainably to ensure some of their apportionment survives – so that later generations also have tree resources. They negotiate with conservation organisations for some set-aside forest to be more intensively managed than conservationists might like.

Yellow cedar and Sitka spruce are currently the most important tree species for the markets. Native logging sites are scattered and remote, accessible mainly by boat or plane. Management is similar to that of the Forest Service, with selective old-growth logging (highgrading) or clear-fell. Subsequent natural regeneration is sometimes managed by brashing and thinning. This opens out the young forest, which encourages herbaceous and shrub vegetation as fodder for Sitka white-tailed deer, hunted as subsistence food. Young trees are thinned down to about 450 stems per ha (1000 per acre) to give optimum timber at harvest time. Locally-collected tree seed has been used to grow small trees and plant them at 2633 per ha (6506 per acre) by hand

Local Tlingit, Haida and Tsimshian men are sometimes contracted to carry out forest work, but most is done by loggers from further south. Logs – roundwood – are exported by ship to Korea and Japan, as well as supplying the sawmill Viking Lumber, on Prince of Wales Island; this can take logs up to 91 cm (3 ft) diameter. Within the constraints of strong economic output, Sealaska Timber Corporation tries to reflect the strong spiritual attachment to land, knowledge of flora, fauna and their use as harvests of its shareholder First Nations people.

Other local or urban corporations, such as Shee Atiká Incorporated, have their own ecological resources and other portfolios such as buildings. But Sitka spruce still figures as a major natural resource in Southeast Alaska. However, highgrading has now turned its attention to scarce Yellow cedar in the hemlock-spruce forests, and to Western red cedar further south.

FIGURE 16.11. Wilderness conservation requires all parts of the rainforest to be sustained, not just the trees; Olympic National Park, WA.

PHOTO: SYD HOUSE, 2010.

The preservation and conservation movements

The Pacific Coast temperate rainforests of western Canada were as important to its economy as were those of the USA. Sitka spruce and its main compatriots, Western red cedar, Western hemlock and Nootka cypress and had been produced from Puget Sound, Vancouver Island and the coastal archipelagos of British Columbia. Canada was the most important exporter of timber to the UK during the Second World War. As well as aeroplane manufacture, Sitka spruce was needed by the UK for pulp: its own 20–25 year old state forests could supply its pit props but not paper pulp.

Forest covered about 48% of Canada and contained about 180 tree species at the start of European settlement. The timber industry was, and is, one of the most important for home supply and export; eastern provinces such as Nova Scotia and Quebec had been denuded of their most valuable old-growth forests by 1900. A survey of British Columbia in 1918 showed that over 60% of once-forested land had been destroyed by fire; the search for gold had felled or burned riparian forests of Yukon Territory even in the nineteenth century.

Along with more noticeable flooding, dust storms, erosion and wildlife loss, visible removal of large forests and their ecological replacement by barrens and muskeg concerned non-foresters. The urban public expected forests to be permanent landscape features; foresters and associated industries saw them as features to be quickly removed and used for human and economic benefit before moving to mature forests elsewhere.

During the later nineteenth century, a 'conservation' ethic had developed amongst some landowners and thinkers in relation to nature, and land

FIGURE 16.12. Sustaining the microclimates of temperate rainforest needs a large area such as the 373,380 ha (922,000 acres) Olympic National Park.

PHOTO: SYD HOUSE, 2010.

management (including forestry) of North America. There were two versions: the total *Preservation of Wilderness*, of whole landscapes, their geology, plants and animals without human intervention of any kind, allowing them to return to a pristine 'natural', 'virgin' or 'wilderness' state. The declaration of National Parks arose from the success of the preservation movement with presidents and politicians.

Nature Conservation management of *plants and animals* by human *intervention* to ensure their future survival (a possibly futile intention now we know about long-term climate changes) was a more pro-active view of land management.

Rapacious, clear-fell logging of old-growth forest – which moves across whole regions until all the old-growth has vanished, and then moves to another locality – conflicts with both these intentions.

During the inter-war years, professional North American foresters developed the theory of conservation management into an ethic of ongoing forest management. The intention was to have 'sustained yield' which would mean continuity of both a forest and its economic productivity: no need to move on elsewhere after logging because any one region would be forested permanently. This would mean changed styles of silvicultural management.

Forestry as timber production was an economic activity which remained paramount; as always, wildlife species either survived or did not under the new regimes. Sustained yield, as carried out by foresters, consisted of regulating each year's timber harvest to an area equivalent to the length of the rotation. By the year the final coup of the rotation was harvested, the first coup had grown up sufficiently to be re-harvested. However, most foresters, mills, and politicians wanted old-growth timber – not young trees – for both income and ease of management. Sustained yield is difficult to organise unless a forest inventory has been taken. This involves an assessment of: *What is there? Where is it? How much of it is there?* Until this was perceived, few forests received the necessary quantitative assessment.

During the eighteenth and nineteenth centuries, native people had suffered from loss of their traditional resources when old-growth forests were logged and waterways polluted by incomers. However, after the ANCSA of 1971, they benefited financially – but as agents of old-growth logging. Their own timber corporations felled old-growth on their granted lands to provide shareholder communities with much-needed income.

Forestry methods on public lands of the Pacific Coast came into conflict with salmon fisheries, nature conservationists, tourism workers and the urban public. By the 1990s, pulp mills were closing, old-growth forests were vanishing and the general public realised – from its knowledge of environmental issues – that *functioning forest ecosystems* which support their natural flora, fauna and soils are perhaps the *real* 'Wealth of Nations'. It was even suggested that the most important forest products are not timber and wood, but their contribution to the planet's weather and water systems, unpolluted rivers, self-perpetuating fish (especially salmon) populations, tree, flora and fauna populations, ability to support human activities such as quiet outdoor recreation and tourism and small-scale, local timber businesses.

Professional foresters started to listen to others' points of view, to negotiate, to re-think their aims and methods and consider a potentially different future for the industry and forested lands. Their colleagues in the Republic of Ireland and the UK were going through a similar process of adjustment.

Holistic forestry

During the later twentieth century both forestry and nature conservation organisations thought their activities should trump other land uses.

Forestry itself expected to trump all other potential uses of trees-covered land, because of its economic and employment importance. Nature conservation expected to trump forestry, farming, human recreation, food and craft harvesting so as to save threatened species and ecosystems.

More pragmatic attitudes and methods have started to prevail more recently. Members of the general public are increasing their understanding of ecology and land use; sophisticated ecological science is modernising the philosophy and practice of land use in the Pacific rainforest region. But I have a weekly sigh when I read in Canada's online *Working Forest Newspaper* of complaints by forestry companies at others' wishes or demands that they amend their logging practices.

Non-specialists, including communities close to temperate rainforest, are consulted by the Forest Service and contribute ideas and information to practice. Research ecologists analyse the status of forest flora and fauna and the factors affecting their populations, before consultation and discussion with other groups. So, relevant data is on hand to support decision-making.

The public and governments seem to want forests to cater for many uses at one time, to provide their paper and construction material, but not to alter forest landscapes. It is rarely possible to accommodate all needs and desires in the same forest.

People want from public forests: many types of quiet recreation on land, fresh and sea water; clean rivers and ocean; hydroelectricity; biomass for fuel; subsistence harvesting and salmon fishing for native people, other residents and visitors; conservation of selected plants and animals; soil conservation; outdoor education; timber for local and international firms.

Tongass Land and Resources Management Plan

Tongass Forest is the largest USA National Forest consisting of 4 million ha (10 million acres) of rainforest and 2.8 million ha (7 million acres) of ice, rock, barrens and muskeg. Designated a Forest Reserve in 1902 and a National Forest in 1907, it has since enlarged to cover most of Southeast Alaska. The area available for old-growth logging is restricted to 12% of what remains; over twenty wilderness areas represent all the habitats and landscapes.

Long-term plans guide the management of National Forests; other, non-Forest Service stakeholders contribute to the plans if they wish. The plans are intended to be flexible and updated when necessary.

The first overall management plan for Tongass Forest was written in 1997 but was superseded in 2008. The primary aim of the document was the management of the Forest for timber and its associated industries, under forestry legislation. Its success as a working document has been under review since 2013, with input from the public. The most important change now being considered is from logging old-growth forest to silviculture based on short rotations. The intention is to harvest existing secondary forests of smaller, much younger trees. This should mean that remaining old-growth forests can be left untouched. However, short-rotation silviculture is more complex, methods will require testing and monitoring. Local and distant markets will need to be persuaded that smaller logs are just as valuable as larger ones.

It is the view of some conservation organisations that the review gives insufficient emphasis to the ending of old-growth logging, to protection of salmon fisheries and tourism. This cry sounds familiar. A new Tongass Advisory Committee, the TAC, has been set up to involve more organisations and communities in official policy and to monitor ongoing multi-use forest management.

Modern management of waterways takes into account not only water, but salmon populations and their relationship with forest ecosystems through which they pass. On Baranof Island, despite 40 years of highgrading especially in alluvial

FIGURE 16.13 *(left)*. Information and education at the Olympic National Forest, a 254,189 ha (628,115 acres) commercial and recreational buffer zone to the National Park.

PHOTO: SYD HOUSE, 2010.

FIGURE 16.14 *(right)*. Information and interest at the Forest History Centre, Toledo, OR.

PHOTO: SYD HOUSE, 2010.

valleys, some intact old-growth rainforest exists higher in river catchments. Salmon are now the most important rainforest species economically. Sitka Conservation Society, which has worked since 1967 to sustain the Tongass Forest, suggests that river valleys are the most appropriate units for forest management. Indeed the Society ensures all habitats and human communities receive its attention, directly and by negotiation with other organisations and people.

Cruise tourists and other visitors probably visit Alaska to see and experience beautiful seascapes and landscapes, trees, nature-based traditional art and crafts. Landscape and seascapes alone might justify management which maintains forest cover, unpolluted waterways, optimum salmon populations and resources for First Nations art and crafts.

Haida Gwaii Forest Stewardship Plan

Only 6% of Canada's forests have been logged. So there is still doubt in the forest industry about the need to set-aside 'no-logging' areas and the requirement to work with other organisations in forest management (see CanBio 2014 below).

Timber and tourism are the main industries of British Columbia. The BC Forest Service recognises that the huge area of Canada's forests is vital to the water cycle and climatic stability of the planet. It regards itself as the steward of public (government) lands and resources including timber and does not only monitor economic forestry on public lands. Staff liaise with local groups, so that forested lands can be used for many purposes including recreation, food, wilderness, and of course logging. They cooperate with other agencies on watershed management, wildlife, fish, heritage, minerals, energy and tourism.

British Columbia's two Pacific Coast rainforest reserves are the Pacific Rim and Gwaii Haanas National Parks. Pacific Rim National Park, inaugurated in 1971, lies on south-west Vancouver Island within reach of big urban areas; it receives 700,000 visitors during seven months each year so that some regulations are needed to ensure rainforest protection.

The islands of Haida Gwaii, close to the Alaska border, now have a range of protected rainforest and other sites. Human access to the Gwaii Haanas National Park is limited and strictly controlled by guide-only access. First Nations have traditional rights on agreed sites; forestry firms can harvest timber from agreed, suitable locations.

The Haida Gwaii Forest Stewardship Plan is overseen and run by the BC Forest Service. It divides the islands into landscape units for forest management purposes. Silvicultural operations require public consultation, must reach high standards and take account of Haida traditionally-used trees and historic sites. There are reserves where nature conservation of species or habitat types takes precedence (for instance marbled murrelet nesting, forest swamps); some geological features such as changing river systems can progress naturally without containment.

The Forest History Society volumes by D. W. MacCleery and K. Drushka give an overview of USA and Canadian forest history.

Sitka spruce education and culture

First Nations

The Haida run an education programme for their children, in which Sitka spruce has its own unit of nine lessons taking places indoors and in forests.

First Nations people of the Pacific Coast demonstrate their wood crafts skills in museums, shops and their own art galleries. Western red cedar rather than Sitka spruce wood is sometimes used even beyond its northern limit, where Sitka spruce is the traditional material.

The University of Alaska runs classes in basket-making using Sitka spruce and other woody roots. Skilled First Nations teachers run these popular classes, which any person may attend. At the Sealaska Heritage Institute in Juneau, First Nations and other personnel carry out research, education and publication on Tlingit and Haida culture, such as art and language.

European

Museums in Alaska, Washington and Europe store and display Tlingit and Haida cultural objects made of Sitka spruce.

The Faculty of Forestry of the University of British Columbia runs an intensive field study semester on Haida Gwaii for its undergraduates in natural resources. Students learn about Haida culture and the complex ecological, social and economic issues facing the Haida communities of the islands.

Many popular and technical books have been and are published on the history and uses (past and present) of Sitka spruce in North America; they encourage an appreciation of this tree species in its native habitats. Sitka spruce has been included in long-term academic research on Native American food plants. Modern commercial products made with Sitka spruce, such as beer and essential oil, enhance public interest in this tree.

Futures

A two-pronged approach to Pacific Coast temperate rainforests futures has been described in this chapter: the 'wilderness preservation' of remaining old-growth forest sites and the complex, 'sustainable' management of naturally-regenerated, 'secondary' forests of rainforest species. New types of forests are likely to emerge according to ecological and social factors when society's needs alter. When climatic change becomes pronounced within the current rainforest range, Sitka spruce may die off in some localities or extend in others, continuing a post-glacial process of landscape migration. Pests and diseases may affect it and other trees differently according to location, regional climates and susceptibility. Sitka spruce is a genetically variable species and likely to maintain itself in North America, with or without humans, or even with another glaciation.

Sitka spruce
this giant spruce of enormous span
one thousand years
in the age of man
carries on with only trust
perfectly planted on a coast remote
away from the pollution of
haunted hosts
and somewhere, sometime
someone said
you will be spared the
long saw of men
a gentle giant among
younger souls
exhaling oxygen
to fill the air
its spirit alive with
cooperation and faith
that it has lived its life
in perfect pace
and all the while
the outer world
is unaware and
rushes on to exist
and I just grin ...
because
after we all
have long been gone
it will still stand
and continue on

Debby J. Rosenberg 2006. American author, poet and friend to trees, describing a Sitka spruce in the Olympic National Park, 58.22 m (191 ft) tall and 5.72 m (18 ft 9 in) diameter.

CHAPTER SEVENTEEN

New Temperate Rainforests: Futures in Ireland and Britain

The 'greening' of these forests springs from changing social attitudes and attests to their cultural malleability. It also shows a certain amount of slippage between concepts of 'nature' and 'artifice'. Does this matter? Certainly not, it seems, for the red squirrels of Kielder. And I'm not sure that it matters too much for us humans... We live in a world where nature and culture are all jumbled up, and arguably always have been for as long as humans have lived on this earth...

Chris Pearson 2011

What now?

Since 1831 Sitka spruce has become the most important tree in the UK and Republic of Ireland for commercial forestry and its markets. It contributes to homes, businesses and fuel. Its forests have developed a patchwork of colours, textures and shapes in upland landscapes and form a matrix in which upland recreation and education take place. It is accumulating unprecedented combinations with native species, especially lower plants and invertebrates. Birds rarely seen in deciduous woodland, such as siskins and Common crossbills, are now common; the mix of forests and ungrazed moorland suit Hen harriers. In Ireland it is the most successful tree ecologically since the Scots pine disappeared in the twelfth century.

In 185 years, the percentage contribution of Sitka spruce to planted forests has increased phenomenally in comparison with other conifer and broadleaved trees. Its numbers and area in British and Irish plantations is greater than of any other tree species (Tables 9 and 10). It has influenced people's lives whether they know so or not. Map 17 shows how the distribution of Sitka spruce increased even between 2000 and 2010, based on records collected by members of the Botanical Society of Britain and Ireland.

Total woodland area	**= 3,154,000 ha**
% total area which is broadleaves	= 49%
% total woodland which is conifers	= 51%
Actual area of conifers	= 1,614,000 ha
% total woodland which is Sitka spruce	= 26%
% conifer area which is Sitka spruce	= 51%
So, actual area of Sitka spruce	= 823,140 ha

TABLE 9. Contribution of Sitka spruce to UK forestry, 2014.

SOURCE: FORESTRY COMMISSION 2015.

English Name	Latin name	% Total forest area
Sitka spruce	*Picea sitchensis*	53.3
Lodgepole pine	*Pinus contorta*	10.3
Willow species	*Salix* spp.	5.2
Birch species	*Betula pendula and B.pubescens*	4.8
Norway spruce	*Picea abies*	4.2
Japanese larch	*Larix kaempferi*	3.7
Ash	*Fraxinus excelsior*	3.1
Hazel	*Corylus avellana*	2.2
Alder	*Alnus glutinosa*	1.9
Douglas fir	*Pseudotsuga menziesii*	1.7
Other broadleaves and conifers	Multiple spp.	9.5

Total forest area in Republic of Ireland = 750,000 ha
Contribution of Sitka spruce = 399,750 ha

TABLE 10. Contribution of Sitka spruce to forests in Republic of Ireland, 2007.

SOURCE: PHILLIPS (2013).

Sitka spruce's attributes suggest that it will continue to have an important role in landscapes and economy, and an increasing role in our culture. But this will depend on our continuing to use its products and on people's attitudes to its commercial value and ecological status, and its beauty. Changing climates this century may affect where Sitka spruce and other trees can grow or be grown, and which species we categorise as diseases and pests.

Potential climate changes

The potential future for commercial forest trees takes into account climate changes forecast for the coming decades and century. The Meteorological Office's computer modelling uses several sources of data to summarise expected climate change at increasing levels of certainty: its document UKCP09 gives details.

The climates of the Republic of Ireland and UK will continue to be oceanic, except possibly in their south-east corners. Mean summer temperatures will probably rise by 2.5°C in north-western Scotland and by 4.2°C in southern England if emissions of 'greenhouse gases' continue rising to a medium extent. Mean winter temperatures will rise less than summer values.

Seasonal rainfall patterns will alter this century. Summers throughout the country will become drier, especially in eastern and south-eastern Scotland and in southern England. Dry summer periods will increase so that droughts could be more frequent and severe, especially in eastern and southern localities, but soil type and topography will determine their extent.

Winters will become considerably wetter. Heavy rain days of over 2.5 mm (0.1 in) per day are likely to double in summer and increase by 2–4 times in winter. Look forward to much wetter winters! Frequent heavy rain is likely to increase flooding, to scour river banks, loosen tree roots and erode soil. Future

MAP 17 *(left). Picea sitchensis* distribution map for Britain and Ireland based on data up to 2000 from the Botanical Society of Britain and Ireland; *(right) Picea sitchensis* distribution map for Britain and Ireland based on data up to 2014 from the Botanical Society of Britain and Ireland.

wind patterns are less understood, but it is thought that there will be higher wind speeds especially in winter. The land area of the UK will decrease because ocean levels, for several reasons, are rising faster than any land uplisft.

In Scotland, Sitka spruce distribution may change deliberately and naturally. It will still fit conditions in large areas of the west, but eastern Scotland may become less suitable. In some eastern areas it already suffers from drought stress: 'drought-cracks' form in its bark making it vulnerable to disease, to breaking and to wind blow. Timber quality and productivity is then reduced. If drought limits Sitka spruce in eastern areas, alternative tree species will be needed: potential applicants include Macedonian pine (*Pinus peuce*), Douglas fir, Western hemlock and Chilean pine (Monkey Puzzle).

In Wales, broadleaved trees are expected to flourish and produce quality timber in central and eastern areas. Wales' suitability for growing Sitka spruce will decline slightly in the eastern half of the country but the south-west and north-west will provide even more suitable areas, including on Ynys Môn.

The growing season for plants in the UK starts *three weeks* earlier than 60 years ago. Sitka spruce trees on forest edges now produce more 'Lammas shoots' from their trunks – Lammas is 1 August, a time when trees show a burst of late-summer growth.

Soils will be subjected to increased drought in summer, standing water in winter, or be lost under the onslaught of rain, wind and drought; operations which break the soil surface will have to be reduced. New soil and climate maps will be needed so as to formulate where Sitka spruce can be planted.

Pests and diseases in future

Deer populations and their deleterious effects on forest trees show no sign of decreasing. Where deer numbers are high, Sitka spruce is an appropriate commercial species because it is the least affected tree due to its prickliness. It is eleventh out of fifteen trees in its desirability to deer for bark stripping and sixth out of eight conifers in desirability for browsing.

It is difficult to predict how wetter winters will affect deer: rain does not cover conifer shoots but snow does, so there could be more food. Earlier plant growth in spring along with drier summers might increase deer's reproductive success. But in the drier south, food availability and nutritious-ness could decrease during summer droughts.

Warmer, longer summers are likely to benefit rabbits because burrows and stops will be drier, potentially improving their reproductive success. I look forward to more rabbit pies! When weather and soils alter, plant life will change not only in timing but in spatial distribution and amounts, so there may be more rabbits in some localities but less in others. Wetter winters and more severe rain episodes could have the opposite effects to summer drought, drowning adults and kits in their burrows. Oh, fewer rabbit pies!

Red and grey squirrel populations may be affected by changing periodicity and quantities of tree-seed production; sugar levels in trees' inner bark could differ from now, altering the frequency and tree species from which squirrels tear bark.

Moorland may not even endure in its current make-up or locations, its suitability for forestry, farming or grouse. Heather and bilberry may relocate (dying *in situ* but colonising elsewhere) forcing hares and Black grouse to migrate for their favoured foods or to change foods as capercaillie have already done.

Sitka spruce is more resistant to insect and fungal diseases than other tree species grown in the UK, an advantage since European legislation prohibited certain insecticide dips for small trees. However, insects and fungi respond differently to climate according to species; those which are currently forest pests may become more or less so. Fungi and insects which are not at present pests to Sitka spruce could find that changing seasons or altered chemical composition of growing wood (for instance) suited their relationship with Sitka spruce, causing damage as yet unknown.

We can attempt predictions according to what we know of plant and animal ecology. For instance, the numbers of Green spruce aphids are usually controlled by low winter temperatures. Warmer winters will allow bigger populations to over-winter and they may grow quicker in the warmer summers. More carbon dioxide should encourage plant growth, so more Sitka spruce foliage will be available as food. Outbreaks of damage by this aphid, by weevils and by bark beetles could be more frequent and serious. But – of course – populations of their predators may also increase. Drought-stressed trees are more liable to insect and fungal diseases, so that possibility should be added into potential scenarios.

Ecological predictions are difficult as each species lives in unique conditions and is one member of a mesh of interrelationships. Computer modelling can enhance levels of certainty of potential outcomes but does not foretell the future. It may be a matter of 'wait and see', even for a Sitka spruce plantation:

> Ecology would be easy, were it not for all the ecosystems – vastly complex and variable as they are. Even the most austere desert or apparently featureless moor is a dense, intricate network of thousands of species of photosynthesisers, predators, prey animals, parasites, detritovores, and decomposers…
>
> It would be useful to have broad patterns and commonalities in ecology. To know how ecosystems will respond to climate change, or to be able to predict the consequences of introducing or reintroducing a species, would make conservation more effective and efficient. But a unified theory of everything is not the only way to gain insight.
>
> Anon 2014

More Sitka spruce?

Sitka spruce planting has declined during the past two decades in the UK. Diseases affecting other trees and expansion of deer populations countrywide mean this policy is to be re-assessed. Resistance to diseases and deer damage makes Sitka spruce a sensible choice for sites where other trees are vulnerable. For these two important reasons, and for commercial conifer forestry to remain viable, foresters in some localities will probably retain or *increase* the content of Sitka spruce in their forests.

In Wales losses of Corsican pine and larches to diseases led the Welsh Assembly to decide that they will no longer be planted in its forests. Wales will reduce dependence on Sitka spruce, with possible susceptibility to new diseases in mind. To ensure that all Welsh eggs are not put in one basket, 22 other conifer and broadleaf species are undergoing tests as potential commercial trees in Wales.

The Republic of Ireland intends to continue with Sitka spruce as the main commercial species but to reduce it from 69% to 65% and then to 60% as the long-term target. Afforestation with broadleaf species will rise from 4% to 10% and monocultures will be restocked with several tree species including Norway spruce.

Producing millions of (better) Sitka spruce

Sources of seeds

Seeds brought back by David Douglas from the Columbia River locality are now tall trees; seed imported in the later nineteenth century originated from places visited by explorers; the source of seed for early UK plantations like Durris Estate is not known; seed sent by the USA and Canadian governments in the twentieth century was collected from known sites; seed bought from J. Rafn's Copenhagen nursery came mainly from Haida Gwaii; seed brought back to the UK for provenance research was collected from throughout the geographic range of Sitka spruce; seed and transplants used in early state forests came from wherever foresters could buy them.

A 1920s Forestry Commission list shows that its Sitka spruce seed at that time originated from rainforests in British Columbia, some from Washington and Oregon and a little from nurseries but of unknown origin. Durris Estate in Scotland gifted 454 g (1 lb) of year 1922 seed. At the start of the Second World War, the bulk of Sitka spruce seed still came from Canada and the USA but home forests provided the seed of other conifers and native broadleaves. In 1942 the small amount of spruce seed available was from home-grown *Norway* spruce with a little imported Sitka spruce seed.

Between 1944 and 1947, medium quantities of Sitka spruce seed from British Columbia and Washington were again imported and small amounts from home forests. But other conifers and many broadleaves contributed the most seed those years.

In 1949, 9 kg (20 lb) of Sitka spruce seed came from the UK; 18 kg (40 lb) from Johannes Rafn's nursery in Denmark; 907 kg (2000 lb) from a USA nursery; and 295 kg (650 lb) from a British Columbia nursery. These 5 million seeds would afforest (at 1500 trees per acre) 146,257 ha (361,408 acres) and could be used for up to ten years before losing viability. Seed from Alaska has rarely contributed to state afforestation, but it was fortunate that early plantings on the Corrour Estate, Inverness-shire used it.

After 1960, large-scale expansion of forests into the very marginal uplands necessitated getting hold of millions of Sitka spruce seeds quickly. How to obtain so much seed when virgin rainforests in the usual North American sites were becoming scarce?

Tree breeding

Although a genetics research section was set up by the Forestry Commission in 1948, it was not until after 1960, when it became the most-needed tree, that Sitka spruce became the focus of genetics research.

What is the point of genetics research? How are the results used in forests?

A tree species is a population of interbreeding individuals which vary in traits such as height, speed of growth, canopy shape, seed weight, their susceptibility to frost, wind and diseases.

Foresters are interested in only some traits, particularly growth rate, timber volume (taller tree with larger girth), straight trunks, and wood density. They wished to 'improve' – change for their own purposes – what evolution has produced. Humans have been doing this sort of thing for millennia with, for instance, dogs, cattle, tomatoes and maize.

The aim was to produce bigger, stronger, straighter, trees with more unblemished timber of acceptable construction strength per unit of land.

The difficulty with breeding trees is their long life-cycle: 25–30 years for Sitka spruce to grow from seed to tree and then seed production. Compare cattle

FIGURE 17.1 *(left).* Collecting Sitka spruce cones from Ledmore Seed Orchard, Perthshire.

PHOTO: GLENN BREARLEY, COURTESY OF FOREST RESEARCH. © CROWN COPYRIGHT.

FIGURE 17.2 *(right).* A 'SuperElite' rooted Sitka spruce grown in a plug from a cutting off a genetically-superior mother tree or stock hedge.

PHOTO: MATT HOMMEL, COURTESY CHRISTIE-ELITE NURSERY, FORRES, 2010.

FIGURE 17.3. After six months in a polytunnel, rooted cuttings are transplanted out to the nursery's fields.

PHOTO: MATT HOMMEL, COURTESY CHRISTIE-ELITE NURSERY, 2012.

FIGURE 17.4. Homes and fuel for the future? Close view of the growing transplants in a nursery field.

PHOTO: MATT HOMMEL, COURTESY CHRISTIE-ELITE NURSERY, 2011.

with 9–22 months from birth to reproduction; dogs six months; sunflowers about twelve months.

Starting in 1963, genetics researchers looked countrywide, chose and marked 1800 individual Sitka spruce 'Plus Trees' which showed traits they wished for. They chose trees of Oregon, Washington, Haida Gwaii and Alaska provenances.

Researchers then collected seed from each Plus Tree: all seeds from an individual Plus Tree have the same mother but different fathers (depending where the pollen blew in from). Progeny of each Plus Tree were compared in field experiments, to compare the genetic quality of their mother Plus Trees. The best 240 of the original 1800 Plus Trees were then selected for a Breeding Programme.

From the 240 Breeding Programme trees, 40 of the *very best* were chosen to be planted in a seed orchard. The 40 trees mated naturally via wind pollination and their resulting seeds were available for sale as 'Improved Sitka spruce'. Another level of improvement was reached by cross-pollinating *by hand* the very best Plus Trees – so that *fathers* as well as mothers of their seeds were known. Their highly selected seeds fetch a high price.

Of course it is necessary to check that 'improved' Sitka spruces really are better than 'original' trees. Researchers travelled countrywide to measure the girth, height and wood density of *improved trees* and *non-improved trees* from Haida Gwaii seed.

Statistical analysis of their data indicated that the improvement was: an increase of about 25% in trunk (timber) volume and up to 20% greater wood density (better for milling and construction). Tree breeders in New Brunswick, Canada, obtained greater increases, but with other spruce species. You can see the results of selection in forests today, because taller, bigger-diameter, less-branched, narrower-crowned Sitka spruce will be growing there.

The next phase of research was to seek ways of producing large numbers of Improved Sitka spruce for commercial use.

Vegetative propagation

Cutting and grafting: Many young shoots from the crown of 'Improved' Sitka spruces are cut off and grafted onto rootstocks of young trees. The scions grow into short plants on the host stock and form an orchard of small, but fertile, clonal trees. They mate with each other and produce seeds which can be collected in commercial numbers at a handy height. The process is then repeated.

Cuttings from offspring of 'Improved' trees can also be rooted directly, to miss out the grafting stage. Christie-Elite Nursery at Forres, Moray takes its cuttings from parent stock in February each year and puts them tiny pots. Five months later they have grown roots and can be transplanted into the field until grown sufficiently for sale to the industry.

Tissue culture: This method of producing 'Improved' trees on a large scale has great potential. It bypasses the need to take cuttings because tissue can be taken directly from high-value trees. Millions of offspring can be produced from one parent tree.

Researchers cut embryos from the seeds of one tree and make a few copies of each in a laboratory. From each seed, some copies are grown into young clonal plants helped by hormones, and some are stored as tissue in liquid nitrogen.

The best clones are chosen after six years in field trials, but they are discarded. Their sibling embryonic tissues are taken out of nitrogen store and multiplied by thousands to grow into rooted plants, each genetically identical.

This method is in the development stage in the UK but in Canada it is used on an industrial scale. J. D. Irving a large firm based in St John, New Brunswick, tested four spruces (including Sitka) and seven pines species. The firm now produces one million small conifers of nursery stock annually by tissue culture.

Effects on Sitka spruce populations: The genetic variability of Sitka spruce populations along the Pacific coast of North America was determined by natural selection and evolution. In contrast, the genetic variability of most Sitka spruce populations in the UK and some in the Republic of Ireland is a mix determined by many humans.

Nearly 100% of the Sitka spruce now planted is 'Improved'. The potential for disease to afflict a whole forest is greater with cloned than with genetically variable trees, but foresters have tried to guard against this by storing a variety of clonal tissues in liquid nitrogen.

Practicalities: seed and plant supply

At Alice Holt, Surrey, the Forestry Commission obtains, processes and stores tree seed. Conifer seed is collected from registered stands and seed orchards, extracted, cleaned and added to the cold store where over *five tonnes* of seed is stored long term. Before being sent on or sold, seed is tested for its viability and registered according to UK *Forest Reproductive Materials Regulations* under an EU directive.

Seed lots and treelets are identifiable from start to finish of the supply chain: provenance, collection site, extractory site, nursery, buyer and user. Christie-Elite private nursery has a 7 acre (2.8 ha) seed orchard at Forres, Moray where it maintains 40 genetically-superior clones of Sitka spruce, so that it can produce large numbers of treelets in its nursery fields.

Three Forestry Commission nurseries produce Sitka spruce, at Delamere (Cheshire), Wykeham (Yorkshire) and Newton (Moray). Treelets are kept long-term in cold stores to extend the tree-planting season. The Newton nursery grows 3–4 million treelets each year and grows-on 160,000 clonal rooted cuttings from Delamere Nursery, which are ready for sale after about fifteen months.

Coillte nurseries grow 25–35 million transplants of many tree species for European and Irish forests; 2.6 million of them are vegetatively-propagated Sitka spruce from Clone Nursery, Co. Wicklow. Plants from Washington and

FIGURE 17.5. Recent afforestation at Jerah, a derelict moorland farm at 400m (1312 ft) in the Ochil Hills, Stirlingshire; temperate rainforest for the future?

PHOTO: SYD HOUSE, 2015.

Oregon seed are most commonly planted in the Republic of Ireland, with Haida Gwaii origin used only on frost-prone sites and altitudes above 300 m (984 ft).

Future roles of Sitka spruce

The Enlightenment plantations of introduced trees were intended to beautify landscapes and increase their productivity. Today, people and governments expect forests to fulfil these two purposes but also sundry others.

Reduce imports

The UK is a small country but the third-largest importer of timber in the world. More Sitka spruce trees will be needed just to supply the continually-rising human population of the UK and Republic of Ireland. All regions of the UK and the Republic of Ireland intend to increase afforestation in coming decades, but many farmers oppose this.

Support home wood-processing industries

In 2010, about 29,000 people were employed in the UK primary wood-processing industries, in firms of 1–600 employees. Sitka spruce is the mainstay of firms in the rural north and west Britain and throughout Ireland.

Rural employment

The UK forest industry employs 14,000 people countrywide in forests, offices and laboratories. In the Republic of Ireland, Coillte employs 1000 people, but many farmers also engage in forestry party-time: 18,000 have planted farm woodlands averaging 8 ha (20 acres) since about 1990. Woodlands bought and run by local communities are popular, usually near towns and villages. Community woodlands are also on the increase in the UK, for instance at Abriachan by Loch Ness and Woodstock, near Oxford. Although residents give their time voluntarily, professional help is usually required for both forest planning and silvicultural work. Communities may not be keen to plant Sitka spruce in their woodlands but it may be necessary sometimes to balance the books.

Energy supply: heating

Sitka spruce logs and chips are now important fuel for wood-burning boilers in cold country castles, mansions, villages, schools and homes. Domestic wood-

FIGURE 17.6. An attractive woodland of Sitka spruce and broadleaved trees developing on Foyne's Island, Co. Limerick.

PHOTO: ANDY TITTENSOR, 2013.

FIGURE 17.7. Wind turbines are one of the most recent uses for Sitka spruce forests: the largest European on-shore windfarm, Whitelee, Ayrshire.

PHOTO: RUTH TITTENSOR, 2008.

burning stoves have become popular and that is likely to continue as other forms of heating are costly.

The upland landscape: water and flooding

Although farmers and home-owners tend to see flood-relief as walls of concrete or heavily-dredged waterways, ecologists and foresters prefer more sustainable options. Expanding the woodland area on appropriate high-level catchments is one approach: Sitka spruce would be suitable. Over time, tree canopies, vegetation and soils will retain more rainwater, releasing it slowly into waterways compared with moorland. If waterside farmland were also planted with open woodland of Sitka spruce, broadleaved trees and shrubs, the new riparian zones would hold floodwater and provide habitats for desired fish and many other flora and fauna.

Amenity, recreation, tourism and education

The population of the UK at the end of 2013 was 63.7 million and is expected to rise. Most people live in cities and need rural holidays. The Republic of Ireland had a population of 4.6 million in 2013, expected to rise to just over 5 million by 2021. Forests can hold thousands of people yet feel 'empty' and 'wild'.

Forests of Sitka spruce are nowadays well-used for outdoor recreation, from dog-carting to orienteering and star-gazing. The bigger forests offer modern visitor facilities and educational activities for schools and informal groups.

Nature conservation

Nature conservationists are currently more interested in *removing* Sitka spruce from deciduous woodlands and afforrested moorland than in conserving it. Yet ideas change: Red squirrels were treated as vermin for their depredations to trees during the nineteenth and early twentieth century. Thousands were killed on private land throughout Britain between 1835 and 1933. The idea of spending money and time *conserving* them would have been considered ludicrous, but that is what happens now. Our evaluation of Sitka spruce may change in future, if, for instance, large, old individuals, or beautiful trees lining well-used trails are destined for felling.

CO_2 sequestration

From way back in geological time trees have absorbed carbon dioxide from the atmosphere, used it in photosynthesis and converted it into plant tissue. Trees and other plants hold this 'fixed' carbon dioxide within their tissues, until it is released into the air when they die and decompose. But, because trees respire, they release carbon dioxide all their lives; the amount of photosynthesis compared with respiration decides how much they 'fix'.

After megaana of trees quietly getting on with photosynthesis, along came humans to measure the CO_2 'fixed' by or 'sunk' semi-permanently in trees and soils. Researchers discovered that Sitka spruce is an immense CO_2 sink. Fast growth is one reason for its high CO_2 uptake. It contains 57% of all the CO_2 held by UK conifer forests. Scots pine and larches are the next biggest sinks at 19% and 14%. Sitka spruce also hold more CO_2 than mixed-deciduous forests: a hectare at 17–21 years old takes up 26 tonnes of CO_2 (7 tonnes of carbon) a year, while a hectare of 72–80 year old deciduous forest takes up 18 tonnes (4.9 tonnes of carbon).

Deciduous forests have no leaves in winter and photosynthesise for about 150 days of the year; during the remaining 215 days, carbon dioxide is not taken in, *only given out by respiration.* Sitka spruce, being evergreen, carries out photosynthesis for about 303 days annually and so takes up CO_2 from the atmosphere for twice as long as deciduous trees.

Soils hold about twice as much carbon dioxide as do trees, but peat soils contain four to six times as much. As digging and ploughing peat releases its CO_2 into the atmosphere, afforestation on deep peat is now discouraged.

Felling trees removes carbon dioxide, in the form of timber and wood, from the forest – CO_2 is taken elsewhere *in the timber*. Felled timber of any kind used for construction, particle board, furniture or aeroplanes, contains carbon until it is burned or thrown away to decompose. When Sitka spruce is used as biomass fuel the CO_2 returns to the atmosphere quickly by burning.

Because Sitka spruce takes in and holds so much CO_2, it would be possible to expand Sitka spruce forests so as to draw off CO_2 from our atmosphere. However, to absorb 10% of the UK's CO_2 emissions it would require afforestation of 23,200 extra hectares (56,824 acres) of land every year from 2016 until 2050.

Food

People of First Nations who live by the Pacific coast of North America traditionally use Sitka spruce for food and drink. British and Irish may wish to try adding some of these nutritious items to their diet.

Silvicultural changes

The more varied silvicultural techniques of the last 25 years are likely to fulfil amenity, nature conservation and public approval wishes, despite changing climate.

Economic timber production, however, is still the primary purpose of forestry. Markets at the moment want young, straight, even-aged, even-sized trees without blemishes for machine conversion. Some Sitka spruce forests will be intensively managed into the foreseeable future, so plant-clearfell-restock systems are likely, though with genetic improvements boosting productivity.

More 'mixed' forestry

'Low Impact' and 'Continuous Cover' systems using Sitka spruce have a variety of objectives: landscape amenity and tourism at Gwydyr; to remove some conifers and return to broadleaved forest at Tintern; to encourage nature conservation with 'old-growth' at Glenbranter; to improve amenity at Kielder; to conserve the European red squirrel and Black grouse populations in Clocaenog Forest. These systems are also intended to increase 'resilience' of forests to climate change.

High productivity, plant-clearfell-restock management is now carried out in smaller and more complex spatial mosaics; low impact and continuous cover systems vary according to site and objectives. Forests now and in future will be very *tightly planned and managed*; yet they will *look more natural* so as to be acceptable to conservationists, visitors and diversity experts.

More 'naturalisation'

Foresters are still under pressure to grow what might be called 'pseudo-natural' commercial forests in which 'natural' or 'old-growth' features are deliberately encouraged. These include an elaborate 3-D structure; many habitats; many tree species; individuals of all age-classes; dead standing and fallen trees; decomposition; natural regeneration in desired places; ecological change and mature soils.

Wind disturbance: Gaps formed by windblow or dying trees are typical of the few natural type woodlands we experience in Europe. Depending upon their size, shape and frequency, gaps provide habitats for many light-needy plants and animals, for vegetation change and tree regeneration.

Sitka spruce plantations are susceptible to wind-blow in some conditions. Whole trees, including their root-plates, usually fall. It might be expected that the resulting gaps would increase ecological diversity, if only from the pond inside each root plate's previous home. Which plants colonise a gap depends on how quickly the fallen trees decompose, on what seeds and spores happen to pass by and drop into the gap, and on seeds in the soil which germinate in the temporary warmth and light.

In a plantation, a gap becomes part of a bigger clear-fell area at harvesting time and is replanted with small trees. So it is difficult to turn windblow gaps into copies of natural forest disturbance, except for short periods. In natural forests, plants and animals succeed each other in gaps until canopy forest of some sort is formed again; or gaps may be maintained by grazing mammals.

Old growth conservation: Another way to increase the naturalness of Sitka spruce plantations is to retain a core of trees or even a whole plantation of 50–100 ha (124–247 acres) in which trees are allowed to grow 100–200 years old. Then they start to show old-growth features such as great height and diameter, boles with bird holes, epiphytes and fungal decomposition, some fallen and standing dead timber, inward migration of other tree species and abundant shrubs.

In long-derelict woodlands such as Białowieża Forest in Poland and The Mens woodland in England, one-fifth of standing trees are dead or decomposing.

FIGURE 17.8. In Clocaenog Forest, Denbighshire, judicious thinning by Continuous Canopy management produces a varied age-structure and many young Sitka spruce under the 61-year-old trees.

PHOTO: ANDY TITTENSOR, 2013.

Mackenzie's Grove and Ritual Grove in Inverliever and Glenbranter Forests respectively, contain Sitka spruce about a century old: they are being spared felling so that old-growth features develop.

Ennerdale Forest, Cumbria: The English Lake District is famed for its upland landscapes and waters, its literary connections and twentieth-century anti-conifer campaigns. Its spruce-dominated Ennerdale Forest, planted between the 1920s and 1970s, is taking forward an experiment in forestry, nature and landscape conservation. The 'Wild Ennerdale Partnership' is a joint venture between four national organisations in which 4300 ha (10,625 acres) of commercial forest, upland sheep and cattle farms, historic and water supply sites are being managed as a whole landscape. The Forest itself is 800 ha (1976 acres) of the whole, with Sitka spruce contributing 42% of the trees.

The aims are first, to allow more land to undergo 'natural' processes, without excessive intervention. Continuous cover silviculture produces a more natural-looking structure; domestic cattle range freely rather like prehistoric wild aurochs; trees regenerate naturally or are planted. The Forest will appear more 'natural' or 'wild' yet be managed.

Second, people are encouraged to participate, increasing their experience of 'wild-ness' by taking on practical tasks, woodland crafts and activities in forest-schools. Outdoor events and sporting activities encourage people with other interests into the open air.

Best of all, Sitka spruce, the forest's timber income, is valued as a legitimate member of the flora, an innovative idea to non-foresters. However, its natural regeneration *is controlled,* while native broadleaved trees are allowed to regenerate

FIGURE 17.9. After clear-felling, seedlings of Sitka spruce make a carpet of natural regeneration in Loch Ard Forest, Stirlingshire.

PHOTO: RUTH TITTENSOR, 2012.

unfettered! The degree of naturalness or wild-ness in Ennerdale Forest locality is thus limited. Perhaps we are a little afraid of both real wilderness and of a natural, 'uncontrolled' Sitka spruce forest?

New homes for Sitka spruce

Riparian zones

During afforestation in the first half of the twentieth century, trees were planted very close to waterways so as to maximise the eventual economic productivity of the land. Bare land was considered wasted land. However, public criticism cited several potentially negative effects on water life: shaded, cooled water; reduced photosynthesis by aquatic plants; water acidification from falling conifer needles, uptake of industrial atmospheric acids by conifers; an increased sediment load after forestry operations.

As a result, 1988 guidelines precluded planting conifers within 20 m (65 ft) of waterways. Foresters removed trees from riparian zones to encourage herbaceous vegetation. Tree-less riparian zones contrast with Scandinavia and north-west Canada where native conifers grow naturally close to rivers and link ecological processes of forest and water (see Chapter Four). Their riparian conifers are important in sustaining fish populations. River flood-plains are one of Sitka spruce's optimum habitats.

It has recently been suggested that new types of riparian zones might be planted in the UK. Sitka spruce or Douglas fir, native Common alder and willows could be planted and managed by low impact methods and infrequent

FIGURE 17.10.
Sitka spruce forest regenerating naturally on high moorland above Glenbranter Forest, Argyll.

PHOTO: BILL MASON, COURTESY OF FOREST RESEARCH, 2005. © CROWN COPYRIGHT

harvesting. Waterways would benefit from a zone with several ages and sizes of tree, partial shade over water, infrequent soil disruption and sediment input. Adjusting the tree species according to soil type would reduce the likelihood of acidification. As yet, possible effects upon aquatic flora and fauna are not known.

Natural regeneration outside forests: where is it going?

In southern Scotland, there are high moorlands above about 400 m (1312 ft) where Sitka spruce is regenerating naturally from seed of adjoining plantations. Trees are appearing beyond forest boundaries, across wide-open heather and grass landscapes. Fewer sheep and less 'muirburn' during the past two decades may explain this.

Foresters and conservationists usually dislike extra-mural Sitka spruce regeneration. They focus on potential loss of moorland and rare 'montane scrub' should Sitka spruce 'parkland' or woodland develop autonomously – but how about *new* montane scrub and woodlands? These could be useful to land-use, important experimentally and grow into n*ew temperate rainforest* in these high rainfall localities of 1200–2000 mm (47–79 in) annual average. Other species will accrue in due course.

On high moorland beyond the upper margin of Kielder Forest, Sitka spruce is also regenerating sparsely where moorland is nowadays lightly grazed by sheep and goats and heather no longer burned. Two such localities are scheduled for nature conservation, so the Sitka spruce is removed every several years. The Border Mires (Chapter 12) also make good places for Sitka spruce to seed onto, but conservationists hope that drain-blocking and a higher water-table will stop that. Extra-mural Sitka spruce regeneration is welcomed only where foresters and conservationists deliberately create a transition zone between forest and moorland with birch seed.

There are a few places in Northern Ireland too, where Sitka spruce is regenerating at low density on open ground next to mature plantations. As elsewhere, it will be removed if it 'threatens' designated conservation areas.

It will be interesting to observe how faunal diversity changes in low density Sitka spruce 'parkland' when it expands above existing plantations.

Sitka spruce in future forests

State forests

Forests in the UK are being prepared for a future in changing climates. Despite Sitka spruce's fast growth, carbon sequestration abilities and useful products, many other tree species are being tested for the future. Species as yet rarely planted, such as Monterey pine, Chilean pine and Southern beeches are under consideration.

Upgrading the infrastructure within forests is in progress, for instance, constructing more-robust drains and bigger culverts under forest roads; extracting directly onto railway wagons at night when passenger trains do not use adjacent railways; building more miles of forest roads to reduce damage by timber lorries to public roads. As well as timber, future forests will provide for education, recreation, health, tourism, nature conservation, social and ecological research, jobs, carbon sequestration, district heating and hydro schemes. Inverliever Forest's 20,000 ha (49,421 acres) has 163 km (90 miles) of vehicle tracks available for walking and cycling, has viewpoints, picnic places, historic sites and many privately-provided recreational facilities. What a change since 1907! In Scotland and Wales more forests will contain wind farms – which necessitate felling large areas to avoid turbulence for turning blades. Sitka spruce regenerates profusely in these opened-up areas.

Ecologists, conservationists and landscape specialists are some of the professionals who give advice and data to foresters when long-term forest plans are being drawn up. Local residents are also encouraged to give their views on draft plans. The widespread dislike of Sitka spruce among non-foresters may affect its contribution to plans and forests, a worry for foresters who will still be required to provide the public with timber products.

Scotland intends to increase its forest area from the current 18% to 25% by 2030; Wales from 15% to about 25%; England from 10% to 15% by 2060; the Republic of Ireland from 11% to 17%. The proportion of Sitka spruce in new forests will depend on public attitudes, the countries' needs, politicians' views and variation in soils and climate. In the Republic of Ireland, Sitka spruce is viewed as the main tree species for the foreseeable future. In Wales its role as main timber producer may be replaced in the east by Douglas fir which can

Canada	34
USA	33
European Union (average)	37
UK (average)	13
Iceland	0.3
Republic of Ireland	11
England	10
Northern Ireland	6
Scotland	18
Wales	15

TABLE 11. Percentage of woodland cover in Europe and North America, 2014.

SOURCE: WORLD BANK 2016.

take advantage of drier summers, and trees such as birch, oak, Coast redwood and Grand fir are potential companions. Table 11 shows that despite tremendous increases in woodland cover during the past century, the Republic of Ireland and UK have some way to go before they reach the average woodland cover of North America and Europe.

Private forests

The private sector holds 60% of Britain's forests. Landowners with commercial forests in the north and west maintain Sitka spruce plantations as their bread and butter: they are constrained by the need to make a living, keep their homes, farms and workforce. Nevertheless, many do grow other tree species, allow for nature conservation, public access and amenity. There is no shortage of investors wishing to afforest land or buy Sitka spruce forests.

Along the River Don in Aberdeenshire, the Grant family has been tree-planting since at least 1720. Sir Archie Grant has decided, that when he fells mature conifer plantations, no more Sitka spruce will be planted: it is vulnerable to frosts and low rainfall in north-east Scotland. Norway spruce is his choice as the main replacement on his dry soils.

At Kyloes Forest in Northumberland, conifers were planted in the first decade of the twentieth century. Some have grown extremely tall. Huge Western hemlock, Western red cedar, Coast redwood and Sitka spruce go for pulp and fencing, a waste of their timber. This difficulty with marketing big conifer timber is a fairly common problem.

The Atholl Estate is now within the new Cairngorms National Park: timber production is still the main aim of forestry, but leisure, amenity and environmental benefits are also vital in a National Park. Small felling coups will be the main way of including these other uses.

On Corrour Estate, there are areas of Sir John Stirling-Maxwell's early plantations, including Sitka spruce, still growing by Loch Ossian. They are managed using continuous cover methods, with amenity, biodiversity and continuity as main aims. Recent commercial Sitka spruce plantations on more accessible estate land will remain in plant-clearfell-restock management, especially as wet peat soils and potential windblow preclude thinning.

In Northern Ireland, private landowners are encouraged to help expand its 6% forest area by small, annual additions. There will be more flexibility in Sitka spruce management if frequency and intensity of high wind increases. Social and recreational use of forests is increasing; strategic consultations with an assortment of forest owner are in progress; groups who rarely visit forests such as young, disabled and elderly are now encouraged to participate in forest planning.

In the Republic of Ireland farmer-landowners will be assisted by Coillte to expand the country's commercial forest area in future, as the state does not intend to buy further afforestation land yet.

FIGURE 17.11. Some of the early Sitka spruce planted by Sir John Stirling-Maxwell in his experimental plots, Corrour Estate, Inverness-shire.

PHOTO: JOHN SUTHERLAND, 2014.

Conservation organisations such as the Woodland Trust usually remove Sitka spruce and other conifers from their woodlands with the aim of returning woodlands to their notional, 'native' vegetation.

The Royal Society for the Protection of Birds fells Sitka spruce and Lodgepole pines on its 10,117 ha (25,000 acres) Forsinard Reserve in Flow Country. They are placed into the long forestry drains where their skeletal remains are chosen by hen harriers and song birds for nesting sites. Earth or plastic dams later raise the water level of the peat gradually. The RSPB intends to encourage peat regeneration and migratory wading birds into the born-again peat landscape.

New temperate rainforests?

Nature conservation has largely ignored ecological *processes* and functioning in woodlands and has concentrated on *age*, *biodiversity* and *naturalness*.

The post-war period has over-emphasised what are called 'ancient' and 'semi-natural' woodlands as the most valuable category for nature conservation and as vital aspects of our heritage. This has not just clouded, but closed, our eyes to other types of woodlands, their ecology, history, beauty and contributions to human lives.

'Ancient woodlands' are defined by continuity with tree cover for several centuries and at least since the time of early reliable maps: 1750 in Scotland and 1600 in England and Wales. They currently make up 1% of England and Wales, 1% of Scotland and 0.08% of Northern Ireland's landscapes. Conservation organisations suggest that some may link with what they call the 'wild-wood' of 8000 years ago.

'Ancient woodlands', then, are at least 266 or 416 years old. The first private Sitka spruce plantations we know of, were initiated 150 years ago (1866) while the early state forests of Inverliever and Hafod Fawr are 109 and 117 years old respectively. On a time-elapsed basis, they are becoming ancient-ish. Of course they do not replicate ancient woodland composition as we see it now. They are gaining their own ecological composition and processes which may (or may not) be valuable to people living in say, 157, 266 or 299 years' time.

Ancient woodland by definition is always stuck in the same place. It is a static concept which gives us the un-ecological idea that woodlands (specifically) and vegetation (generally) stay in the same place. If we lived for a millennium,

FIGURE 17.12. Sitka spruce seedlings on a nurse log in a grove of his original trees, behaving as if they are at home on the Pacific coast.

PHOTO: ANDY TITTENSOR, 2015.

FIGURE 17.13 *(above left)*. Informing visitors about one of the oldest Sitka spruce in the UK; Glenalmond Arboretum, Perthshire.

PHOTO: SYD HOUSE, 2005.

FIGURE 17.14 *(above right)*. A family enjoying outdoor fun in Hamsterley Forest, Durham.

PHOTO: ISOBEL CAMERON, COURTESY FORESTRY COMMISSION. © CROWN COPYRIGHT.

FIGURE 17.15 *(left)*. Winter reflections of Sitka spruce in Loch Ossian, Corrour Estate, Inverness-shire, a remote area but well-used by walkers and cyclists in all seasons.

PHOTO: ANDY TITTENSOR, 2015.

we would know that tree species (and therefore woodlands) move around the landscape with landslip, flood, windblow, grazing, fire and tree death. Vegetation, including woodlands, is static only if held in place by humans who enclose, manage and try to preserve it.

The Woodland Trust's web site tells us that 'ancient woodland' is our richest habitat, containing abundant fungi and supporting rare and vulnerable species. We now know that some 'modern forests' of Sitka spruce are developing rich habitats which also contain hundreds of fungus species, as well as rarities such as red squirrels and black grouse. Sitka spruce forests have an advantage over ancient woodlands as their huge size allows for species needing large territories or which move long distances.

We are told that 'ancient woodlands are a delight to visit'. So are Sitka spruce forests – unless the 850 000 annual visitors to Galloway Forest Park (2004 figures) are disappointed by the surroundings.

We are told that 'ancient woodlands' are 'living history books'. So are *all* woodlands. I have observed signs of the past even in one-acre woodlands; I have observed historic features in all the woodlands I have ever studied.

Historian Richard Oram, in *The Curious Case of the Missing History* was shocked that residents of Kielder Forest area are somehow severed from their long, rich (but not obvious) history. He pointed out that visitors to Kielder are given little information about its past, as if there is none. Modern Kielder Forest has 'history', but most is not recognised locally while the rest is seen as lacking in merit: no large monuments!

To sum up: it has been generally assumed that Sitka spruce forests lack natural history *and* history, perhaps a result of the twentieth century's intense negative perception of conifers usurping romantic, wild landscapes.

I argue that Sitka spruce and its forests have social histories, heritage value, ecological features and futures equal to their alternatives in Britain and Ireland. They are gradually becoming acceptable in upland landscapes, although nature conservationists have difficulty in assessing them impartially, while foresters still need affirmation that their 'improvements' are creditable.

As time moves on, Sitka spruce plantations will be recognised as 'traditional' countryside features with cultural and ecological identity comparable with hunting parks, Dorset heathland, rabbit warrens, watered meadows and the

FIGURE 17.16. Sitka spruce plantation in Heiðmörk, a popular recreational forest near Reykjavík, Iceland, lit by the Aurora Borealis and city lights.

PHOTO: EDDA SIGURDÍS ODDSDÓTTIR, 2014, COURTESY SKOGRAEKT RIKISINS.

Norfolk Broads. I hope, however, that some naturally-regenerated Sitka spruce tracts in the high-rainfall uplands will progress towards new-style temperate rainforests without expectations (or interference) on our part as to their structure and processes.

But:

> ...I noticed, emerging from the heath on the far side of the site, two trees. I fixed my binoculars on them then swore out loud: Sitka spruce! The seed must have blown in from the great plantations across the mountains.
>
> George Monbiot 2014

> We almost worship it here.
>
> Richard Carstensen, Alaska ecologist, via email, January 2015

Bibliography

Chapter 1. The Most Hated Tree?

Bennett, K. D. (1986) The rate of spread and population increase of forest trees during the postglacial. *Philosophical Transactions of the Royal Society of London* B 314, 523–31.

Brimble, L. J. F. (1948) *Trees in Britain: Wild, Ornamental and Economic.* London: Macmillan.

DellaSala, D. A. (2011) *Temperate and Boreal Rainforests of the World: Ecology and Conservation.* Washington DC: Island Press.

Edlin, H. L. (1949) *British Woodland Trees.* 3rd edn. London: Batsford.

Ellis, C. J. and Hope, J. (2012) *Lichen Epiphyte Dynamics in Scottish Atlantic Oakwoods: The Effect of Tree Age and Historical Continuity.* Edinburgh: Scottish Natural Heritage Commissioned Report 426.

Flora of Northern Ireland (2010) http://www.habitas.org.uk/flora/species.asp?item=3811.

Hall, V. (2011) *The Making of Ireland's Landscape Since the Ice Age.* Cork: Collins Press.

Harris, E., Harris, J. and James, N. D. G (2003/2010) *Oak: A British History.* Oxford: Windgather Press.

Hickie, D. and O'Toole, M. (2002) *Native Trees and Forests of Ireland.* Dublin: Gill & Macmillan.

Mitchell, A. (1974) *A Field Guide to the Trees of Britain and Northern Europe.* London: Collins.

More, D. and White, J. (2003) *Cassell's Trees of Britain and Northern Europe.* London: Cassell.

Nash, R. (1967/1992) *Wilderness and the American Mind.* New Haven, CT and London: Yale University Press.

Native Woodland Trust (2014) http://www.nativewoodlandtrust.ie/en/learn/irish-trees.

Parducci, L. *et al.* (2012) Glacial survival of boreal trees in northern Scandinavia. *Science* 335(6072), 1083–86.

Peterson, E. B., Peterson, N. M., Weetman, G. F. and Martin, P. J. (1997) *Ecology and Management of Sitka Spruce, Emphasising its Natural Range in British Columbia.* Vancouver: UBC Press.

Pryor, F. (2004) *Britain BC: Life in Britain and Ireland Before the Romans.* London: Harper Perennial.

Schneeweiss, G. M. and Schönswetter, P. (2011) A re-appraisal of nunatak survival in arctic-alpine phylogeography. *Molecular Ecology* 20(2), 190–92.

Simmons, I. G. (2001) *An Environmental History of Great Britain from 10,000 Years Ago to the Present.* Edinburgh: Edinburgh University Press.

Stace, C. (2010) *New Flora of the British Isles.* 3rd edn. Cambridge: Cambridge University Press.

Tittensor, R. M. (1980) Ecological history of the yew (*Taxus baccata* L.) in southern England. *Biological Conservation* 17, 243–65.

Unwin, A. H. (1905) *Future Forest Trees.* London: Fisher Unwin.

Walton, J. (1967) *Argyll Forest Park Guide.* Edinburgh: Forestry Commission.

Wickham-Jones, C. (2010) *Fear of Farming.* Oxford: Oxbow Books.

Chapter 2. The Tree From Sitka

Anon (1950) *Native Trees of Canada.* Canada Department of Resources and Development, Forestry Branch Bulletin 61. Ottawa: Edmond Cloutier.

Bongard, [M.] (1831/1833) Observations sur la végétation de l'Ile de Sitcha. *Mémoires de L'Academie Imperilae des Sciences de St.-Pétersbourg* 6(2), 119–77.

Brockman, C. F. and Merrilees, R. (1979) *Trees of North America. A Guide to Field Identification.* New York: Golden Press.

Copes, D. L. and Beckwith, R. C. (1977) Hybrid swarm Sitka spruce and White spruce, Alaska. *Botanical Gazette* 138(4), 512–21.

Dallimore, W. and Jackson, A. (1961) *A Handbook of Coniferae.* rev. edn. London: Edward Arnold.

Ellyson, W. J. T. and Sillett, S. C. (2003) Epiphyte communities on Sitka spruce in an old-growth redwood forest, *Bryologist* 106, 197–211.

Farrar, J. L. (2005) *Trees in Canada.* Markham, Ontario: Fitzhenry & Whiteside and Canadian Forest Service

Faulkner, R. (1987) Genetics and breeding of Sitka spruce. In Henderson, D. M. and Faulkner, R. (1987) Sitka spruce. *Proceedings of the Royal Society of Edinburgh,* Section B, 93(1–2), 41–50.

Friis, E. M., Crane, P. R. and Pederson, K. R. (2011) *Early Flowers and Angiosperm Evolution.* Cambridge: Cambridge University Press.

Government of British Columbia, Ministry of Forests,

Lands and Natural Resource Operations (post-2000) *Compendium of Tree Species: Engelmann Spruce.* https://www.for.gov.bc.ca/hfp/silviculture/Compendium/EngelmannSpruce.htm.

Government of British Columbia, Ministry of Forests, Lands and Natural Resource Operations (post-2000) *Compendium of Tree Species: Sitka Spruce.* https://www.for.gov.bc.ca/hfp/silviculture/Compendium/SitkaSpruce.htm.

Johnson, O. and More, D. (2004) *Collins Tree Guide.* London: HarperCollins.

Lindsay, A. and House, S. (1999/2005) *The Tree Collector: The Life and Explorations of David Douglas.* London: Aurum Press.

Little, W., Fowler, H. W. and Coulson, J. (1967) *The Shorter Oxford Dictionary of Historical Principles.* 3rd edn. Oxford: Clarendon Press.

Missouri Botanical Garden (2012) http://www.tropicos.org/Name/40008450; http://www.tropicos.org/Name/24900172; http://www.tropicos.org/Name/50231643.

Portland State University (2009) *Sitka Spruce.* http://www.oregonencyclopedia.org/article/sitka_spruce/.

Rana, J.-H., Weia, X. and Wanga, X.-Q. (2006) Molecular phylogeny and biogeography of *Picea* (Pinaceae): implications for phylogeographical studies using cytoplasmic haplotypes. *Molecular Phylogenetics and Evolution* 41(2), 405–19.

Seaside and Cannon Beach, Oregon (2012) *Welcome to Klootchy Creek Park, Oregon, Home to the Giant Sitka Spruce.* http://www.lewisandclarktrail.com/section4/orcities/seaside/sitkaspruce.htm.

Sharples, A. W. (1938/1958) *Alaska Wild Flowers.* Stanford, CA: Stanford University Press.

Sillett, S. (2012) *Photo Tour: Sitka Spruce* Picea sitchensis http://www.humboldt.edu/redwoods/photos/spruce.php. Accessed 2014.

Sigurgeirsson, A. and Szmidt, A. (1993) Phylogenetic and biogeographic implications of chloroplast variation in *Picea. Nordic Journal of Botany* 13(3), 233–46.

Sterry, P. (2007) *Collins Complete British Trees.* London: HarperCollins.

Chapter 3. Origin, Migration and Survival on the Edge

Ager, T. A. and Rosenbaum, J. G. (2009) Late glacial–Holocene pollen-based vegetation history from Pass Lake, Prince of Wales Island, southeastern Alaska. In Haeussler, P. J. and Galloway, J. P. (eds), *Studies by the US Geological Survey in Alaska, 2007.* US Geological Survey Professional Paper 1760-G, https://pubs.usgs.gov/pp/1760.

Brodribb, T. and& Field, T. (2009) *How Angiosperms Took Over the World* http://news.sciencemag.org/sciencenow/2009/12/08-04.html.

Chaisurisri, K., Mitton, J. B. and El-Kassaby, Y. A. (1994) Variation in the mating system of Sitka spruce (*Picea sitchensis*): evidence for partial assortative mating. *American Journal of Botany* 81(11), 1410–15.

Chunjing, Z., Weduo, X., Shimizu, H. and Kaiyun, W. (2012) Overview of spruce forests in China. In Oteng-Amoako, A. A. (ed.) *New Advances and Contributions to Forestry Research*, 205–24. Published online by Intech.

Copes, D. L. and Beckwith, R. C. (1977) Hybrid swarm Sitka spruce and White spruce, Alaska. *Botanical Gazette* 138(4), 512–21.

Page, C. N. and Hollands, R. C. (1987) The taxonomic and biogeographic position of Sitka spruce. In Henderson, D. M. and Faulkner, R. (1987) Sitka Spruce. *Proceedings of the Royal Society of Edinburgh, Section* B, 93(1–2), 13–24.

Rana, J.-H., Weia, X. and Wanga, X.-Q. (2006) Molecular phylogeny and biogeography of *Picea* (Pinaceae): Implications for phylogeographical studies using cytoplasmic haplotypes. *Molecular Phylogenetics and Evolution* 41(2), 405–19.

Ritchie, J. C. (1987/2003) *Postglacial Vegetation of Canada.* Cambridge: Cambridge University Press.

Schmidt, P. (2003) The diversity, phytogeography and ecology of spruces (*Picea*: Pinaceae) in Eurasia. In Mill, R. R. (ed.), *Proceedings of the Fourth International Conifer Conference. Acta Horticulturae* 615, 189–201.

Sigurgeirsson, A. and Szmidt, A. (1993) Phylogenetic and biogeographic implications of chloroplast variation in *Picea. Nordic Journal of Botany* 13(3), 233–46.

Simmons, I., Sirkin, L. and Tuthill, S. J. (1987) Late Pleistocene and Holocene deglaciation and environments of the southern Chugach Mountains, Alaska. *Geological Society of America Bulletin* 99(3), 376–384. www.bulletin.geoscienceworld.org/.

Taylor, R. J. (1993) 5. *Picea* and 7. *Picea sitchensis.* In Flora of North America Editorial Committee (eds), *Flora of North America North of Mexico* Vol. 2. New York and Oxford: Oxford University Press; http://www.efloras.org/florataxon.aspx?flora_id=1&taxon_id=125375; http://www.efloras.org/florataxon.aspx?flora_id=1&taxon_id=233500914.

Thomas, A. (2009) *Vegetation Development on Heceta island, Southeastern Alaska during the Late Glacial and Holocene.* http://gsa.confex.com/gsa/2007CD/finalprogram/abstract_120853.htm.

University of California Museum of Palaeontology (2011) *The Mesozoic Era.* http://www.ucmp.berkeley.edu/mesozoic/mesozoic.php.

University of California Museum of Palaeontology (2011) *The Palaeozoic Era.* http://www.ucmp.berkeley.edu/paleozoic/paleozoic.php.

Whitlock, C. (1992) Vegetational and climatic history of the Pacific Northwest during the last 20,000 years: implications for understanding present-day biodiversity. *Northwest Environmental Journal* 8, 5–28.

Woodward, J. (2014) *The Ice Age: A Very Short Introduction.* Oxford: Oxford University Press.

Wynn, G. (2007) *Canada and Arctic North America: An Environmental History.* Santa Barbara CA: ABC-CLIO.

Chapter 4. At Home in North American Rainforests

Ager, T. A. and Rosenbaum, J. G. (2009) Late Glacial–Holocene pollen-based vegetation history from Pass Lake, Prince of Wales Island, southeastern Alaska. In Haeussler, P. J. and Galloway, J. P. (eds), *Studies by the US Geological Survey in Alaska, 2007.* USGS Professional Paper 1760-G. http://pubs.usgs.gov/pp/1760.

BBC (2012) *Secrets of Our Living Planet: The Magical Forest.* BBC Earth DVD, Disc 2. http://bbc.co.uk/programmes/b01k73zy. Presenter Chris Packham.

Carstensen, R., Armstrong, R. H. and O'Clair, R. M. (2014) *The Nature of Southeast Alaska: A Guide to Plants, Animals and Habitats.* 3rd edn. Portland, OR: Alaska Northwest Books.

DellaSala, D. A. (2011) *Temperate and Boreal Rainforests of the World: Ecology and Conservation.* Washington DC and London: Island Press.

Ecotrust and Conservation International (2012) *Coastal Temperate Rainforests: Ecological Characteristics, Status and Distribution Worldwide.* Portland, OR: Ecotrust.

Gavin, D. G., Brubaker, L. B. and Lertzman, K. P. (2003) Holocene fire history of a coastal temperate rainforest based on soil charcoal radiocarbon dates. *Ecology* 84(1), 186–201.

Ground Truth Trekking (2007) *Temperate Rainforests of the Northern Pacific Coast.* http://www.groundtruthtrekking.org/.

Heinrich, B. (2013) *Life Everlasting: The Animal Way of Death.* New York: Mariner Books.

Kauffmann, M. E. (2014) *Conifers of the Pacific Slope. A Field Guide to the Conifers of California, Oregon and Washington.* Kneeland, CA: Backcountry Press.

Klinkenberg, B. (ed.) (2013) *E-Flora BC: Electronic Atlas of the Flora of British Columbia* Vancouver: Laboratory for Advanced Spatial Analysis, Department of Geography, University of British Columbia. http://ibis.geog.ubc.ca/biodiversity/eflora.

Kricher, J. C. (1993) *A Field Guide to the Ecology of Western Forests.* Boston MA: Houghton Mifflin.

McAllister, I. and McAllister, K. (1997) *The Great Bear Rainforest: Canada's Forgotten Coast.* San Francisco CA: Sierra Club Books.

Millman, L. (1998) Pacific traditions. In *Exploring Canada's Spectacular National Parks*, 18–45. Washington DC: National Geographic Society.

National Parks of Canada (2009) *Pacific Rim National Park Reserve of Canada.* http://www.pc.gc.ca/pn-np/bc/pacificrim/natcul/natcul1.aspx.

Romano-Lax, A. (2011) *Tongass National Forest: A Temperate Rainforest in Transition.* Anchorage: Alaska Geographic Association.

Schoonmaker, P. K., von Hagen, B. and Wolf, E. C. (1997) *The Rain Forests of Home: Profile of a North American Bioregion.* Washington DC: Island Press.

Sillett, S. C. (2012) *Institute for Redwood Ecology.* http://www2.humboldt.edu/redwoods/.

Chapter 5. Sitka Spruce in the Lives of First Nations

Betts, R. C. (1998) The Montana Creek fish trap 1: archaeological investigations in southeast Alaska. In Bernick, K. (ed.) *Hidden Dimensions: The Cultural Significance of Wetland Archaeology*, 239–51. Vancouver: UBC Press.

Blackman, M. B (2012) *Haida.* http://www.encyclopedia.com/topic/haida.aspx.

Bongard, [M.] (1831/1833) Observations sur la végétation de l'Ile de Sitcha. *Mémoires de L'Academie Imperilae des Sciences de St.-Pétersbourg* 6(2), 119–77.

Busby, S. (2003) *Spruce Root Basketry of the Haida and Tlingit.* Seattle: Marquand Books and University of Washington Press.

Cohen, J. M. (2014) *Paleoethnobotany of Kilgii Gwaay: A 10,000-year-old Ancestral Haida Archaeological Wet Site.* MA thesis, University of Victoria.

De Laguna, F. (2009) *Under Mount Saint Elias: The History and Culture of the Yakutat Tlingit.* Haverford and Ottawa: Frederica de Laguna Northern Books.

Ecotrust, Pacific GIS and Conservation International (1995–2012) *The Rain Forests of Home: An Atlas of People and Place*. http://www.inforain.org/rainforestatlas/.

Emmons, G. T. (1991) ed. with additions by De Laguna, F. *The Tlingit Indians*. Seattle WA: University of Washington Press.

Erlandson, J. M. *et. al.* (2007) The Kelp Highway hypothesis: marine ecology, the coastal migration theory, and the peopling of the Americas. *Journal of Island and Coastal Archaeology* 2, 161–74.

Fifield, T. E. (1998) *Alaska Archaeology Week Poster*. Sponsored by the Public Education Committee of the Alaska Anthropological Association.

Frankenstein, E. (2015) *Tracing Roots: A film by Ellen Frankenstein with Delores Churchill*. http://www/tracingrootsfilm.com for trailer, http://www.newday.com to watch the film.

Gulick, A. and Troll, R. (2010) *Salmon in the Trees: Life in Alaska's Tongass Rain Forest*. Seattle WA: Mountaineers Books.

HighBeam Research (2002) *Haida*. http://www.highbeam.com/doc/1G2-3048800091.html.

Holm, W. (1965) *Northwest Coast Indian Art: An Analysis of Form*. Vancouver: Douglas & McIntyre.

Loring, J. M. (1992) *Montana Creek Fish Trap, Juneau, Alaska: Site Overview*. 19th Annual Alaska Anthropological Association Meeting. Juneau. Loring Research.

Moerman, D. E. (2010) *Native American Food Plants: An Ethnobotanical Dictionary*. Portland: Timber Press.

Östlund, L., Bergman, I. and Zackrisson, O. (2004) Trees for food: a 3000-year record of subarctic plant use. *Antiquity* 78(300), 278–86.

Raghaven, M., Skoglund, P., Graft, K. E. *et al.* (2014) Upper Palaeolithic Siberian genome reveals dual ancestry of Native Americans. *Nature* 50, 87–91.

Russell, P. N. (2011) *Alutiig Plantlore: An Ethnobotany of the Peoples of Nanwalek and Port Graham, Kenai Peninsula, Alaska*. Fairbanks: University of Alaska Fairbanks, Centre for Cross-Cultural Studies.

Sierra Club (2014) *Temperate Rainforest Ecology: An Introduction*. http://www.sierraclub.bc.ca/education/teachers/educator-resources-by-topic/#Rainforest.

Thomas, M. *et al.* (2008) DNA from pre-Clovis human coprolites in Oregon, North America. *Science* 320(5877), 786–89.

Thornton, T. F. (2008) *Being and Place Among the Tlingit*. Seattle WA: University of Washington Press.

Treuer, A., Wood, K. and Fitzhugh, W. W. (2010) *Indian Nations of North America*. Washington DC: National Geographic.

Turner, N. J. (1998/2007) *Plant Technology of First Peoples in British Columbia*. Victoria: Museum Handbook, Royal British Columbia Museum,

Turner, N. J (1995/2010) *Food Plants of Coastal First Peoples*. Victoria: Museum Handbook, Royal British Columbia Museum.

Waters, M. R. *et al.* (2011) Pre-Clovis Mastodon hunting 13,800 years ago at the Manis site, Washington. *Science* 334 (6054), 351–53.

Wonders, K. (2008) *First Nations: Land Rights and Environmentalism in British Columbia*. http://www.firstnations.de/fisheries/kwakwakawakw-quatsino.htm.

Worl, R. Lanáat'tlaa, Shaadoo'tlaa and Gregory, D. Héendei (2014) *A Basic Guide to Northwest Coast Formline Art*. Juneau: Sealaska Heritage Institute.

Chapter 6. Prehistoric Lives and Woodlands in Britain and Ireland

Allen, M. J. and Scaife, R. (2007) A new downland prehistory: long-term environmental change on the southern English chalklands. In Fleming and Hingley (eds) (2007), 16–32.

Anon (2011) Bronze Age eel trap may be oldest known. *British Archaeology* 120/121, 7.

Bennett, K. D. and Provan, J. (2008) What do we mean by 'refugia'? *Quaternary Science Reviews* 27, 2449–55.

Bunting, J. (2015) Mesolithic Orkney, *Archaeology Scotland* 24, 8–9.

Coles, J. (1985) *Somerset Levels Papers 15*. Exeter: Department of History and Archaeology, University of Exeter.

Coles, B. and Coles, J. (1986) *Sweet Track to Glastonbury: The Somerset Levels in Prehistory*. London: Thames & Hudson.

Coppins, S. and Coppins, B. (2012) *Atlantic Hazel: Scotland's Special Woodlands*. Kilmartin, Argyll: Atlantic Hazel Action Group.

Corbet, G. B. (1975) *Finding and Identifying Mammals in Britain*. London: British Museum Natural History.

Dalton, C. P. D. and Battarbee, R. W. *et al.* (2001) Climate history through the Holocene at Lochnagar, Scotland. *Proceedings of the Royal Irish Academy* 101B (1–2), 159.

Davies, A. N. (1999) *High Spatial Resolution Holocene Vegetation and Land-Use History in West Glen Affric and Kintail, Northern Scotland*. PhD thesis, University of Stirling. Available on-line via EthOS: http://www.ethos.bl.uk/.

Dixon, N. (2007) *The Crannogs of Perthshire: A Guide*. Perth: Perth and Kinross Heritage Trust.

Finlayson, B. (1998) *Wild Harvesters: The First People in Scotland.* Edinburgh: Birlinn and Historic Scotland.

Fleming, A. and Hingley, R. (eds) (2007) *Prehistoric and Roman Landscapes.* Oxford: Windgather Press.

Gaffney, V., Fitch, S. and Smith D. (2009) *Europe's Lost World: The Rediscovery of Doggerland.* York: Council for British Archaeology Research Report 160.

Hall, V. (2011) *The Making of Ireland's Landscape Since the Ice Age.* Cork: Collins Press.

Hampshire and Wight Trust for Maritime Archaeology (2012) *Bouldnor Cliff.* http://www.maritimearchaeologytrust.org/bouldnor.

Lee, D. and Thomas, A. (2012) *Orkney's First Farmers.* http://www.archaeology.co.uk/features/orkneys-first-farmers.htm.

Lewis, B. (2009) *Hunting in Britain from the Ice Age to the Present.* Stroud: History Press.

Lillie, M. and Ellis, S. (2007) *Wetland Archaeology and Environments: Regional Issues, Global Perspectives.* Oxford: Oxbow Books.

Lindsay, R. (1995) *The Ecology, Classification and Conservation of Ombrotrophic Mires.* Perth: Scottish Natural Heritage.

Linnard, W. (2000) *Welsh Woods and Forests: A History.* Landysul: Gomer.

McQuade, M. and O'Donnell, L. (2007) Late Mesolithic fish traps from the Liffey estuary, Dublin, Ireland. *Antiquity* 81(313), 569–84.

Mithen, S. (2003) *After The Ice: A Global Human History 20,000–5,000 BC.* London: Weidenfeld & Nicolson.

Mithen, S. and Wicks, K. (2015) The voyagers of Rubha Port ant-Seilich: the discovery of Late Glacial hunters on Islay. *British Archaeology* 145, 24–29.

Momber, G., Stachell, J. and Gillespie, J. (2011) Bouldnor Cliff. *British Archaeology* 121, 31–34.

Parducci, L. *et al.* (2012) *Glacial Survival of Boreal Trees in Northern Scandinavia.* http://www.sciencemag.org/content/335/6072/1083.abstract?.

Pryor, F. (2001) *Seahenge: A Quest for Life and Death in Bronze Age Britain.* London: Trafalgar Square Publishing.

Pryor, F. (2004) *Britain BC: Life in Britain and Ireland Before the Romans.* London: HarperPerennial.

Reynolds, P. (1979) *Iron-Age Farm: The Butser Experiment.* London: Colonnade Books.

Strachan, D. (2010) *The Carpow Logboat: A Bronze Age Vessel Brought to Life.* Perth: Perth and Kinross Heritage Trust.

Thomas, A. (2011) The Braes of Ha'Breck, Wyre. In Milburn, P. (ed.) *Discovery and Excavation in Scotland, New Series* 12, 134–36. Musselburgh: Archaeology Scotland.

Tipping, R. (2003) Living in the past: woods and people in prehistory to 1000 BC. In Smout, T. C. (ed.) *People and Woods in Scotland: A History.* Edinburgh: Edinburgh University Press.

Tipping, R. (2008) Blanket peat in the Scottish Highlands: timing, cause, spread and the myth of environmental determinism. *Biodiversity Conservation* 17, 2097–2113.

Tittensor, R. M. (1991) *West Dean: A History of Conservation on a Sussex Estate.* Bognor Regis: Felpham Press.

Tittensor, R. M. (2010) *What's so Special about Peat? Its Importance to Scotland and the World.* Darvel, Ayrshire: Leaflet, Countryside Management Consultancy.

Twigger, S. N. and Haslam, C. J. (1991) Environmental changes in Shropshire during the last 13,000 years. *Field Studies* 7, 743–758.

Van de Noort, R. and Fletcher, W. (2000) Bronze Age human dynamics in the Humber Estuary. In Bailly. G., Charles, R. and Winder, N. (eds), *Human Ecodynamics: Proceedings of the Association for Environmental Archaeology Conference*, 47–54. Oxford: Oxbow Books.

Warren, G. (2005) *Mesolithic Lives in Scotland.* Stroud: History Press.

Wells, S. (2011) *Pandora's Seed: Why the Hunter-Gatherer Holds the Key to Our Survival.* Harmondsworth: Penguin.

Whitehead, P. F. (2007) The alluvial archaeobiota of the Worcestershire River Avon. *Worcestershire Record* 20, 42–42.

Willerslev, E. (2012) *Sturdy Scandinavian Conifers Survived Ice Age.* http://news.ku.dk/all_news/2012/2012.3/sturdy-scandinavian-conifers-survived-ice-age.

Yalden, D. W. (1999) *The History of British Mammals.* London: T. & A. D. Poyser.

Chapter 7. Woodland History and Britain's Need for Sitka Spruce

Coillte (2007) *Bringing the Bogs Back to LIFE.* DVD. Available: Newtonmountkennedy, Co. Wicklow, Eire; info@coillte.ie.

Constable, J. (1815) *Boat-building near Flatford Mill.* Painting held by Victoria and Albert Museum: London. Prints available.

DiPietro, M. (2014) *In Terms of Sacredness: Tree Laws and Status in Medieval Ireland.* https://www.academia.

edu/4403496/In_Terms_of_Sacredness_Tree_Laws_and_Status_in_Medieval_Ireland.

Evans, J. (ed.) (2001) *The Forests Handbook. Vol.2: Applying Forest Science for Sustainable Management.* Chichester: Wiley.

Forest Service, Department of Agriculture, Fisheries and Food (2008) *Irish Forests: A Brief History.* https://www.agriculture.gov.ie/media/migration/forestry/forestservicegeneralinformation/abouttheforestservice/IrishForestryAbriefhistory200810.pdf.

Gautier, A. (2007) Game parks in Sussex and the Godwinesons. *Anglo-Saxon Studies* 29, 51–64.

Harmer, R. and Howe, J. (2003) *The Silviculture and Management of Coppice Woodlands.* Edinburgh: The Forestry Commissison.

Length of British coastline: Ordnance Survey data using digital analysis of MHWM. http://www.cartography.org.uk/default.asp?contentID=749.

Lewis, B. (2009) *Hunting in Britain from the Ice Age to the Present.* Stroud. History Press.

Mackie, C. (1819/1821) *The History of the Abbey, Palace, and Chapel-Royal of Holyrood House.* Edinburgh (digitised by New York Public Library)

Manwood, J. (1598) *Treatise and Discourse of the Lawes of the Forrest.* London: Thomas Wight & Bonham Norton.

Mills, C. M. and Crone, A. (2012) Dendrochronological evidence for Scotland's native timber resources over the past 1000 years. *Scottish Forestry* 66(1), 18–33.

Moray, Earl of (2010) Darnaway: the agony and ecstasy of managing a Scottish forest. In Mills, C. (ed.) *Woods as Working and Cultural Landscapes, Past and Present.* Native Woodland Discussion Group Woodland History Conference: Notes 15, 11–14.

Oram, R. (2014) Between a rock and a hard place. Climate, weather and the rise of the lordship of the Isles. In Oram, R. D. (ed.) *The Lordship of the Isles,* 40–61. Leiden: Brill.

Phillips, M. T. T. (1998) Historical information about the ancient forest of Darnaway Estate, Morayshire. *Scottish Forestry* 52(1), 42–46.

Simmons, I. (2000) *An Environmental History of Great Britain: From 10,000 Years Ago to the Present.* Edinburgh: Edinburgh University Press.

Swedish Phytogeographical Society (1965) *The Plant Cover of Sweden.* Acta Phytogeographica Suecica 50. Uppsala: Almqvist & Wiksells.

Smout, T. C. (ed.) (1997) *Scottish Woodland History.* Edinburgh: Scottish Cultural Press.

Smout, T. C., MacDonald, A. R. and Watson, F. (2005/2007) *A History of the Native Woodlands of Scotland, 1500–1920.* Edinburgh: Edinburgh University Press.

Tipping, R. (2004) Palaeoecology and political history: evaluating driving forces in historic landscape change in southern Scotland. In Whyte and Winchester (eds) (2004), 11–20.

Tittensor, R. M. (1970) History of the Loch Lomond Oakwoods. *Scottish Forestry* 24(2), 100–118.

Tittensor, R. M. (1978) A History of The Mens: A Sussex Woodland Common. *Sussex Archaeological Collections* 116, 347–74.

Wells, S. (2011) *Pandora's Seed: Why the Hunter-gatherer Holds the Key to Our Survival.* London: Penguin.

Whittemore, C. (2012) *A Living from the Edinburghshire Countryside.* Edinburgh: Whittemore Books, colin.whittemore@btinternet.com.

Whyte, I. D. and Winchester, A. J. L. (2004) *Society, Landscape and Environment in Upland Britain.* London: Society for Landscape Studies: supplementary series 2.

Witney, K. P. (1976) *The Jutish Forest: A Study of the Weald of Kent from 450 to 1380 AD.* London: Athlone Press.

Yalden, D. W. and Albarella, U. (2009) *The History of British Birds.* Oxford: Oxford University Press.

Chapter 8. Realisation: New Trees for New Woodlands

Anderson, J. (2006) *Atholl Estates.* Perth: Perth and Kinross Heritage Trust. http://www.pkht.org.uk.

Barrett, J. R. (2011) Regular Reforestation: Woodland Resources in Moray, 1720–840. In *Native Woodlands Discussion Group Conference Notes XVI, (2011)*: Community, woodlands and perceptions of ownership: a historical perspective. Available from Institute for Environmental History, University of St Andrews.

Benn, A. W. (2011) *The Levellers and the Tradition of Dissent.* http://www.bbc.co.uk/history/british/civil_war_/benn_levellers_01.shtml.

C[hurche], R[ooke] (1612) *An Olde Thrift Newly Revived.* London: Richard Moore. (7 copies in British and 4 in N. American institutions). Abstracts online *e.g.* http://books.google.co.uk/books/about/An_Olde_Thrift_Newly_Revived.html?id=fVk8PAAACAAJ&redir_esc=y.

DellaSala, D. A. (2011) *Temperate and Boreal Rainforests of the World: Ecology and Conservation.* Washington DC and London: Island Press.

Devine, T. M. (1999/2006/2012) *The Scottish Nation: A Modern History*. Harmondsworth: Penguin.

Dyer, C. (2003) *Making a Living in the Middle Ages: The People of Britain 850–1520*. Harmondsworth: Penguin.

Evelyn, J. (1664) *Sylva; or, A Discourse on Forest-Trees and the Propagation of Timber in His Majesty's Dominions*. London: John Martyn.

Hall, V. (2011) *The Making of Ireland's Landscape since the Ice Age*. Cork: Collins Press.

House, S. and Dingwall, C. (2003) 'A Nation of Planters': introducing the new trees, 1650–1900. In Smout, T. C. (2003) *People and Woods in Scotland: A History*, 128–57. Edinburgh: Edinburgh University Press.

Hunter, T. (1883) *Woods, Forests, and Estates of Perthshire: With Sketches of the Principal Families in the County*. Perth: Henderson, Robertson & Hunter.

Levett, C. (1618) *An Abstract of Timber-Measures*. London: MS Book, William Jones. http://www.worldcat.org/identities/lccn-n88-32230.

Linnard, W. (1970) Thomas Johnes of Hafod – pioneer of upland afforestation in Wales. *Ceredigion* 6(1–4), 309–319.

Linnard, W. (1979) A forester's manual of the thirteenth century. *Quarterly Journal of Forestry* 73(2), 95–99.

Linnard, W. (2000) *Welsh Woods and Forests: A History*. Landysul, Ceredigion: Gomer.

Mackenzie, S. and Blackett, S. (2009) Royal Scottish Forestry Society Spring 2009 excursion to Aberdeenshire. *Scottish Forestry* 63(3), 37–46.

Mills, C. M. and Crone, A. (2012) Dendrochronological evidence for Scotland's native timber resources over the past 1000 years. *Scottish Forestry* 66(1), 18–33.

Monteath, R. (1820) *The Forester's Guide: Or, a Practical Treatise on the Training and Pruning of Forest Trees, Etc.* BiblioBazaar Reprint Books Online (2010).

Peterken, G. F. and Welch, R. C. (eds) (1975) *Bedford Purlieus: Its History, Ecology and Management*. Huntingdon: Institute of Terrestrial Ecology/Monks Wood Experimental Station Symposium 7.

Reid, J. (1683) *The Scots Gardener*. Edinburgh: text available on-line, http://www.archive.org/stream/scotsgardnertogeooreidrich/scotsgardnertogeooreidrich_djvu.txt.

Scott, P. J. (1974) *Conifers of Newfoundland*. St. John's: Memorial University of Newfoundland.

Smout, T. C., MacDonald, A. R. and Watson, F. (2005/2007) *A History of the Native Woodlands of Scotland, 1500–1920*. Edinburgh: Edinburgh University Press.

Tubbs, C. R. (2001) *The New Forest: History, Ecology and Conservation*. Newton Abott: David & Charles.

University of Waterloo Library (2009) *Scholarly Societies Project; Society of Improvers in the Knowledge of Agriculture in Scotland*. http://www.scholarly-societies.org/history/1723sikas.html.

Walker, G. J. and Kirby, K. J. (1987) An historical approach to woodland conservation in Scotland. *Scottish Forestry* 41(2), 87–98.

Wilson, J. J. (1891) *The Annals of Penicuik. Being a History of the Parish and of the Village*. Facsimile reprint published by SPA Books, Stevenage (1985).

Chapter 9. Ships, Surveyors, Scurvy and Spruces

Anon (2014) *Plant Hunters in the Northeast*. http://nynjctbotany.org/plnthunt/plnthunt.html.

Anon (2013) *General Amherst's Spruce Beer (from his 1758–1763 Journal)*. http://www.canadahistory.com/sections/documents/arts/sprucebeer.htm.

Banks, Sir J. (2006) *The Endeavour Journal of Sir Joseph Banks*. Teddington: Echo Library.

Cook, J. [1795] *Captain Cook's Third and Last Voyage to the Pacific Ocean, in the Years 1776, 1777, 1778, 1779 and 1780*. London: John Fielding & John Stockdale (modern facsimile: Filiquarian Publishing, LL/Qontro).

Dictionary of Canadian Biography Online (2014) Banks, Sir Joseph. http://www.biographi.ca/009004-11901-e.php?BioId=36373.

Dictionary of Canadian Biography Online (2014) Menzies, Archibald. http://www.biographi.ca/en/bio.php?id_nbr=3557.

Durzan, D. J. (2009) Arginine, scurvy and Cartier's 'tree of life'. *Journal of Ethnobiology and Ethnomedicine* 2(5), 5.

Edwards, P. (1999/2003) *James Cook: The Journals*. (Chosen from J. C. Beaglehole's Selection for the Hakluyt Society 1955–67). London: Penguin Books.

Frost, O. W. (ed.) (1988) *Georg Wilhelm Steller: Journal of a Voyage with Bering, 1741–1742*. Trans. Engle, M. E. and Frost, O. W. Stanford: Stanford University Press.

Golder, F. A. (1925) *Steller's Journal of the Sea Voyage from Kamchatka to America and Return on the Second Expedition, 1741–1742*. Trans. L. Stejneger. American Geographical Society Research Series No. 2, New York. https://archive.org/stream/beringsvoyagesaco2gold/beringsvoyagessaco2gold_djvu.txt.

Hough, R. (1994/1995) *Captain James Cook*. London: Hodder & Stoughton.

Justice, C. (2000) *Mr Menzies' Garden Legacy: Plant*

Collecting on the Northwest Coast. Vancouver: Cavendish Books Inc.

Kalm, P. (1752) The method of making spruce beer practiced in the north of America, from the letters of P. Kalm sent to the Swedish Academy. *Gentleman's Magazine and Historical Chronicle* 22, 399–400.

Kelpius Society (2010) *Johannes Kelpius*. http://kelpius.org/aboutus.html.

Lindsay, A. (2008) *Seeds of Blood and Beauty: Scottish Plant Explorers*. Edinburgh: Birlinn.

Littlepage, D. (2006) *Steller's Island: Adventures of a Pioneer Naturalist in Alaska*. Seattle: The Mountaineers Books.

Lockett, J. (2010) *James Cook in Atlantic Canada: The Adventurer and Map Maker's Formative Years*. Halifax, NS: Formac Publishing.

Lysaght, A. M. (1971) *Joseph Banks in Newfoundland and Labrador, 1766: His Diary, Manuscripts and Collections*. London and Berkely CA: University of California Press.

Martini, E. (2002) Jacques Cartier witnesses a treatment for scurvy. *Vesalius* 8(1), 2–6.

Newcombe, C. V. (ed.) (1923) *Menzies' Journal of Vancouver's Voyage, April to October, 1792*. Full text. Archives of British Columbia, Digital Library. http://archive.org/stream/menziesjournalof1792menz/menziesjournalof1792menz_djvu.txt; http://www.openlibrary.org/books/ OL6661787M/Menzies'_journal_of_Vancouver's_voyage_April_to_October_1792.

Nisbet, J. (2009) *The Collector: David Douglas and the Natural History of the Northwest*. Seattle WA: Sasquatch Books.

O'Brian, P. (1970) *Master and Commander*. London: Collins.

O'Brian, P. (1987/1997) *Joseph Banks*. London: Harvill Press.

Radall, T. H. (1948/1993) *Halifax: Warden of the North*. Halifax, NS: Nimbus.

Saunders, G. (1970) *Trees of Nova Scotia: A Guide to the Native and Exotic Species*. Nova Scotia: Department of Lands and Forests.

Scott, P. J. (1974) *Conifers of Newfoundland*. St John's: Memorial University of Newfoundland.

Sörlin, S. (2008) *Globalizing Linnaeus: Economic Botany and Travelling Disciples*. TijdSchrift voor Skandinavistiek, 29(1–2), 117–43.

Spons, E. and Spons, F. (1879) *Spons' Encyclopaedia of the Industrial Arts, Manufactures, and Commercial Products*. London and New York: E. & F. Spons.

Victoria University of Wellington Library (2014) Captain Cook's method of making Spruce beer. In *Cook's Second Voyage Towards the South Pole*. 4th edn, I, 99 and 101. http://www.nzetc.victoria.ac.nz/tm/scholarly/tei-ShoSout-t1-back-d1-d6.html.

Chapter 10. Journeys and Experiments for Seeds and People

Alden, J. and Bruce, D. (1989) *Growth of Historical Sitka Spruce Plantations at Unalaska Bay, Alaska*. USDA, Forest Service, General Technical Report PNW-GTR-236.

Barclay, D. (1887) On the plantations on the Estate of Sorn, in the County of Ayr, N.B. *Transactions of the Scottish Arboricultural Society* 11, 29–35.

Boylan, C. (2010) *Champion and Heritage Trees of Ireland*. Paper presented at IFPRA World Congress 22, Hong Kong.

Donald, J., Wood, R. F., Edwards, M. V. and Aldhous, J. R. (1964) *Exotic Forest Trees in Great Britain*. London: Forestry Commission, HMSO.

Douglas, D. (1914/2011) *Journal Kept by David Douglas During his Travels in North America 1823–1827: Together with a Particular Description of Thirty-three Species of American Oaks and Eighteen Species of* Pinus. Cambridge: Cambridge University Press.

Dunn, M. (1891) The value in the British Islands of introduced conifers. *Journal of the Royal Horticultural Society* 14, 73–102.

France, C. S. (1869) Remarks upon Coniferae grown at Powerscourt, Co. Wicklow, Ireland. *Transactions of the Scottish Arboricultural Society* 5, 83–85.

Hetherington, J. (2004) *The Forestry Department at Bangor 1904 to 2004: A Personal Account*. Bangor: University of Wales, School of Agricultural and Forest Sciences.

James, N. D. G. (1982) *A Forest Centenary: A History of the Royal Forestry Society of England, Wales and Northern Ireland*. Oxford: Blackwell.

Jefferies, R. (1945) *The Wood From the Trees*. London: Pilot Press.

Joyce, P. M. and OCarroll, N. (2002) *Sitka Spruce in Ireland*. Dublin: National Council for Forest Research & Development (COFORD).

Lindsay, S. (2008) *Seeds of Blood and Beauty: Scottish Plant Explorers*. Edinburgh: Birlinn.

Linnard, W. (1970) Thomas Johnes of Hafod – Pioneer of Upland Afforestation in Wales. *Ceredigion* 6(1–4), 309–319. http://welshjournals.llgc.org.uk/browse/viewpage/llgc-id:1093205/llgc-id:1095169/llgc-id:1095508/get650.

Linnard, W. (1975) Scots foresters in Wales in the 19th century. *Scottish Forestry* 29(4), 268–73.

Linnard, W. (1981) The Hafod Fawr experiment. *Journal of the Merioneth Historical and Record Society* 9(1), 89–96.

Linnard, W. (1985) Angus Duncan Webster: a Scottish forester at Penrhyn Castle, North Wales. *Scottish Forestry* 39, 265–74.

Linnard, W. (1988) Historical planting at Cawdor. *Scottish Forestry* 42(1), 181–84.

Linnard, W. (2000) *Welsh Woods and Forests: A History*. Landysul, Ceredigion: Gomer.

Lutz, H. J. (1963) *History of Sitka Spruce Planted in 1805 at Unalaska Island by the Russians*. Juneau: Northern Forest Experiment Station, Forest Service, USDA.

Mackay Brown, G. (1992) *Vinland*. Edinburgh: Polygon.

Magnusson, M. and Pálsson, H. (1965) *The Vínland Sagas: The Norse Discovery of America*. Harmondsworth: Penguin.

M'Laren, J. (1884) Forest and ornamental trees of recent introduction which might be generally cultivated in Scotland. *Transactions of the Royal Scottish Arboricultural Society* 10, 209–222.

Niall, I. (1972) *The Forester*. London: Heinemann.

Nisbet, J. (1900) *Our Forests and Woodlands*. London: Dent.

Oliver, S. (2009) Planting the nation's 'waste lands': Walter Scott, forestry and the cultivation of Scotland's wilderness. *Literature Compass* 6(3), 585–98.

Oregon Cultural Heritage Commission. (2012) *Finding David Douglas*. DVD from http://www.FindingDavidDouglas.org.

Ravenscroft, E. J. (1863–1884). *The Pinetum Britannicum, A Descriptive Account of Hardy Coniferous Trees Cultivated in Great Britain*. Edinburgh: Ballantyne.

Roberts, C. (2007) *The Unnatural History of the Sea: The Past and Future of Humanity and Fishing*. London: Octopus.

Smith, S (2001) *The National Inventory of Woodland and Trees: England*. http://www.foresty.gov.uk/pdf/frnationalinventory0001.pdf

Smout, T. C. (ed.) (2003) *People and Woods in Scotland: A History*. Edinburgh: University Press.

Stirling-Maxwell, J. (1929) *Loch Ossian Plantations: An Essay in Afforesting High Moorland*. London: privately printed. Available from Library, Forest Research, Alice Holt Lodge, Surrey.

Tree Council of Ireland (2013) *The Heritage Tree Database. Abbeyleix Estate, Abbeyleix, Co. Laois*. Picea sitchensis. http://www.treecouncil.ie.

Veniaminov, I. (1840) *Notes on the Islands of the Unalaska District*. St Petersburg. Trans. Black, L. T. and Geoghegan, R. H. (1984). Kingston, Ontario: Limestone Press.

Woods, P. and Woods, J. (pre-2004) *The Oregon Expedition 1850–1854: John Jeffrey and his Conifers*. Edinburgh: Manuscript, Royal Botanic Garden Edinburgh.

Chapter 11. From Rare Ornamental to Upland Carpet

Anderson, J. (2007) *Atholl Estates: A Brief History*. Perth: Perth and Kinross Heritage Trust. http://www.pkht.org.uk.

Anon (2009) *Hatfield in World War II*. http://www.hatfield-herts.co.uk/features/wwII.html.

Butter, R. (2011) *Kilmartin, Scotland's Richest Prehistoric Landscape*. 2nd edn. Kilmartin: Kilmartin House Trust.

Crozier, J. D. (1910) The Sitka spruce as a tree for hill planting and general afforestation. *Transactions of the Royal Scottish Arboricultural Society* 23, 7–16.

Davies, E. J. M. (1967) Silviculture of the spruces in West Scotland. *Forestry* 40, 37–46.

Davies, E. J. M. (1972) History and background of Sitka spruce in the United Kingdom to the present day. In Malcolm, D. C. *Proceedings of Joint Study Group on Sitka Spruce. Scottish Forestry*, 26(1), 61–68.

Department of the Environment, Community and Local Government, Ireland. (2014) *Augustine Henry (1857–1930) Botanical Explorer, Dendrologist*. http://www.askaboutireland.ie/reading-room/life-society/science-technology/irish-scientists/henry-augustine.

Dunn, M. (1892) The value in the British Islands of introduced conifers. *Journal of the Royal Horticultural Society* 14, 73–102.

Edlin, H. L. (1970) *Forestry in Scotland*. Edinburgh: Forestry Commission.

Foot, D. (2012) Roy Robinson: a story of research and development. *Scottish Forestry* 66(4), 31–35.

Forestry Commission (1948) *Britain's Forests: Forest of Ae*. London: HMSO.

Forestry Commission (1950) *Britain's Forests: Tintern*. London: HMSO.

Forestry Commission (1954) *Snowdonia National Forest Park Guide*. London: HMSO.

Forestry Commission (1970) *Forestry in Scotland*. London: HMSO.

Forestry Commission (1974) *British Forestry*. London: HMSO.

Hansen, H. (2004) The role of Sitka spruce, part

2. *Northwest Native Plant Journal*. http://www.nwplants.com/information/emag/vol2-4.pdf.

Hansard (1803–2005) *The Estate of Interliever.* Lords Sitting of 10 July 1918. London: HL Deb 10 July 1918. Vol. 30 cc 744–49.

Hatt, D. E. (1919) *Sitka Spruce: Songs of Queen Charlotte Islands.* Vancouver: R. P. Latta & Company. http://archive.org/details/cihm_74437 (Microfilm held by Archives University of Alberta).

House, S. and Dingwall, C. (2003) 'A Nation of Planters': introducing the new trees, 1650–1900. In Smout, T. C. *People and Woods in Scotland: A History*. Edinburgh: Edinburgh University Press.

Icelandic Forest Research (2013) *Forestry in a Treeless Land.* http://www.skogur.is/english/forestry-in-a-treeless-land/.

Joyce, P. M. and OCarroll, N. (2002) *Sitka Spruce in Ireland.* Dublin: National Council for Forest Research and Development (COFORD).

Kilpatrick, C. S. (1987) *Northern Ireland Forest Service: A History.* Belfast: Forest Service, Dept of Agriculture and Rural Development, Ballymiscaw; dardhelpline@dardni.gov.uk.

Linnard, W. (1984) Angus Duncan Webster: a Scottish forester at Penrhyn Castle, north Wales. *Scottish Forestry* 39, 265–274.

McEwan, J. (1998) A *Life in Forestry*. Perth: Perth and Kinross Libraries.

MacDonald, J. (1931) Sitka spruce in Great Britain. Its growth, production and thinning. *Forestry* 6, 100–107.

Mears, G. (2013) *Sitka Spruce and the Mosquito Aircraft.* Unpublished typescript, Mosquito Aircraft Museum. http://www.dehavillandmuseum.co.uk/.

Mitchell, A. F. (1972) *Conifers in the British Isles: A Descriptive Handbook*. London: Forestry Commission Booklet 33.

Niall, I. (1972) *The Forester*: The *Story of James Lymburn Shaw, Forester of Killearn, Tintern, and Gwydyr.* London: Heinemann.

North, F. J., Campbell, B. and Scott, R. (1949) *Snowdonia: The National Park of North Wales*. London: Collins New Naturalist Series 13.

Peterken, G. (2008) *The Wye Valley*. London: HarperCollins New Naturalist Series 105.

RAF Wyton (2012) *De Havilland Mosquito.* http://www.raf.mod.uk/rafbramptonwyton/history/dehavilandmosquito.cfm.

Robinson, R. (1931) Use of Sitka spruce in British afforestation. *Forestry* 5(2), 93–95.

Robinson, R. (1951) *History of Inverliever Forest 1907–1951*. Farnham: Forestry Commission. Unpublished typescript and 1935 compartment map.

Rooke, D. B. (ed.) (1974) *British Forestry.* Edinburgh: Forestry Commission.

Shaw, D. L. (1971) *Gwydyr Forest in Snowdonia: A History*. London: Forestry Commission Booklet 28.

Stewart, M. (2007) *Smell of the Rosin – Noise of the Saw: The Story of Forestry in Mid-Argyll in the 20th Century.* Edinburgh: Forestry Commission.

Stirling-Maxwell, J. (1929/1951) *Loch Ossian Plantations: An Essay in Afforesting High Moorland.* Farnham: Forest Research, unpublished (also 1913 edn).

Stirling-Maxwell, J. (1931) Sitka spruce on poor soils and at high elevations. *Forestry* 6, 96–99.

Sutherland, W. (2009) The push for suitable afforestation land. In Tittensor, R. M. (2009), 78–79.

Taylor, C. M. A. and Tabbush, P. (1990) *Nitrogen Deficiency in Sitka Spruce Plantations*. London: Forestry Commission Bulletin 89.

Taylor, W. L. (1946) *Forests and Forestry in Great Britain*. London: Crosby Lockwood & Son.

Tittensor, A. M. and Tittensor, R. M. (1985) The rabbit warren at West Dean near Chichester. *Sussex Archaeological Collections* 123, 151–85.

Tittensor, R. M. (2009) *From Peat Bog to Conifer Forest: An Oral History of Whitelee, its Community and Landscape*. Chichester: Packard Publishing and Darvel: Countryside Management Consultancy. http://www.ruthtittensor.co.uk.

Walton, J. (ed.) (1967) *Forest Park Guides: Argyll.* 4th edn. Edinburgh: HMSO.

Weir, E. M. K. (2003) German submarine blockade, overseas imports, and British Military production in World War II. *Journal of Military and Strategic Studies* 6(1), 1–42.

White, J. (1995) *Forest and Woodland Trees in Britain.* Oxford: Oxford University Press.

Wonders, W. C. (1991) *'The Sawdust Fusiliers': The Canadian Forestry Corps in the Scottish Highlands in World War Two.* Montreal: Canadian Pulp and Paper Association.

Zehetmayr, J. W. L. (1953) Problems of moorland afforestation. *Quarterly Journal of Forestry* 57, 32–40.

Zehetmayr, J. W. L. (1954) *Experiments in Tree Planting on Peat.* London: Forestry Commission Bulletin 22.

Chapter 12. Peat: The Final Frontier

Carey, N. G. (1978) *A Guide to the Queen Charlotte Islands.* Anchorage: Alaska Northwest Publishing.

Cowie, G. (2009) The push for suitable afforestation land. In Tittensor (2009), 78.

Crozier, J. D. (1910) The Sitka spruce as a tree for hill planting and general afforestation. *Transactions of the Royal Scottish Arboricultural Society* 23, 7–16.

Day, W. R. (1957) *Sitka Spruce in British Columbia.* London: Forestry Commission Bulletin 28.

Edlin, H. L. (1970) *Forestry in Scotland.* Edinburgh: Forestry Commission.

Farrelly, N., Dhubhain, A. N., Nieuwenhuis, M. and Grant, J. (2009) The distribution and productivity of Sitka spruce (*Picea sitchensis*) in Ireland in relation to site, soil and climatic factors. *Irish Forestry* 67, 51–73.

Forestry Commission (1954) *Snowdonia National Forest Park Guide*. London: HMSO.

Forestry Commission (1965) *Seed Identification Numbers.* London: Forestry Commission Research Branch Paper 29.

Fraser, D. (2004) *Highland Reflections: The Reminiscences of a Forester.* Coupar Angus: self-published.

Hartley, M. and Ingilby, J. (1975) *Life in the Moorlands of North-East Yorkshire*. London: Dent.

Innes, P. (2009) Why Whitelee? Buying land for the New Forest. In Tittensor (2009), 79.

International Union of Forestry Research Organisations (1976) *Sitka Spruce* Picea sitchensis *(Bong.) Carr. International Ten Provenance Experiment: Nursery Stage Results.* Dublin: Department of Lands, Forest and Wildlife Service.

Johannes Rafn & Son (1937–8) *The Scandinavian Tree Seed Establishment*. Copehagen: self-published.

Joyce, P. M. and OCarroll, N. (2002) *Sitka Spruce in Ireland*. Dublin: National Council for Forest Research & Development (COFORD).

McEwan, J. (1998) A *Life in Forestry*. Perth: Perth and Kinross Libraries.

Neustein, S. A. (1976) A history of plough development in British forestry. *Scottish Forestry* 30(1), 89–111.

North, F. J., Campbell, B. and Scott, R. (1949) *Snowdonia: The National Park of North Wales.* London: Collins. New Naturalist 13.

Samuel, C. J. A., Fletcher, A. M. and Lines, R. (2007) *Choice of Sitka Spruce Seed Origins for Use in British Forests.* Edinburgh: Forestry Commission Bulletin 127.

Shaw, D. L. (1971) *Gwydyr Forest in Snowdonia: A History*. London: Forestry Commission Booklet 28.

Stewart, M. (2007) *'Smell of the Rosin – Noise of the Saw': the Story of Forestry in Mid-Argyll in the 20th Century.* Edinburgh: Forestry Commission.

Stirling-Maxwell, J. (1929/1951) *An Essay in Afforesting High Moorland.* Farnham: Forestry Research, Forestry Commission.

Tittensor, R. M. (2009) *From Peat Bog to Conifer Forest: An Oral History of Whitelee, its Community and Landscape*. Chichester: Packard Publishing and Darvel: Countryside Management Consultancy. http://www.ruthtittensor.co.uk.

Tittensor, A. M. and Lloyd, H. G. (1983) *Rabbits.* London: Forestry Commission Forest Record 125.

Walton, J. (ed.) (1962) *National Forest Park Guides: The Border.* 2nd edn. London: HMSO.

Walton, J. (ed.) (1967) *Forest Park Guides: Argyll.* 4th edn. Edinburgh: HMSO.

Webster, A G. (1905) *Webster's Practical Forestry: A Popular Handbook on the Growth of Trees for Profit or Ornament.* 4th edn. London: W. Rider & Son.

Wilson, K. and Leathart, S. (eds) (1982) *The Kielder Forests: A Forestry Commission Guide.* Edinburgh: Forestry Commission.

Chapter 13. Perceptions

Anon (1987) *The Lednock Letter*. Glen Lednock, Perthshire: Typescript Newsletter.

BBC News (2013) *Fingle Woods: National Trust and Woodland Trust Team Up*. http://www.bbc.co.uk/news/uk-england-devon-23540859. 2 August 2013.

Bunce, R. G. H., Smart, S. M. and Wood, C. M. (2012) *The Impact of Afforestation on the British Uplands. Proceedings of Conference, Landscape Ecology: Linking Environment and Society*. Edinburgh: NERC Open Research Archive. http://nora.nerc.ac.uk/id/eprint/500912.

Corbett, J. (2004) *Castles in the Air: The Restoration Adventures of two Young Optimists and a Crumbling Old Mansion (Gwydyr Castle)*. London: Ebury.

Cowie, G. (2009) The push for suitable afforestation land. In Tittensor (2009), 78.

Crowther, R. E. and Low, A. J. (1986) *Advice on Establishment and Tending of Trees*. Reprinted from Forestry Commission Bulletin 14. 1987 edn. London: HMSO.

Foot, D. (2010) *Woods and People: Putting Forests on the Map*. Stroud: Gloucestershire: The History Press.

Forestry Commission Scotland (2015) *Cowal and Trossachs Forest District Strategic Plan 2007–2017*. http://scotlandforestry.gov.uk/images/corporate/pdf/CowalTrossachsDsp2014-17.pdf.

Fraser Darling, F. and Morton Boyd, J. (1964) *The Highlands and Islands*. London: Collins New Naturalist Series 6.

Harvey, L. A. and St Leger-Gordon, D. (1953) *Dartmoor*. London: Collins New Naturalist Series 27.

MacDonald, A. (1999) *Fire in the Uplands: A Historical Perspective*. Edinburgh: Scottish Natural Heritage Information and Advisory Note 108.

Mason, W. L. (2007) Changes in the management of British forests between 1945 and 2000 and possible future trends. *Ibis* 149 (suppl. 2), 41–52.

Mason, W. L. and O'Kane, S. (2014) The species composition and structure of the old conifer plantations at Loch Ossian, Corrour, north Scotland a century after their establishment. *Scottish Forestry* 68(3), 8–16.

McVean, D. N. and Ratcliffe, D.A. (1962) *Plant Communities of the Scottish Highlands*. London: Nature Conservancy.

Milner, N., Taylor, B., Conneller, C. and Schadla-Hall, T. (2013) *Star Carr: Life in Britain After the Ice Ages*. York: Council for British Archaeology.

Nature Conservancy Council (1986) *Nature Conservation and Afforestation in Britain*. Peterborough: Nature Conservancy Council.

Proctor, M. (2013) *The Vegetation of Britain*. London: HarperCollins New Naturalist Library 122.

Ratcliffe, D. (2007) *Galloway and the Borders*. London: HarperCollins New Naturalist Series 101.

Sheail, J. (1998) *Nature Conservation in Britain: The Formative Years*. London: HMSO.

Shoard, M. (1980) *The Theft of the Countryside*. London: Maurice Temple.

Simmons, I. G. (2003) *The Moorlands of England and Wales. An Environmental History 8000 BC–AD 2000*. Edinburgh: Edinburgh University Press.

Scott, M. (2008) The Flow Country revisited. *British Wildlife* 19(4), 229–39.

Smout, T. C. (2000) *Nature Contested: Environmental History in Scotland and Northern England*. Edinburgh: Edinburgh University Press.

Stroud, D. A., Reed, T. M., Pienkowski, M. W. and Lindsay, R. A. (1988) *Birds, Bogs and Forestry: The Peatlands of Caithness and Sutherland*. In D. A. Ratcliffe and P. H. Oswald (eds). Peterborough: Nature Conservancy Council. http://jncc.defra.gov.uk/page-4322.

Thomas, A. S (1975) *The Follies of Conservation*. Ilfracombe: Arthur Stockwell.

Thomas, K. (1983) *Man and the Natural World: Changing Attitudes in England 1500–1800*. London: Allan Lane.

Thomas, R. S. (1963) *The Bread of Truth*. London. Rupert Hart-Davis. Currently published as: *Collected Poems 1945–1990 by R. S. Thomas*. London: Orion.

Tipping, R. (2008) Blanket peat in the Scottish Highlands: timing, cause, spread and the myth of environmental determinism. *Biodiversity Conservation* 17, 2097–2113.

Tittensor, R. M. (1970) History of the Loch Lomond oakwoods. *Scottish Forestry* 24 (2), 100–118.

Tittensor, R. M. (1981) *A Sideways Look at Nature Conservation in Britain*, 1–45. London: University College Discussion Paper in Conservation 29.

Tittensor, R. M. (1985) Conservation of our historic landscape heritage. *Folk Life* 23, 5–20.

Tomkins, S. (1989) *Forestry in Crisis: The Battle for the Hills*. London: Christopher Helm.

Chapter 14. Contribution to Modern Societies

Anon (2013) *Spruce Production Division*. http://en.wikipedia.org/wiki/Spruce_Production_Division

Coast Forest Products Association (2004) *Wood Species and Products from the Coast Regions of British Columbia*. http://www.coastforest.org/wp-content/uploads/2012/03/sitka-spruce.pdf.

Falconer, J. and Rivas, B. (2013) *De Havilland Mosquito: Owners' Workshop Manual*. Yeovil: Haynes.

Forestry Commission (2012a) *UK Wood Production and Trade: 2011 Provisional Figures*. Edinburgh: Forestry Commission.

Forestry Commission (2012b) *Timber Utilisation Statistics 2010 and 2011. Estimates: and updated study for the Forestry Commisssion*. Nicholas Moore of Timbertrends. Forestry Commission Contract CFS 03/09. Edinburgh: Forestry Commission.

Forestry Commission Scotland (2013) *Full Steam Ahead for New Timber Pier on Mull*. News Release 16127, 16 December 2013, Forestry Commission. Edinburgh: Forestry Commission.

James Jones & Sons (2013) *Products and Services*. http://www.jamesjones.co.uk.

James Jones & Sons (2013) *JJI-Joists: Superior I-Joists from the First Purpose-built Production Line in the UK*. http://www.jamesjones.co.uk/assets/downloads/JJI%20Joists%20Leaflet.pdf.

Koster, J. (1995) *Wood in Early American Keyboard Instruments as Evidence of Origins*. 1995 WAG Postprints–St Paul, Minnesota. http://cool.conservation-us.org/coolaic/sg/wag/1995 /WAG_95_koster.pdf.

Lakin, G. (2013) *The de Havilland Aircraft Heritage Centre: The Birthplace of the Wooden Wonder*. http://www.dehavillandmuseum.co.uk/aircraft/de-havilland-dh98-mosquito-b-mk-35/.

Mackovjack, J. (2010) *Tongass Timber: A History of Logging and Timber Utilization in Southeast Alaska.* Durham, NC: Forest History Society.

Moore, J. (2011) *Wood Properties and Uses of Sitka Spruce in Britain.* Edinburgh: Forestry Commission Research Report.

Musical Forests (2013) *Tonewood in the Making.* http://www.tonewood.ca/servicepage.htm.

Norbord (2013) *Norbord Make it Better. Norbord Products Timber Species.* Leaflet available from Norbord, Cowie, FK77 7BQ.

RateBeer (2013) *Baranof Island Sitka Spruce Tip.* http://www.ratebeer.com/beer/baranof-island-sitka-spruce-tip/163961/.

Saer Coed Iago (2013) *SCI Joinery: Timber Frame Construction.* http://www.saercoediago.com/timbers_sitka_spruce.html (accessed 2014).

Sharp, C. M. and Bowyer, M. J. F. (1967, 1995) *Mosquito.* Manchester: Crécy.

Steinway Pianos (2013) *The Diaphragmatic Soundboard: The Heart of the Steinway Tone, Color and Richness.* http://www.steinway.com/news/articles/the-diaphragmatic-soundboard-the-heart-of-the-steinway-tone-color-and-richness/.

Sutton, A. and Black, D. (2011) *Cross-Laminated Timber: An Introduction to Low-Impact Building Materials.* http://www.bre.co.uk/filelibrary/pdf/projects/low_impact_materials/IP17_11.pdf.

Taylor Guitars (2008) *Taylor Guitars Honored for Environmental Efforts.* https://www.taylorguitars.com/news/2008/11/17/taylor-guitars-honored-environmental-efforts.

Tonsfeldt, Ward (2013) *The US Army Spruce Production Division at Vancouver Barracks, Washington, 1917–1919.* Vancouver: Fort Vancouver National Historic Site, National Park Service. Contract P11PX842-51. http://www.nps.gov/fova/historyculture/upload/Tonsfeldt-SPD-Study-2013-2.pdf.

US Department of Agriculture Forest Service (1984) *Sitka Spruce: An American Wood.* FS-265. http:// www.fpl.fs.fed.us/documnts/usda/amwood/265sitka.pdf.

Wagner, J. (2013) *Fort Vancouver Spruce Mill.* http://storify.com/JoshuaWagner/comjour-333-reporting-across-platforms-spruce-mill#.

Webster, A. D. (1905) *Webster's Practical Forestry: A Popular Handbook on the Rearing and Growth of Trees for Profit and Ornament.* London: W. Rider & Son.

Wetzel, S. Duchesne, L. C. and Laporte, M. F. (2006) *Bioproducts from Canada's Forests: New Partnerships in the Bioeconomy.* Dordrecht: Springer.

Wheeler, R. and Alix, C. (post-2003) *Economic and Cultural Significance of Driftwood in Coastal Communities of Southwest Alaska.* http://www.uaf.edu/files/aqc/Wheeler%20and%20Alix.pdf.

Chapter 15. Plantation Ecology: Plants and Animals Re-assemble

Adamson, R. S. (1912) An ecological study of a Cambridgeshire woodland. *Journal of the Linnaean Society, Botany* 40, 339–87.

Anderson, M. L. (1950, 1961) *The Selection of Tree Species: An Ecological Basis of Site Classification for Conditions Found in Great Britain and Ireland.* Edinburgh: Oliver & Boyd.

Armstrong, H., Gill, R., Mayle, B. and Trout, R. (2003) *Protecting Trees from Deer: An Overview of Current Knowledge and Future Work*, 28–39. http://www.forestry.gov.uk/pdf/FR0102deer.pdf/$FILE/FR0102deer.pdf.

Balmer, D. E., Gillings, S., Caffrey, B. J., Swann, R. L., Downie, I. S. and Fuller, R. J. (2013) *Bird Atlas 2007–2011: The Breeding and Wintering Birds of Britain and Ireland.* Thetford: BTO Books.

Bryce, J., Cartmel, S. and Quine, C. P. (2005) *Habitat Use by Red and Grey Squirrels: Results of Two Recent Studies and Implications for Management.* Information Note. Edinburgh: Forestry Commission.

Burnett, J. H. (ed.) (1964) *The Vegetation of Scotland.* Edinburgh: Oliver & Boyd.

Butterfield, J., Luff, M. L., Baines, M. and Eyre, M. D. (1995) Carabid beetle communities as indicators of conservation potential in upland forests. *Forest Ecology and Management* 79(2), 63–77.

Coed y Gororau Forest District, Welshpool (now Natural Resources Wales) (2006) *Clocaenog LISS Management Plan including Red Squirrel Management.* Typed Report available from Coed y Gororau Froest District, Powells Lane, Welshpool SY21 7JY.

Coillte (2005) *Deer Damage in Farm Forestry.* Teagasc Forestry Development Unit Farm Forestry Series 9. http://www.teagasc.ie/forestry/docs/advice/teagascdeerdamage9.pdf.

Condry, W. (1967) The *Snowdonia National Park.* London: Collins New Naturalist 47. 2nd edn.

Condry, W. (1974/1976) *Woodlands.* London: Collins.

Condry, W. (1981/1982) *The Natural History of Wales.* London. Collins New Naturalist 66.

Condry, W. (1987) *Snowdonia.* Newton Abbot: David & Charles.

Davies, F. H. (1952–54) Natural regeneration at 1,200

feet above sea level at Glasfynydd Forest. *Journal of the Forestry Commission* 23, 55–56.

Day, W. R. (1957) *Sitka Spruce in British Columbia: A Study in Forest Relationships*. London: HMSO/ Forestry Commission Bulletin 28.

Douglass, J. (2002, 2004–2006) *The Lichens of Whitelee Forest*. 4 Reports in: Whitelee Forest Oral History Project Archives (2009) at National Museum of Scotland, Edinburgh, and Forestry Commission, Alice Holt Lodge, Farnham, Surrey.

Forestry Commission (1954) *Snowdonia: National Forest Park Guide*. London: HMSO.

Forest Service of Republic of Ireland (2015) *Sitka spruce (*Picea sitchensis *(Bong.) Carr.) Sprús Sitceach*. Sheet 13. https://www.agriculture.gov.ie/media/migration/forestry/publications/SitkaSpruce_low.pdf.

Fraser Darling, F. (1947) *Natural History in the Highlands and Islands*. London. Collins New Naturalist Series 6.

Fuller, L., Irwin, S., Kelly, T., O'Halloran, J. and Oxbrough, A. (2013) The importance of young plantation forest habitat and forest road-verges for ground-dwelling spider diversity. *Proceedings of the Royal Irish Academy* 113B(3), 1–13.

Haworth, P. F. and Fielding, A, H. (2009) *An Assessment of Woodland Habitat Utilisation by Breeding Hen Harriers*. Battleby: Scottish Natural Heritage Project No. 24069.

Haworth, P. F. and Fielding, A, H. (2013) *Golden Eagles and Woodland*. Unpublished report to Forestry Commission Scotland and Scottish Natural Heritage. Project no. 24069. Isle of Mull: Haworth Conservation and Battleby: Scottish Natural Heritage.

Hill, M. O. (1986) Ground flora and succession in commercial forests. In Jenkins, D. (ed.), 71–78.

Humphrey, J., Ferris, R. and Quine, C. P. (2003) *Biodiversity in Britain's Planted Forests: Results from the Forestry Commission's Biodiversity Assessment Project*. Edinburgh: Forestry Commission.

Jenkins, D. (ed.) (1986) *Trees and Wildlife in the Scottish Uplands*. Abbots Ripton, Huntingdon: Symposium 17 of the Institute of Terrestrial Ecology. http://nora.nerc.ac.uk/5296/.

Joyce, P. M. and OCarroll, N. (2002) *Sitka Spruce in Ireland*. Dublin: National Council for Forest Research and Development (COFORD).

Kimmins, J. P. (1999) Biodiversity, beauty and the 'beast': are beautiful forests sustainable, are sustainable forests beautiful, and is 'small' always ecologically desirable? *Forestry Chronicle* 75(6), 955–60.

Little, B., Davison, M. and Jardine, D. (1995) Merlins (*Falco columbarius*) in Kielder Forest: influences of habitat on breeding performance. *Forest Ecology and Management* 79(1–2), 147–52.

Lloyd Praeger, R. (1950, 1972) *Natural History of Ireland*. Wakefield: EP Publishing.

Mason, P. A. and Last, F. T. (1986) Are the occurrences of sheathing mycorrhizal fungi in new and regenerating forests and woodlands in Scotland predictable? In Jenkins, D. (ed.), 63–73.

McVean, D. and Ratcliffe, D. A. (1962) *Plant Communities of the Scottish Highlands*. London: HMSO.

North, F. J., Campbell, B. and Scott, R. (1949) *Snowdonia: The National Park of North Wales*. London: Collins New Naturalist 13.

Oxbrough, A., Irwin, S., Kelly, T. C. and O'Halloran, J. (2010) Ground-dwelling Invertebrates in Reforested Conifer Plantations. *Forest Ecology and Management* 259(10), 2111–21.

Petty, S. J. *et al.* (2000) Spatial synchrony in field vole (*Microtus agrestis*) abundance in a coniferous forest in northern England: the role of vole-eating raptors. *Journal of Applied Ecology* 37 (suppl. 1), 136–47.

Picozzi, N., Moss, R. and Catt, D. C. (1996) Capercaillie habitat, diet, and management in a Sitka spruce plantation in central Scotland. *Forestry* 69(4), 373–88.

Proctor, M. (2013) *Vegetation of Britain and Ireland*. London: HarperCollins New Naturalist Library 122.

Purser, P., Wilson, F. and Carden, R. (2009) *Deer and Forestry in Ireland: A Review of Current Status and Management Requirements*. Report to Woodlands of Ireland (Coillearnacha Dúchasacha). http://www.woodlandsofireland.com/sites/default/files/DeerStrategy.pdf.

Quine, C. P. (2001) A preliminary survey of regeneration of Sitka spruce in wind-formed gaps in British planted forests. *Forest Ecology and Management* 151, 37–42.

Quine, C. P. and Humphrey, J. W. (2009) Plantations of exotic tree species in Britain; irrelevant for biodiversity or novel habitat for native species? *Biodiversity Conservation* 19, 1503–12.

Rackham, R. (1986) *The History of the Countryside*. London: Dent.

Ratcliffe, D. (2007) *Galloway and the Borders*. London: HarperCollins New Naturalist Series 101.

Ratcliffe, P. R. and Peterken, G. F. (1995) The potential for biodiversity in British upland spruce forests. *Forest Ecology and Management* 79(1), 153–60.

Salisbury, E. J. (1916, 1918) The oak-hornbean woods of Hertfordshire, parts 1–4. *Journal of Ecology* 4, 83–117; 6, 14–52.

Shorten, M. (1962) *Squirrels and their Control.* London: HMSO, MAFF Bulletin 184.

Simpson, B. (2006) *A Field-notebook of the Whitelee Hills 1960–2005* and *Recording and Transcript.* Whitelee Forest Oral History Project Archives (2009), available at National Museum of Scotland, Edinburgh and Forestry Commission, Alice Holt Lodge, Farnham, Surrey.

Smith, J. J. (1952–54) Rabbit clearance in King's Forest 1947 to 1951. *Journal of the Forestry Commission* 23, 70–71.

Tansley, A. G. (1939) *The British Islands and Their Vegetation.* Cambridge: Cambridge University Press.

Wainhouse, D. and Inward, D. J. G. (2016) The influence of climate change on forest insect pests in Britain. *Forestry Commission Research Note.* http://www.forestry.gov.uk/pdf/FDRN021.pdf/$FILE/FCRN021.pdf.

Ward, L. K. (1973) Conservation of juniper 1. Present status of juniper in southern England. *Journal of Applied Ecology* 10, 165–88.

Watt, A. S. (1926) Yew communities of the South Downs. *Journal of Ecology* 14, 282–316.

Webster, A. D. (1890) The plantations on the Penrhyn Estate, North Wales. *Transactions of the Royal Scottish Arboricultural Society 12*, 165–80. http://archive.org/stream/transactionsofro1888roya/transactionsofro1888roya_djvu.txt.

Wilson, K. and Leathart, S. (eds) (1982) *The Kielder Forests: A Forestry Commission Guide.* Edinburgh: Forestry Commission.

Chapter 16. Sustainability in North America

Anon (n.d.) *Why Forestry in BC Must Change.* http://www.pacificfringe.net/sustainedyield/change.htm.

Anon (1971) *Alaska Native Claims Settlement Act.* http://en.wikipedia.org/wiki/Alaska_Native_Claims_Settlement_Act.

Anon (2014) *Tongass Young Growth Management Strategy 2014.* Available from USDA Forest Service, Juneau, Southeast Alaska.

Andrews, R. W. (1954) *'This Was Logging': Selected Photographs of Darius Kingsley.* Seattle WA: Superior Publishing Company.

Andrews, R. W. (1968) *Timber: Toil and Trouble in the Big Woods.* Seattle WA: Superior Publishing Company.

British Columbia Forest Service (2011) *Haida Gwaii Forest Stewardship Plan.* https://www.for.gov.B.C..ca/ftp/tch/external/!publish/FSP/QCI/BCTS_QCI_FSP.pdf.

CanBio (2014) *The Working Forest Newspaper.* http://www.workingforest.com/.

Cary, N. L. (1922) *Sitka Spruce: Its Uses, Growth and Management.* Washington: United States Department of Agriculture. Bulletin no. 1060.

Coastal First Nations (2014) *EBM Forestry.* http://www.coastalfirstnations.ca/programs/ebm-forestry.

Cooper, F. (2013) *Riders of the Tides.* Minneapolis MN: Langdon Street Press.

Drushka, K. (2003) *Canada's Forests: A History.* Montreal: Forest History Society/McGill-Queen's University Press.

Durbin, K. (2005) *Tongass: Pulp Politics and the Fight for the Alaska Rain Forest.* Corvallis OR: Oregon State University Press.

Grinev, A. V. (1991) *The Tlingit Indians in Russian America, 1741–1867.* Available in English translation (2005). Lincoln NE: University of Nebraska Press.

Hak, G. (2000) *Turning Trees into Dollars: The British Columbia Coastal Lumber Industry, 1858–1913.* Toronto: University of Toronto Press.

Harris, A. S. (1974) Clear-cutting, reforestation and stand development on Alaska's Tongass National Forest. *Journal of Forestry* 72, 330–37.

Litwin, T. S. (ed.) (2005) *The Harriman Alaska Expedition Retraced: A Century of Change, 1899–2001.* New Brunswick: NJ Rutgers University Press.

MacCleery, D. W. (1992/2011) *American Forests: A History of Resiliency and Recovery.* Durham NC: Forest History Society.

Mackovjak, J. (2010) *Tongass Timber: A History of Logging and Timber Utilization in Southeast Alaska.* Durham NC: Forest History Society.

Morgan, M. (1955) *The Last Wilderness.* Seattle WA: University of Washington Press.

Natural Resources Canada (2014) *Aboriginal Forestry Initiative.* http://www.nrcan.gc.ca/forests/federal-programs/13125.

Rosenberg, D. J. (2006) *Sitka Spruce.* http://www.authorsden.com/visit/viewpoetry.asp?&id=168635.

Sealaska Timber Corporation (2016) http://www.sealaska.com/what-we-do/sustainable-natural-resources/sealaska-timber-company.

Shee AtikǍ Incorporated (2014) *About Shee AtikǍ Incorporated.* http://www.sheeatika.com/about-shee-atika.

Taylor, G. W. (1975) *Timber: History of the Forest Industry in BC.* Vancouver: J. J. Douglas.

The Wildlife Society (2012) *Conservation and Management of Old-Growth Forest on the Pacific Coast of North America.*

http://wildlife.org/wp-content/uploads/2016/04/PS_ConservationandMgmtofOldGrowth.pdf.

USDA Forest Service (2014) *Tongass National Forest: Land and Resources Management*. http://www.fs.usda.gov/land/tongass/landmanagement.

Chapter 17. New Temperate Rainforests? Futures in Ireland and Britain.

Anon (2008) *European Forests: Ecosystem Conditions and Sustainable Use*. Copenhagen: European Environment Agency Report 3.

Anon (2014) Editorial: an elegant chaos. *Nature* 507, 139–140.

Atholl Estate (2013) *Atholl Estates Newsletter Winter 2012/2013*. http://www.atholl-estates.co.uk/File/Newsletters/Newsletter_Winter_2012-13.pdf.

Brown, N. (1997) Re-defining native woodland. *Forestry* 70 (3), 191–98.

Browning, G. (2012) Wild forest – wilder valley (Ennerdale). In Grace, J and Morison, J. I. L. (eds) *Managing Forests for Ecosystem Services: Can Spruce Forests Show the Way?* IUFRO Conference, Roslin, Edinburgh 2012. http://www.forestry.gov.uk/fr/iufro2012.

Butterfly Conservation (2014) *Mabie Forest Nature Reserve Leaflet*. http://www.southwestscotland-butterflies.org.uk/mabie_forest_butterfly_conservation_reserve.shtml.

Emery, M., Martin, S. and Dyke, A. (2006) *Wild Harvests from Scottish Woodlands*. Edinburgh: Forestry Commission.

Evans, J. (ed.) (2001) *The Forests Handbook 2: Applying Forest Science for Sustainable Management*. Chichester: Wiley.

Fenning, T. (ed.) (2014) *Challenges and Opportunities in Forestry*. Houten: Springer.

Forestry Commission (2007) *Forest Reproductive Material: Regulations Controlling Seed, Cuttings and Planting Stock for Forestry in Great Britain*. Edinburgh: Forestry Commission.

Forestry Commission (2014) *Forestry Facts and Figures 2014*. Edinburgh: Forestry Commission leaflet.

Forestry Commission (2015) *Forestry Statistics 2015*. http://www.forestry.gov.uk/pdf/ForestryStatistics2015.pdf/$FILE/ForestryStatistics2015pdf.

Forestry Commission Scotland (2009a) *Cowal and Trossachs Forest District Strategic Plan 2009–2013*. Edinburgh: Forestry Commission.

Forestry Commission Scotland (2009b) *Climate Change Information Pack: What Will Climate Change Look Like?* http://www.forestry.gov.uk/pdf/3_what_will_climate_change_look_like.pdf/$FILE/3_what_will_climate_change_look_like.pdf.

Forest Research (2014) *How Might Climate Change Affect Insect Pest Outbreaks?* http://www.forestry.gov.uk/pdf/FCRN021.pdf/$FILE/FCRN021.pdf.

Grace, J. and Morison, J. I. L. (eds) (2012) Sitka spruce as a sink for carbon dioxide. In *Managing Forests for Ecosystem Services: Can Spruce Forests Show the Way?* IUFRO Conference, Roslin, Edinburgh 2012. http://www.forestry.gov.uk/pdf/IUFRO_spruce_2012_Grace.pdf/$FILE/IUFRO_spruce_2012_Grace.pdf.

Humphrey, J. (2005) Benefits to biodiversity from developing old-growth conditions in British upland spruce plantations: a review and recommendations. *Forestry* 78(1), 33–53.

Mackenzie, S. and Blackett, S. (2009) RSFS Spring excursion to Aberdeenshire. *Scottish Forestry* 63(3), 37–46.

Mason, W. and Perks, M. P. (2011) Sitka spruce (*Picea sitchensis*) forests in Atlantic Europe: Changes in forest management and possible consequences for carbon sequestration. *Scandinavian Journal of Forest Research* 26, S11 http://www.tandfonline.com/doi/full/10.1080/02827581.2011.564383.

McKellar, K. and McKellar, C. (2012) *Scottish Venison: An Industry Review, 2010*. Edinburgh: Scottish Natural Heritage.

Meteorological Office (2012) *Using Climate Projections (UKCP09)*. http://ukclimateprojections.metoffice.gov.uk/21678.

Monbiot, G. (2014) *Feral*. London: Penguin.

Morison, J. *et al.* (2012) *Understanding the Carbon and Greenhouse Gas Balance of Forests in Britain*. Edinburgh: Forestry Commission Research Report.

Oram, R. (2016) The curious case of the missing history at Keilder. In Coates, P., Moon, D. and Warde, P. *Local Places, Global Processes: Histories of Environmental Change in Britain and Beyond*. Oxford: Windgather Press.

Park, Y.-S. and Adams, G. (2009) *Industrial Implementation of Multi-Varietal Forestry for Spruces in New Brunswick, Canada*. St John, New Brunswick: Canadian Wood Fibre Centre of Canadian Forest Service & J. D. Irving Ltd.

Peterken, G. F., Ausherman, D., Buchenau, M. and Forman, R. T. T. (1992) Old-growth conservation in British upland conifer plantations. *Forestry* 65(2), 127–44.

Pearson, C. (2016) The 'nature' of 'artificial' forests. In

Coates, P., Moon, D. and Warde, P. *Local Places, Global Processes: Histories of Environmental Change in Britain and Beyond.* Oxford: Windgather Press.

Phillips, H. (2013) *Sitka Spruce and Ireland's Afforestation Programme.* Paper presented at Nordic Forest History Conference, Iceland. Available from Forest Solutions: hrphillips@gmail.com.

Quine, C. P., Humphrey, J. W. and Ferris, R. (1999) Should wind disturbance patterns observed in natural forests be mimicked in planted forests in the British uplands? *Forestry* 72(4), 337–58.

Ray, D. (2008a) *Impacts of Climate Change on Forestry in Wales.* Forestry Commission Wales Research Note. Available: Forest Research, Roslin, Midlothian EH25 9SY and http://www.forestry.gov.uk/pdf/FCRN301.pdf/$FILE/FCRN301.pdf.

Ray, D. (2008b) *Impacts of Climate Change on Forests and Forestry in Scotland.* Report for Forestry Commission Scotland. Available as above and http://www.forestry.gov.uk/pdf/scottish_climate_change_final_report.pdf/$FILE/scottish_climate_change_final_report.pdf.

Shorten, M. (1962) *Squirrels.* London. HMSO: MAFF Bulletin 184.

Smout, C. (2016) Birds and squirrels as history. In Coates, P., Moon, D. and Warde, P. *Local Places, Global Processes: Histories of Environmental Change in Britain and Beyond.* Oxford: Windgather Press.

Soutar, R. (2012) Mountain woodland in the Galloway Forest Park. *Native Woodlands Discussion Group Newsletter* 37(2), 6–11.

Wilson, S. McG. (2011) Productive conifers in the riparian buffer zone: any case for the defence? *Scottish Forestry* 65(4), 14–18.

Woodland Trust (2014) *What is Ancient Woodland?* http://www.woodlandtrust.org.uk/learn/threats-to-our-woodland/human-impact/what-is-ancient-woodland/.

World Bank (2016) *Data: Forest area (% of land area) 2011–2015.* http://data.worldbank.org/indicator/AG.LND.FRST.ZS.

The Maps

Carstensen, R., Armstrong, R. H. and O'Clair, R. M. (2014) *The Nature of Southeast Alaska: A Guide to Plants, Animals and Habitats.* 3rd edn. Portland, OR: Alaska Northwest Books.

Coles, B. J. (1998) Doggerland: a speculative survey. *Proceedings of the Prehistoric Society* 64, 45–81.

Kauffmann, M. E. (2014) *Conifers of the Pacific Slope. A Field Guide to the Conifers of California, Oregon and Washington.* Kneeland, CA: Backcountry Press.

Klinkenberg, B. (ed.) (2013) *E-Flora BC: Electronic Atlas of the Flora of British Columbia.* Vancouver: Laboratory for Advanced Spatial Analysis, Department of Geography, University of British Columbia. http://ibis.geog.ubc.ca/biodiversity/eflora/.

Ritchie, J. C. (1987/2003) *Postglacial Vegetation of Canada.* Cambridge: Cambridge University Press.

Samuel, C. J. A., Fletcher, A. M. and Lines, R. (2007) *Choice of Sitka Spruce Seed Origins for Use in British Forests.* Edinburgh: Forestry Commission Bulletin 127.

Wynn, G. (2007) *Canada and Arctic North America: An Environmental History*. Santa Barbara CA: ABC-CLIO.

Glossary

AFFORESTATION To plant trees on tree-less land.

AGE-CLASS All the individuals of the same age in a population.

ANGIOSPERMS The most highly-evolved, flowering plants, with seeds enclosed in an ovary.

ANGLO-SAXON a) Germanic peoples who colonised and governed parts of England between *c.*AD 400 and 1100; b) the language they spoke (also known as Old English).

ARBORICULTURE Cultivation of trees and shrubs for ornament.

ASSART Forest land where trees and bushes were removed to allow cultivation.

ATLANTIC A period of the post-glacial (*c.*7500–6500 BP) with warm, wet climate, also called the 'Climatic Optimum'.

BACK-CROSS When a hybrid organism mates with one of its parents to produce offspring.

BECBRETHA An early Irish treatise on beekeeping.

BESPOKE Something designed or made to suit a customer or situation; custom-made.

BERINGIAN LAND BRIDGE A land mass which joined Asia and North America during several periods between 35,000 BP and 15,500 BP due to lower sea levels.

BITUMEN Natural mineral pitch, a thick hydrocarbon liquid.

BOLE Cylindrical trunk of a tree.

BOREAL a) Northern; b) a cold, dry period of the post-glacial *c.*10,300–7500 BP.

BOTTOMLANDS Low-lying land adjoining a river or stream.

BREAST HEIGHT The height above ground at which the girth of a tree is measured: 1.4 m (4ft 7in) in the USA, 1.3 m (4ft 3in) in Britain and Ireland.

BROADLEAVED A tree or shrub with wide, flat leaves.

BRONZE AGE In Britain, the period *c.*4300–2700 BP during which civilisations smelted copper and tin to form bronze and developed metalworking.

BRYOPHYTE Small plant with a haploid dominant generation, usually without water-conducting tissues: mosses, liverworts and hornworts.

BRYTHONIC a) Celtic languages spoken in lowland Scotland, northern England, Wales, Cornwall, and Brittany; b) the people who spoke these languages until *c.*10th century AD.

CANT a) A squared log; b) a division of woodland.

CARPEL The female reproductive apparatus of an angiosperm, which becomes the fruit.

CELLULOSE A complex hydrocarbon of which plant cell-walls are made.

CHECK When young trees stop growing but do not die: usually due to N, P, or K deficiency.

CHIPS Very small pieces of wood produced by shredding ('chipping') timber.

CHLOROPHYLL A plant pigment which absorbs sunlight (except the green wavelength) and uses the energy for photosynthesis.

CLEANING Removal of competing trees of the same or other species.

COPPICE A woodland of trees and shrubs cut on short rotation to stimulate regrowth of multiple stems.

CORACLE Very small, oval boat of wickerwork covered with hide, typically constructed in Ireland and Wales.

CORDILLERAN ICE SHEET The smaller ice-sheet of the Wisconsin Glaciation in western North America.

COTYLEDON The first or 'seed leaves' of an angiosperm.

CUNNINGARE a) An area of land used for rearing rabbits; b) a rabbit warren.

DECIDUOUS A tree or shrub which sheds its leaves seasonally.

DECOMPOSER An organism which breaks down the macro-structure of a dead plant or animal into smaller components.

DEER STALKER Someone trained to locate wild deer and to shoot one dead with a rifle bullet which hits the body in the optimum position.

DENN A clearing or glade in a woodland or forest (Anglo-Saxon).
DEVENSIAN The UK name for the most recent ice-age.
DNA A complex molecule which carries an organism's genetic information.
DOGGERLAND Now submerged, former landmass occupying part of the North Sea between Germany, Denmark and the UK.
DONKEY ENGINE A small steam-engine for lifting and pulling.
DOWCATT/DOOCAT A dove or pigeon house with roosting and nesting sites inside.
DRUMLIN A ridge-shaped hillock left by a retreating glacier.

ECOLOGY The branch of biology which studies plants, animals and their relationships with each other and environments; the study of energy flow through ecosystems.
ECOTONE A boundary or transition between two ecosystems.
ECOSYSTEM An assemblage of plants and animals with ecological and energy interrelationships.
ECOTYPE A population of genetically and geographically distinct individuals in a species.
ELVER Very small juvenile eel.
EMMER A wheat (*Triticum dicoccum*) domesticated in prehistory.
ENVIRONMENT The several factors which contribute to an organism's surroundings.

FELLING Cutting down a tree from near the base of the trunk.
FIBRES Lignified plant cells which can be extracted as threads.
FIRST NATIONS Descendants of people from East Asia, who colonised North America during and soon after the latest ice age.
FLITCH A slice of a tree trunk cut lengthways.
FLOWER Leaves of an angiosperm modified into reproductive apparatus.
FLUME An artificial channel containing a stream of water to transport objects.
FOREST (English) a) An area of land subject to Forest Law; b) a large area of land covered with trees; c) land growing timber.
FOREST (American) a) large area of land covered with trees and other plants which grow close together; b) an ecosystem dominated by trees and other woody vegetation; c) trees growing close together so that their crowns overlap and form 60% to 100% cover.

GAMETE A male or female sex cell with one set of chromosomes.
GENES Lengths of DNA which code for proteins.
GENUS A group of similar species.
GINPOLE A pole, held by guy ropes, for lifting heavy weights with a pulley.
GLACIER A river of ice formed by consolidation of snow in a mountain valley.
GUM a) Secretion of trees and shrubs which hardens when dry but is soluble in water; b) hard sweetmeat of gelatine.

HAG (PEAT HAG) An area of peat bog set aside for digging out as fuel.
HAINING The enclosure of land by hedge, fence or wall.
HAPLOTYPE A situation where several genes/DNA variations are grouped and inherited together from one parent.
HAPLOID Where the cells of an organism contain one set of chromosomes.
HEARTWOOD The central, older core of a tree which is dead and resistant to decay.
HERB/HERBACEOUS A vascular plant which dies back to ground level during winter.
HIGH FOREST A woodland of tall, one-stemmed trees (in contrast to coppice).
HIGHGRADING Selecting and harvesting the best, most marketable, timber trees in a forest, while leaving individuals or species of lesser value or poor shape.
HOLOCENE The geological period after the latest ice sheets melted.
HOMININ Modern and extinct human species and their ancestors.
HUMUS The dark-coloured component of soil which consists of decomposing plant and animal matter.
HYBRIDISATION When two unrelated organisms mate.
HYDROCARBON Complex organic molecules consisting of hydrogen and carbon atoms.

ICE AGE A period of time when surface ice extends out from Earth's Poles to lower latitudes.

INTERGLACAL The warmer time between ice ages when ice recedes, flora and fauna colonise.
INSPISSATE To thicken liquid ready for drinking or use in cooking.
INTROGRESSIVE HYBRIDISATION When hybrid organisms backcross (mate) with both parents so that genes are shared between species.
INVERTEBRATES Animals, usually small, without backbones.
IRON AGE In mainland Britain, the period *c.*2700 BP to AD 43 (before the Roman Conquest) when iron metallurgy was widespread and cereal farming intensive; in Ireland, where there was no Roman invasion, *c.*2500 BP–AD 400.

JET STREAM High altitude, fast moving, west to east flowing, wind currents which have major effects on Earth's climate.
JUVENILE A young plant or animal.

KARST A landscape of limestone (a form of calcium carbonate) rock which is dissolved by rainwater to form caves, swallow holes, pavements and underground rivers.
KERNEL A seed which is a pip, grain or nut.

LAURENTIDE ICE SHEET the larger of the two ice sheets of the Wisconsin Glaciation in North America.
LICHEN A symbiotic association of alga and fungus functioning as one organism.
LIGNIN A complex, hard cellulose in the cell walls of plants, which provides stiffening.
LITTORAL The zone between low and high tide, the seashore.
LOGGING Cutting down or felling trees.
LOWLANDS that part of Britain below 300 m altitude, with good soils, rainfall below about 90 cm and amenable climate.
LUMBER a) Felled tree trunks for milling; b) products of milling.

MEDIEVAL The period of west European history *c.*AD 400–1400.
MEGAANUS A time interval of one million years.
MERK Small unit of Scots currency, now obsolete.
MESOLITHIC The hunter-gatherer period of European prehistory *c.*12,000–7500 BP.
MILLING Converting timber into planks, spars or chips.
MUIRBURN A practice started in the mid-nineteenth century UK where moorland is burned in patches on rotation, to encourage young shots of heather and grass for grouse and sheep fodder.
MYCORRIZA Fungi which live in symbiotic association with higher plants.

NATIVE A species or individual living in the region in which it evolved or colonised itself.
NATURAL REGENERATION When a tree propogates itself rather than being planted by humans.
NATURAL, SEMI-NATURAL A habitat or ecosystem of mainly native species.
NEANDERTHAL A Eurasian species of hominin (*Homo neanderthalensis*) which became extinct by *c.*39,000 BP.
NEOLITHIC In the UK and Ireland, the first farming period of prehistory *c.*7500–4300 BP.
NUNATAK Ice-free mountain peak or coast in an ice-covered landscape.

OLD-GROWTH Forest of great age in which the life-cycles of organisms are not attenuated or affected by human activity.
OUTER BARK The woody material encasing the trunk, branches and twigs of a tree.
OVARY Another name for carpel.
OVULE The cells within a plant ovary which contain the female gamete.

PALAEO-INDIAN Ancient, native people of the American continent.
PALAEOLITHIC The period of prehistory when hominins made large stone tools ('the Old Stone Age'); in the UK *c.*700,000–12,000 BP.
PARK a) Deer-park; b) enclosed field (Scotland).
PEAT Dark 'soil' made of incompletely decomposed organic matter in a matrix of water.
PECK An old measure: 4 UK pints or half a UK gallon or one quarter of a bushel; equivalent to 2.2 litres.
PHLOEM Inner bark, the living cells of plants which transport sugars in solution.
PHOTOSYNTHESIS The process whereby plants convert carbon dioxide and water into carbohydrates using solar energy 'captured' by chlorophyll.
PINNACE A ship's tender with two masts; schooner-rigged or with eight oars.

PITCH a) English: a brown or black mineral liquid made from tar; b) American: plant resin.

PIT PROPS Small timbers supporting the walls and roofs of mines.

PLANTABLE/UNPLANTABLE Land on which trees can be planted and grown/Land which cannot be afforested due to poor soil, exposure or lack of technology.

PLANTS Living organisms usually containing chlorophyll and carrying out photosynthesis.

PODSOL A layered, acid soil where rainwater leaches minerals from upper levels and deposits them below.

POLICY or POLICIES Ornamental landscape with trees and woodland adjoining Scottish country mansions.

POLLARDING A method of obtaining woody material by removing a tree canopy to leave the clean-cut trunk at head height; many shoots later sprout from the cut surfaces.

POLLEN or POLLEN GRAIN The mobile plant organ which contains the male gametes.

POSTGLACIAL The period after the latest ice age.

POTLATCH A ceremony amongst Northwest Pacific Coast people with feasting, dancing, singing and distribution of gifts.

PUMPKIN In American forestry this describes a giant tree.

RAINFOREST A forest, rich in species, growing where rainfall, mist and fog are considerable and occur at all seasons.

RESIN A thick liquid which exudes from resin canals of coniferous trees.

RIDGE AND FURROW Corrugated land surface formed by ploughing land for growing crops or for afforestation.

RIPARIAN The zone of vegetation and water situated between a waterway and land.

ROMANS Natives of the city and state of Rome and their conquered peoples; in the UK, Roman occupation lasted from 43–410 AD.

RUN-RIG A Scottish system of growing crops on hand-dug ridges.

SALLOW Shrub member of the genus *Salix*.

SAP The sugary liquid transported in a plant's phloem.

SAPWOOD The still-living woody cells of a tree which form an outer cylinder to the central heartwood.

SCREEFING Turning over a piece of turf to plant a tree.

SCRUB Vegetation consisting of bushes and shrubs.

SESSILE Without a stalk.

SETT A shoot or young plant put into the ground by hand.

SHREDDING A method of obtaining branchwood, twigs and foliage by cutting branches off at successive heights close to the trunk; many shoots later sprout from the cut surfaces.

SHRUB Small woody plant with many stems.

SILVICULTURE The management of woodlands and forests.

SNEDDING Cutting branches off a felled tree trunk.

SPECIES A group of closely-related organisms which can breed with each other.

SPELT *Triticum spelta*, an early wheat now frequent in southern Europe.

STALKER *see* Deer Stalker.

STANK Straightened stream or ditch; a moat or swamp.

STOMA or STOMATA Those surface cells of a leaf through which gaseous exchange takes place.

STUMPERS Big trees felled directly into the ocean or sea.

SUCCESSION Changes in vegetation through time, when existing plants give way to a new assemblage.

SUMMER WOOD The cells which form anew in the woody tissue of a tree during summer.

SYMBIOSIS When different species live closely together to the benefit of both.

TAIGA Coniferous forest of arctic regions.

TAR A thick, black liquid made by destructive distillation of coal or wood.

TEMPERATE Mild, equable climate.

THINNING Removing poor quality trees from a commercial forest.

TIMBER a) The trunk and large branches of a tree when growing or when felled and converted into structural material; b) the trees (collectively) growing on an area of land.

TREE Woody plant with one main stem.

TUNDRA Treeless land in a cold climate with dwarf vegetation.

TURPENTINE a) Thick resin which exudes from coniferous trees; b) a liquid made by distilling the resin.

UNDERTAKERS Plants and animals whose lives involve recycling dead organisms into soil/nutrients.

UNPLANTABLE *see* Plantable.

UPLANDS Land above *c.*300 m (984 ft) in the UK and Ireland.

VEGETATION A collective noun describing plants growing in physical proximity to form a mosaic, blanket or mass.
VERDERER A judicial officer of a Royal Forest.
VERST A now-obsolete Russian unit of length approximately equal to 1.07 km (⅔ mile).
VERT Green vegetation in woodland and/or forest suitable for deer forage.
VIRGIN Woodland or forest not affected by human activity.

WATERGATE a) A door or gate giving access to a waterway; b) palings inserted across a waterway to catch sediment, vegetation, rubbish and dead animals.
WATTLE A wooden panel or fence made by weaving flexible rods horizontally between upright stakes.
WILDING a) Crab apple tree; b) young sapling of any tree.
WISCONSIN GLACIATION The most recent ice age of North America.
WOOD a) Small woody material; b) branches or coppice shoots.
WOODLAND (American) Land with trees or other woody vegetation covering 25–60% of the area.
WOODLAND (English) Land covered, or mainly covered, with trees.
WOOD FIBRES Those cells of a tree or shrub which are heavily stiffened by lignin.
WORT A plant used for food or medicine.

YARD the enclosed area around a dwelling for growing vegetables and keeping household animals.
YAWL A ship's small boat with sails and four or six oars.

Latin Names

Domestic Animals	
Cattle	*Bos taurus*
Dog	*Canis familiaris*
Goat	*Capra hircus*
Pig	*Sus domestica*
Sheep	*Ovis aries*

Mammals	
American beaver	*Castor canadensis*
American mink	*Neovison vison*
Archipelago wolf	*Canis lupus ligoni*
Aurochs	*Bos primigenius*
Badger (European)	*Meles meles*
Bank vole	*Clethrionomys glareolus*
Black bear	*Ursus americanus*
Black rat	*Rattus rattus*
Black-tailed deer	*Odocoileus hemionus sitkensis*
Blue hare (European)	*Lepus timidus*
Brown bear	*Ursus arctos*
Brown hare	*Lepus europaeus*
Caribou	*Rangifer tarandus*
Chinese water deer	*Hydropotes inermis*
Dormouse, Hazel	*Muscardinus avellanarius*
Douglas' squirrel	*Tamiasciurus douglasii*
Eastern gray squirrel	*Sciurus carolinensis*
Elk (American)	*Cervus canadensis*
Elk (European)	*Alces alces*
Ermine	*Mustela erminea*
European beaver	*Castor fiber*
Fallow deer	*Dama dama*
Field vole	*Microtus agrestis*
Grey wolf	*Canis lupus*
Grey squirrel	*Sciurus carolinensis*
Grizzly bear	*Ursus arctos horribilis*
Haida ermine	*Mustela erminea haidarum*
Least weasel	*Mustela nivalis*
Lemming, Norway	*Lemmus lemmus*
Long-tailed weasel	*Mustela frenata*
Lynx	*Felis lynx*
Mastodon	*Mammut* spp.
Modern human	*Homo sapiens*
Mole (European)	*Talpa europaea*
Mountain hare	*Lepus timidus*
Mule deer	*Odocoileus hemionus*
Muntjac, Reeves'	*Muntiacus reevesi*
Neanderthal human	*Homo neanderthalensis*
Northern flying squirrel	*Glaucomys sabrinus*
Otter (European)	*Lutra lutra*
Pine marten (European)	*Martes martes*
Polecat	*Mustela putorius*
Porcupine	*Erethizon dorsatum*
Prince of Wales flying squirrel	*Glaucomys sabrinus griseifrons*
Pygmy shrew (European)	*Sorex minutus*
Rabbit (European)	*Oryctolagus cuniculus*
Red deer	*Cervus elaphus*
Red fox	*Vulpes vulpes*
Red squirrel (American)	*Tamiasciurus hudsonicus*
Red squirrel (European)	*Sciurus vulgaris*
Reindeer	*Rangifer tarandus*
Roe deer	*Capreolus capreolus*
Sea otter	*Enhydra lutris*
Short-tailed weasel	*Mustela erminea*
Sika deer	*Cervus nippon*
Stoat	*Mustela erminea*
Townsend's chipmunk	*Tamias townsendii*
Wapiti	*Cervus canadensis*
Water vole (European)	*Arvicola terrestris*
Weasel (European)	*Mustela nivalis*
Wild boar (European)	*Sus scrofa*
Wild horse	*Equus ferus*

Birds

American crow	*Corvus brachyrhynchos*
Arctic loon	*Gavia arctica*
Bald eagle	*Haliaeetus leucocephalus*
Bittern (European)	*Botaurus stellaris*
Black grouse	*Tetrao tetrix*
Blackbird (European)	*Turdus merula*
Black-throated diver	*Gavia arctica*
Bullfinch (European)	*Pyrrhula pyrrhula*
Buzzard, Common	*Buteo buteo*
Capercaillie	*Tetrao urogallus*
Carrion crow	*Corvus corone*
Chaffinch, Common	*Fringilla coelebs*
Chestnut-backed chickadee	*Poecile rufescens*
Coal tit	*Parus ater*
Common crossbill	*Loxia curvirostra*
Common raven	*Corvus corax principalis*
Crested tit	*Lophophanes cristata*
Curlew (European)	*Numenius arquata*
Dark-eyed junco	*Junco hyemalis*
Dunnock	*Prunella modularis*
Garden warbler	*Sylvia borin*
Goldcrest	*Regulus regulus*
Golden eagle	*Aquila chrysaetos*
Goshawk, Northern	*Accipiter gentilis*
Grasshopper warbler	*Locustella naevia*
Great spotted woodpecker	*Dendrocopus major*
Grebe	*Podiceps* spp.
Green woodpecker	*Picus viridis*
Hen harrier	*Circus cyaneus*
Jay (European)	*Garrulus glandarus*
Kestrel (European)	*Falco tinnunculus*
Lapwing, Northern	*Vanellus vanellus*
Long-eared owl	*Asio otus*
Marbled murrelet	*Brachyramphus marmoratus*
Merlin	*Falco columbarius*
Osprey	*Pandion haliaetus*
Peregrine falcon	*Falco peregrinus*
Pheasant, Ring-necked	*Phasianus colchicus*
Pied flycatcher	*Ficedula hypoleuca*
Pileated woodpecker	*Dryocopus pileatus*
Red crossbill	*Loxia curvirostra*
Red grouse	*Lagopus lagopus scotica*
Redpoll, Common	*Carduelis flammea*
Ring ouzel	*Turdus torquatus*
Scottish crossbill	*Loxia scotica*
Short-eared owl	*Asio flammeus*
Siskin (European)	*Carduelis spinus*
Slate-colored junco	*Junco hyemalis*
Snipe, Common	*Gallinago gallinago*
Song thrush	*Turdus philomelos*
Sparrowhawk	*Accipiter nisus*
Spotted flycatcher	*Muscicapa striata*
Spotted owl	*Strix occidentalis*
Spruce grouse	*Falcipennis canadensis*
Tawny owl	*Strix aluco*
Townsend's warbler	*Setophaga townsendi*
Tree creeper (European)	*Certhia familiaris*
Tree pipit	*Anthus trivialis*
Whinchat	*Saxicola rubetra*
Willow grouse	*Lagopus lagopus*
Willow warbler	*Phylloscopus trochilus*
Wood pigeon	*Columba palumbus*
Wood warbler	*Phylloscopus sibilatrix*
Yellowhammer	*Emberiza citrinella*

Reptiles and Amphibians

Adder	*Vipera berus*
Common frog (European)	*Rana temporaria*
Common toad (European)	*Bufo bufo*

Fish

Chinook salmon	*Oncorhynchus tshawytscha*
Coho salmon	*Oncorhynchus kisutch*
Dolly Varden	*Salvelinus malma*
Eel (European)	*Anguilla anguilla*
Eulachon	*Thaleichthys pacificus*
King salmon	*Oncorhynchus tshawytscha*
Pacific herring	*Clupea pallasii*
Smelt (European)	*Osmerus eparlanus*
Trout	*Oncorhynchus*, *Salmo*, *Salvelinus* spp.

Invertebrates	
Atlantic stream crayfish	*Austropotamobius pallipes*
Common mussel	*Mytilus* spp.
Common limpet	*Patella vulgata*
Common whelk	*Buccinum undatum*
Edible periwinkle	*Littorina* spp.
Freshwater mussel	*Anodonta, Margaritifera, Unio* spp.
Flat oyster	*Ostrea edulis*
Great spruce bark-beetle	*Dendroctonous micans*
Green spruce aphid	*Elatobium abietinum*
Honeybee (European)	*Apis mellifera*
Large pine weevil	*Hylobius abietis*
Western honeybee	*Apis mellifera*
White-clawed crayfish	*Austropotamobius pallipes*

Flowering Plants	
'American' alder	*Alnus rubra* or *A. viridis*
'American' ash	*Fraxinus* spp.
'American' chestnut	*Castanea* spp.
Apple	*Malus domestica*
Arbor vitae	*Thuja occidentalis*
Ash, European	*Fraxinus excelsior*
Aspen	*Populus tremula*
Bald cypress	*Taxodium* spp.
Balsam fir	*Abies balsamea*
Balsam poplar	*Populus balsamifera*
Barley	*Hordeum vulgare*
Beech	*Fagus sylvatica*
Bigleaf maple	*Acer macrophyllum*
Bilberry	*Vaccinium myrtillus*
Black cottonwood	*Populus trichocarpa*
Black poplar	*Populus nigra* ssp. *betulifolia*
Black spruce	*Picea mariana*
Bluebell	*Hyacinthus non-scriptus*
Bogbean	*Menyanthes trifoliata*
Bog cranberry	*Vaccinium oxycoccus*
Bramble	*Rubus fruticosus* agg.
Bristlecone pine	*Pinus aristata*
Broom	*Cytisus scoparius*
Bulrush	*Typha latifolia*
Bunchberry	*Cornus canadensis*
Buttercup	*Ranunculus* spp.
Cedar of Lebanon	*Cedrus libani*
Chilean pine	*Araucaria araucana*
Cloudberry	*Rubus chamaemorus*
Clover	*Trifolium* spp.
Coast Douglas fir	*Pseudostuga menziesii* var. *menziesii*
Coast redwood	*Sequoia sempervirens*
Common alder	*Alnus glutinosa*
Common elder	*Sambucus nigra*
Common hawthorn	*Crataegus monogyna*
'Common' hazel	*Corylus cornuta*
Common oak	*Quercus robur*
Corsican pine	*Pinus nigra* ssp. *laricio*
Cotton grass	*Eriophorum* spp.
Cottonwood	*Populus trichocarpa, P. balsamifera*
Cowberry	*Vaccinium vitis-idaea*
Crab apple	*Malus sylvestris*
Crowberrry	*Empetrum nigrum*
Cycad	*Cycas* spp.
Dawn redwood	*Metasequoia glyptostroboides*
Deciduous cypress	*Taxodium distichum*
Dock	*Rumex* spp.
Douglas fir	*Pseudotsuga menziesii*
Douglas maple	*Acer glabrum*
Downy birch	*Betula pubescens*
Dwarf birch	*Betula nana*
Eared willow	*Salix aurita*
Early purple orchid	*Orchis mascula*
Eastern hemlock	*Tsuga canadensis*
Emmer wheat	*Triticum dicoccon*
Engelmann spruce	*Picea engelmannii*
English elm	*Ulmus minor* var. *vulgaris*
European larch	*Larix decidua*
European silver fir	*Abies alba*
False acacia	*Robinia pseudoacacia*
Field maple	*Acer campestre*
Foxglove	*Digitalis purpurea*
Foxtail pine	*Pinus balfouriana*
Gean	*Prunus avium*
Giant redwood	*Sequoiadendron giganteum*
Gingko	*Gingko biloba*
Goat willow	*Salix caprea*
Gorse	*Ulex europea*
Grand fir	*Abies grandis*
Grey alder	*Alnus incana*
Harestail cotton-grass	*Eriophorum vaginatum*
Hazel	*Corylus avellana*
Heath bedstraw	*Galium saxatile*
Heather	*Calluna vulgaris*

Hickory	*Carya* spp.
Holly	*Ilex aquifolium*
Hornbeam	*Carpinus betulus*
Horse chestnut	*Aesculus hippocastanum*
Hybrid larch	*Larix* x *marschlinsii*
Incense cedar	*Calocedrus* spp.
Jack pine	*Pinus banksiana*
Japanese larch	*Larix kaempferi*
Juniper	*Juniperus communis*
Kahikatea	*Dacrycarpus dacrydioides*
Labrador tea	*Ledum palustre*
Large-leaved lime	*Tilia platyphyllos*
Lawson's cypress	*Chamaecyparis lawsoniana*
Least willow	*Salix herbacea*
Lemon	*Citrus limonum*
Lodgepole pine	*Pinus contorta*
Lutz spruce	*Picea* x *lutzii*
Macedonian pine	*Pinus peuce*
Maize	*Zea mays*
Maritime pine	*Pinus pinaster*
Matai	*Prumnopitys taxifolia*
Mat-grass	*Nardus stricta*
Meadow rue	*Thalictrum* spp.
Mediterranean cypress	*Cupressus sempervirens*
Medlar	*Mespilus germanica*
Monkey puzzle	*Araucaria araucana*
Monterey pine	*Pinus radiata*
Mountain hemlock	*Tsuga mertensia*
Mountain maple	*Acer spicatum*
Mountain pine	*Pinus mugo*
Mulberry	*Morus nigra*
Netted willow	*Salix reticulata*
Noble fir	*Abies procera*
Nootka cypress	*Chamaecyparis nootkatensis*
Northern bilberry	*Vaccinium uliginosum*
Norway spruce	*Picea abies*
Oriental arbute	*Arbutus menziesii*
Pacific rhododendron	*Rhododenrdon macrophyllum*
Pacific Silver fir	*Abies amabilis*
Pacific yew	*Taxus brevifolia*
Pedunculate oak	*Quercus robur*
Pensylvanian maple	*Acer pensylvanicum*
'Pinus canadensis' (=Eastern hemlock)	*Tsuga canadensis*
Port Orford cedar	*Chamaecyparis lawsoniana*
Primrose	*Primula vulgaris*
Purple moor-grass	*Molinia caerulea*
Quaking aspen	*Populus tremuloides*
Quickthorn	*Crataegus monogyna*
Red alder	*Alnus rubra*
Red columbine	*Aquilegia canadensis*
Red elderberry	*Sambucus racemosa*
Red spruce	*Picea rubra*
Rimu	*Dacrydium cupressinum*
Rowan	*Sorbus aucuparia*
Salmonberry	*Rubus spectabilis*
Scots pine	*Pinus sylvestris* var. *sylvestris*
Scurvy grass	*Cochlearia* spp.
Sea buckthorn	*Hippophae rhamnoides*
Sea pink	*Armeria maritima*
Serbian spruce	*Picea omorika*
Sessile oak	*Quercus petraea*
Siberian larch	*Larix sibirica*
Siberian spruce	*Picea obovata*
Silver birch	*Betula pendula*
Sitka spruce	*Picea sitchensis*
Sloe	*Prunus spinosa*
Small-leaved lime	*Tilia cordata*
Snowy mespil	*Amelanchier lamarckii*
Southern beech	*Nothofagus* spp.
Spike false oat	*Trisetum spicatum*
Stiff sedge	*Carex bigelowii*
Sugar maple	*Acer sacharrum*
Sugar pine	*Pinus lambertiana*
Sweet chestnut	*Castanea sativa*
Sycamore	*Acer pseudoplatanus*
Tacamahaca	*Populus balsamifera*
Tomato	*Solanum lycopersicum*
Trail plant	*Adenocaulon bicolor*
Tufted hairgrass	*Deschampsia cespitosa*
Tufted saxifrage	*Saxifraga cespitosa*
Tulip tree	*Liriodendron tulipifera*
Twinflower	*Linnaea borealis*
Vine maple	*Acer circinatum*
Walnut	*Juglans regia*
Western gorse	*Ulex gallii*
Western hemlock	*Tsuga heterophylla*
Western larch	*Larix occidentalis*
Western red cedar	*Thuja plicata*
Western white pine	*Pinus monticola*
Western yew	*Taxus breviofolia*

Weymouth pine	*Pinus strobus*
White cedar	*Thuja occidentalis*
White fir	*Abies concolor*
White pine	*Pinus strobus*
White spruce	*Picea glauca*
White waterlily	*Nymphaea alba*
Wild cherry	*Prunus avium*
Wild currant	*Ribes* spp.
Wild raspberry	*Rubus idaeus*
Woolly willow	*Salix lanata*
Wych elm	*Ulmus glabra*
Yellow archangel	*Lamium galeobdolon*
Yellow cedar	*Chamaecyparis nootkatensis*
Yellow mountain avens	*Dryas drumondii*
Yellow waterlily	*Nuphar lutea*
Yew	*Taxus baccata*

Ferns

Deer fern	*Blechnum spicant*
Licorice fern	*Polypodium glycyrrhiza*

Fungi and Lichens

Butt rot fungus	*Heterobasidium annosum*
Deceiver	*Laccaria* spp.
Deer 'moss'	*Cladonia rangifera*
Rufous milkcap	*Lactarius rufus*
Saffron webcap	*Cortinarius croceus*
The Sickener	*Russula emetica*
Witch's hair lichen	*Alectoria fuscescens*
Woolly fibrecap	*Inocybe longicystis*

Bryophytes (Mosses and Liverworts)

Alpine haircap	*Polytrichum alpinum*
Bog moss	*Sphagnum* spp.
Common haircap moss	*Polytrichum commune*
Cypress-leaved plait-moss	*Hypnum cupressiforme*
Fingered cowlwort	*Colura calyptrifolia*
Fringe moss	*Racomitrium* spp.
Irish Daltonia	*Daltonia splachnoides*
Stair-step moss	*Hylocomium splendens*

Index